Wildlife of the Mid-Atlantic

Wildlife of the Mid-Atlantic

A Complete Reference Manual

JOHN H. RAPPOLE

University of Pennsylvania Press

PHILADELPHIA

10   9   8   7   6   5   4   3   2   1

Published by
University of Pennsylvania Press
Philadelphia, Pennsylvania 19104-4112

Library of Congress Cataloging-in-Publication Data

Rappole, John H.
  Wildlife of the Mid-Atlantic / John H. Rappole.
    p. cm.
  ISBN-13: 978-0-8122-3982-9 (alk. paper)
  ISBN-10: 0-8122-3982-2 (alk. paper)
  Includes bibliographical references and index.
  1. Animals—Middle Atlantic States. I. Title
QL157.M52R37   2007
591.975—dc22   2006053081

To My Siblings

Francesca, Bert, Robert, and Rosemary

Great people doing wonderful things

CONTENTS

FIGURES

Figures

Table

PREFACE

This book is designed to serve as a basic reference for a course in natural history, wildlife, or conservation biology, as well as for the practicing field biologist or interested amateur enthusiast. "In nature's infinite book of secrecy a little I can read." So says the soothsayer to Cleopatra in Shakespeare's drama, and that is the ambition this book is designed to serve. Two types of people have this ambition: those who need to know about the composition of the natural communities of the region for professional purposes, and those who just want to know. This volume should provide both groups with basic information on all the terrestrial vertebrates found regularly anywhere in the Mid-Atlantic region, serving as a ready reference for details on appearance, abundance, distribution, habitat use, seasonal occurrence, conservation status, reproduction, range, and sources for additional information. My intention is for it to follow in the tradition established by the superb *New England Wildlife* by Dick DeGraaf and Mariko Yamasaki, published by University Press of New England, which made obvious the utility of treatments summarizing life history for all the terrestrial vertebrates of a region.

In vertebrate zoology, people who study the various taxonomic groups are generally quite separate and distinct, with their own hierarchies, professional societies, journals, and traditions. In fact there are stereotypes that are extremely biased, unfair, and generally disreputable for students of each of the major groups: ichthyologists are of two types—fishermen with a studious bent, and statistical modelers; ornithologists are effete intellectuals with more research funds and recognition than is good for them; mammalogists are hard-working party folk who like to run trap lines and skin mammals all night; herpetologists are the Evel Knievels of field biology, living life on the edge, grabbing anything that moves and worrying about how hard it bites or what it injects you with later. Though the sample with which I have had extensive personal experience is small, essentially the vertebrate inventory group organized by Josh Laerm at the Georgia Museum of Natural History in the late 1970s, I think the members fit these taxon caricatures reasonably well. All except for the ornithologist, whom I see more as the rugged, outdoorsy type.

In any event, I recognize that, as an ornithologist, I am guilty of taxonomic trespass in having the temerity to treat the disparate groups in a single volume, and readily acknowledge the debt I owe to my fellow vertebrate zoologists. As is clearly reflected in the bibliography, I have leaned heavily on their expertise in providing basic information on the terrestrial vertebrates of the Mid-Atlantic. I hope these colleagues and their students, as well as others with an interest in regional natural history, will find the result useful in their field studies.

It has been more than thirty years since Robert MacArthur advised the world that it was time to move beyond descriptive natural history in his book *Geographical Ecology*: "To do science is to search for repeated patterns, not simply to accumulate facts.... But not all naturalists want to do science; many take refuge in nature's complexity as a justification to oppose any search for patterns" (MacArthur 1972: 1). This statement, condemning, and indeed discounting, the intellectual efforts of a large portion of practicing biologists at the time as non-science, seems more like the result of a fit of pique, or a tactical maneuver for use in battles for funds, resources, and faculty positions with entrenched dinosaurs in the Biology Department, rather than a reasoned assessment of the scientific value of descriptive biology. Nevertheless, this perspective became the prevailing attitude in academic institutions in a very short time, signaling the elimination of many organismal biology programs, and the replacement of herpetologists, botanists, ichthyologists, mammalogists, and ornithologists with community, population, and evolutionary ecologists.

No one can discount the importance of, and need for, the kind of synthesis called for by MacArthur, or his emphasis on hypothesis-driven testing of ideas. Without such work, we would never be able to understand the organizational principles of the natural world. However, his dismissal of descriptive biology as non-science was wrong for several reasons. First, any definition of science recognizes the "cataloging of facts" as a key step in building explanations for phenomena. Second, the work of the catalogers has not been completed. Myriads of animals, plants, and microorganisms have yet to be described as species, let alone provided with the details of their life history. Third, the natural world is not static. What was described two decades ago as the life history pattern for a species may not be applicable now, as will be abundantly clear from even a brief perusal of the contents of this volume. Fourth, and perhaps most important, human society has had immense effects on the flora and fauna of the Mid-Atlantic region, mostly negative.

Understanding and redressing these effects, the province of wildlife and conservation biology, requires a thorough knowledge of natural communities that cannot be obtained solely from synthesis of facts gathered in the past, even the recent past.

In writing this summary of the life histories of the terrestrial vertebrates of our region, it has become obvious to me that we need a new generation of scientists with sound training in descriptive biology. Such training not only provides the necessary intellectual foundation for synthesis, but also serves as the basis for understanding the conservation requirements of our natural communities, many of which are seriously threatened. I am hopeful that this book will help to stimulate students to find out what is known, as well as what is not known, about the organisms living around them, and to help build on that knowledge so that our understanding can keep pace with the rather dizzying rates of anthropogenic change documented in this work.

INTRODUCTION

The Mid-Atlantic region, for the purposes of this work, includes Pennsylvania, Virginia, West Virginia, New Jersey, Delaware, Maryland, and the District of Columbia (Figure 1). This area encompasses an impressive array of habitats and topography, ranging from the sandy coastal beaches and blackwater swamps of southeastern Virginia to the boreal bogs and spruce-fir forests of northern Pennsylvania and the highest peaks of West Virginia's Appalachian Mountains. In line with this diversity, the region has an extraordinary richness of terrestrial vertebrate fauna: 72 amphibian species; 62 reptiles; 81 mammals; and 331 birds. "Terrestrial" in this case means those that spend most or all of their life cycle on land, and thus excludes most pelagic vertebrates, e.g., seabirds, whales, and sea turtles, as well as fish. Some anomalous species are included, like the Hellbender and sirens, which are aquatic in all life cycle phases, because to exclude them would be to fall just short of complete coverage for the Amphibia.

An additional purpose of this book is to summarize the conservation status and major problems confronting vertebrate populations in the Mid-Atlantic. This part of the country has experienced anthropogenic effects as radical as any on earth over the four centuries since European settlement began, and these changes have had profound effects on the area's wildlife. For instance, moose, lynx, bison, elk, cougars, wolves, Trumpeter Swans, Whooping Cranes, Passenger Pigeons, and Carolina Parakeets were part of the terrestrial vertebrate fauna when Lederer first pushed his westward explorations to the peaks of the Blue Ridge Mountains in the late 1600s (Talbot 1672). Thomas Lewis surveyed northern Virginia's Shenandoah-Rockingham county line in the mid-1700s, recording not only the large mammals and birds observed, but the dominant tree species as well, providing a glimpse of what the primeval forest must have been like in the Appalachians (Lewis 1746). Now, in the early twenty-first

Century, few old growth stands of any habitat type remain in the region: Cook's Forest in northwestern Pennsylvania and Swallow River Falls State Park in the Cumberland of western Maryland (Figure 2) serving as two notable exceptions. Structurally, at least, it makes quite a difference whether a forest is composed of trees 2–3 m (7–10 ft) in diameter at breast height (dbh) or even larger (Figure 3), as was the case in pre-settlement times, versus today in which few trees exceed a meter in dbh. Unfortunately, we can only surmise what difference these structural dissimilarities might make in terms of the vertebrate communities.

Most early explorers and settlers had things on their mind other than the extraordinary natural diversity surrounding them, and made few notes on animals or plants that weren't edible or dangerous. For instance, John Lederer, a German explorer, made three trips to the Blue Ridge Mountains in the late 1600s, in which he catalogued the presence of deer, elk, buffalo, wolves, and mountain lions, but little else in the way of wildlife (Talbot 1672). Mark

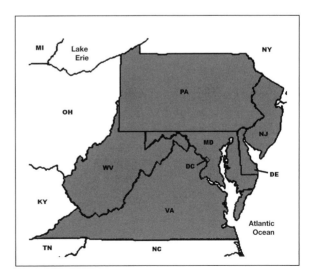

Figure 1. The Mid-Atlantic region.

Catesby's (1731–1743) superb *The Natural History of Carolina, Florida and the Bahama Islands* provided the first extensive treatment of terrestrial vertebrate species for North America, which included not only paintings of each animal and plant discussed, but a brief natural history account as well. His volumes include accounts for 103 birds, 8 mammals, 3 amphibians, and 28 reptiles. Although Catesby's volumes focused on areas south of the Mid-Atlantic, he lived for some time in Virginia, and most of his accounts are relevant for species from the Mid-Atlantic region.

Another relatively early study of Mid-Atlantic natural history is that provided by Thomas Jefferson (1784). Although comprised largely of lists, Jefferson's book catalogued some of the major flora and fauna of Virginia, which included West Virginia at the time of his writing. Works like that of Catesby, Jefferson, William Bartram (1791), and other early naturalists (cf Linzey 1998, Johnston 2003) allow us to establish a baseline of at least some of what existed in temperate eastern North America at that time, nearly two centuries after initial colonization.

The first detailed life history accounts of terrestrial, Mid-Atlantic vertebrates were done on the birds by Alexander Wilson (1808–1831), assisted by George Ord and Charles Lucien Bonaparte. Audubon's work (1840–1844) also focused mainly on birds, and included information from many different parts of North America beyond the Mid-Atlantic. His *Quadrupeds of North America* provided paintings and detailed life history accounts for 37 species of mammals, most of which are, or were, found in the Mid-Atlantic (Audubon and Bachman 1846–1854). David Johnston (2003) has provided an excellent summary of historical natural history liter-

Figure 2. Old growth forest at Swallow Falls State Park, Maryland.

Figure 3. Giant sycamore in southern Illinois during the early 1870s; the great Smithsonian ornithologist Robert Ridgway is seated on the left.

ature for the region. He too focused on birds, and the State of Virginia, but he identifies many of the main historical natural history references pertinent for the entire region. His work, and that of other studies summarizing historical literature for other major taxa (e.g., Linzey's 1998 *The Mammals of Virginia* and Green and Pauley's 1987 *Amphibians and Reptiles in West Virginia*) make clear that systematic efforts to catalog all of the terrestrial vertebrates within the Mid-Atlantic, and to provide the details of the life history for each species, are really a twentieth century phenomenon. Only recently have detailed life history accounts been completed for most of the terrestrial vertebrates of North America. For birds, such accounts became available through the *Birds of North America* project, which began with publication of the Barn Owl account in 1992 and was completed with the Dark-eyed Junco account (#716) in 2002 (Poole and Gill 1992–2002). The American Society of Mammalogy's (1969–2004) "Mammalian Species" was begun in 1969, with publication of the account for Waterhouse's Leaf-nosed Bat, and currently stands at 760 accounts, covering

most North American mammals as well as many from elsewhere in the world. The national herpetological societies have no comparable program, but Petranka's (1998) *Salamanders of the United States and Canada* presents an outstandingly thorough treatment for all members of that group.

Up-to-date life history summaries for members of the major taxa of terrestrial vertebrates (i.e., birds, mammals, reptiles, and amphibians) do not exist for all the states in the region. Those that do exist, however, are excellent. These include Green and Pauley's (1987) *Amphibians and Reptiles in West Virginia*; Martof et al.'s (1980) *Amphibians and Reptiles of the Carolinas and Virginia*; Linzey's (1998) *The Mammals of Virginia*; White and White's (2002) *Amphibians and Reptiles of Delmarva*; Merritt's (1987) *Guide to the Mammals of Pennsylvania*; Webster et al.'s (1985) *Mammals of the Carolinas, Virginia, and Maryland*; and Hulse et al.'s (2001) *Amphibians and Reptiles of Pennsylvania and the Northeast*. *Wildlife of the Mid-Atlantic* is an attempt to take the next step, pulling together summaries from the detailed life history accounts along with distributional and conservation

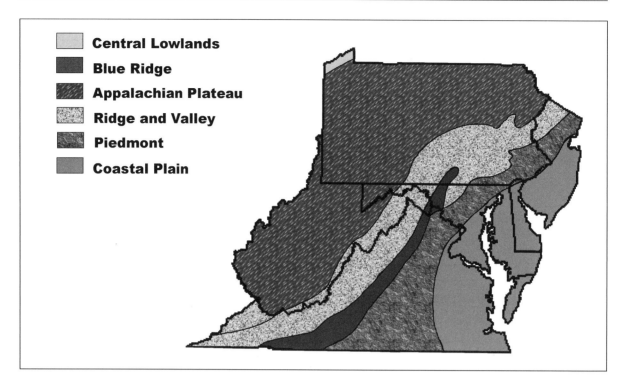

Figure 4. Geologic provinces of the Mid-Atlantic region.

information from state treatments. No publication effort prior to this book presents detailed information on all the terrestrial vertebrates in the Mid-Atlantic region in a single volume, including life history summaries, distribution maps and depictions (photo or drawing), conservation information, and a source in the literature where students can turn for additional information.

Physical Environment

Six major geologic provinces are found in the region: 1) Coastal Plain, 2) Piedmont, 3) Blue Ridge, 4) Ridge and Valley, 5) Appalachian Plateau, and 6) Central Lowlands (Figure 4). The Coastal Plain stretches 3,500 km (2,200 mi) along the edge of the continent from Massachusetts to the Mexican border, forming the boundary between land and ocean. In the Mid-Atlantic, the southern end of this province is about 200 km (120 mi) wide at the Virginia-North Carolina border, although it narrows

sharply and disappears at its northern end in central New Jersey where the Piedmont extends to the ocean shore. Several of the great cities of the Mid-Atlantic, such as Richmond, Washington, Baltimore, and Philadelphia, are located along the boundary between the Coastal Plain and Piedmont provinces, known as the "Fall Line." Here rivers like the James, Delaware, and Potomac flow out of the hard bedrock of the rolling Piedmont onto the sediments of the level Coastal Plain, and waterfalls mark the head of navigation.

The Appalachian Mountains are the predominant physiographic characteristic of the Mid-Atlantic, and the Piedmont Province forms the easternmost of the 4 provinces that are essentially subcategories of this single geological feature (Piedmont, Blue Ridge, Ridge and Valley, and Appalachian Plateau). The hills of the Piedmont extend westward from the Fall Line to the base of the Blue Ridge Mountains in Virginia, Maryland, and southern Pennsylvania, and to the mountains of the Ridge and Valley of Pennsylvania and New Jersey. The Blue Ridge is a long, narrow province consisting of a single ridge or series of parallel ridges and valleys running generally from southwest to northeast 885 km (550 mi) from Carlisle, Pennsylvania, to north Georgia. North and west of the Blue Ridge is a wider series (up to 130 km) (80 mi) of higher ridges and deeper valleys that also run southwest to northeast; these constitute the Ridge and Valley Province, which extends nearly 2,000 km (1,200 mi) from the St. Lawrence River at the Canadian border to central Alabama. West and north of this province are the steep hills of the Appalachian Plateau, covering most of West Virginia and western Pennsylvania. A sixth province, the Central Lowlands, occurs in only a small portion of our region, namely that part of Pennsylvania bordering Lake Erie, where Presque Isle and the city of Erie are located.

Four major climatic zones occur in the Mid-Atlantic: 1) the Austral is characterized by hot summers, mild winters, and moderate precipitation, and is found only in southeastern Virginia; 2) the Carolinian is characterized by hot summers, cool winters, and moderate precipitation, and covers the Piedmont region and valleys of the southern Appalachians; 3) the Alleghenian is characterized by warm summers, cold winters, and moderate precipitation, and covers most of the Appalachian Highlands; and 4) the Canadian is characterized by cool summers, harsh winters, and moderate precipitation, and is found only at the highest elevations in our region.

Living Environment

Life Zones and Biotic Provinces

The varied landforms and climates of the Mid-Atlantic create the basis for the diversity of plant and animal communities or "habitats" that occur here. Specific associations or communities of plants and animals constitute a habitat for a given species, and, with experience, one can learn to recognize these communities and know which organisms to

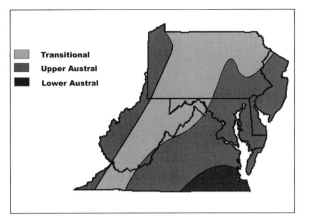

Figure 5. Merriam's life zones for the Mid-Atlantic.

expect in them. However, there are differences among the classifications for the living environment provided by biologists. For instance, C. Hart Merriam (1894) recognized 3 major "life zones" for the Mid-Atlantic based on temperature (isotherms for the mean temperature of the 6 hottest weeks of the year): Lower Austral, Upper Austral, and Transitional (Figure 5). Dice (1943) modified this system, replacing the "life zone" classification with "biotic provinces." In doing so, he used characteristics of temperature, topography, and soil to define "province" boundaries, similar in many cases to those drawn by Merriam. Thus his "Austroriparian Province" roughly matches Merriam's "Lower Austral"; his "Carolinian" approximates Merriam's "Upper Austral"; and his "Canadian" is similar to that of Merriam's "Transitional." These and similar attempts to apply subjective taxonomic rules lumping large numbers of ecological communities together retain interest because they reflect reality at a gross level. For instance, there is no doubt that something of considerable ecological significance occurs in southeastern Virginia, where many species of animals reach the northern extremity of their distribution along the boundary of Dice's "Austroriparian Biotic Province." For this reason, terms from the various attempts at community classification at continental scales, like "Austroriparian" and "Transitional," remain in common natural history parlance, long after their theoretical basis has been more or less discredited. The fundamental problem with these various organizational efforts is that no single principle or set of principles can explain the distributions of all organisms, because the distribution of each species is the result not only of factors affecting all other species in its community, e.g., changes in climate and topography, but also of factors unique to itself, e.g., the length of time that it has existed as a species, and the various chance mutations through which it has evolved over that period. To cite extreme examples, one must take into consideration the breakup of Pangaea 200 million years ago to understand the current distribution of Hellbenders (Cryptobranchidae), which are found only in China, Japan, and the eastern United States, while the breeding distributions of more than 150 species of birds have changed in a matter of decades in response to global warming (Mathews et al. 2004).

To avoid these complications, while recognizing some utility in terms of community classification,

ecologists have reverted to strictly descriptive mapping of the major plant communities. Like the higher-level groupings discussed above, and for the same reasons, these plant community classifications are not always indicative of the animal communities that they contain. Nevertheless, they have the virtue of being largely objective in that they reflect mapping exercises of major plant associations. Perhaps the most widely used of these classifications is that developed by Küchler (1975) (Figure 6). His classification includes 9 principal habitat types for the Mid-Atlantic region: Northern Cordgrass Prairie; Northeastern

Spruce-Fir Forest; Beech-Maple Forest; Mixed Mesophytic Forest; Appalachian Oak Forest; Northern Hardwoods; Northeastern Oak-Pine Forest; Oak-Hickory-Pine Forest; and Southern Floodplain Forest. It is important to note that Küchler's habitat map is titled, *Potential Natural Vegetation of the United States* (Küchler 1975). Thus he shows what he believes *would* be present in the region, if humans

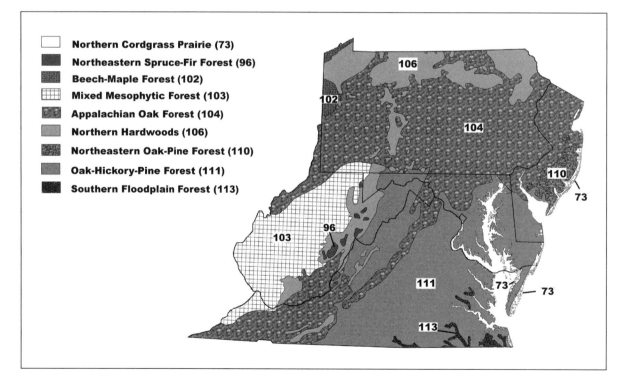

Northern Cordgrass Prairie (73)
Northeastern Spruce-Fir Forest (96)
Beech-Maple Forest (102)
Mixed Mesophytic Forest (103)
Appalachian Oak Forest (104)
Northern Hardwoods (106)
Northeastern Oak-Pine Forest (110)
Oak-Hickory-Pine Forest (111)
Southern Floodplain Forest (113)

Figure 6. Küchler's (1975) major plant communities of the Mid-Atlantic.

Figure 7. Coastal habitat on Assateague Island, Virginia.

had not modified the habitats. Of course, humans have modified every habitat, so what you find at any given site could range anywhere from a hundred-year-old oak forest to a parking lot. Almost all sites are in some seral stage short of the climax vegetation communities whose distribution is hypothesized in Küchler's map. Also, global warming has caused rapid change in the distribution of some communities, a process that is in progress at a rate scientists are only beginning to understand.

Habitats

In defining a habitat classification for the Mid-Atlantic region, I follow a modification of Küchler's system. I add three major habitats that his classification does not address: Coastal Waters and Shoreline, Agricultural and Residential, and Freshwater Wetlands; also, I combine several of his deciduous forest classifications into a single broadleaf forest category.

Coastal Waters and Shoreline (Figure 7): The coastal marine habitat can be broken into three major subdivisions, each with its own characteristic group of species: 1) Pelagic (open ocean) and Bays; 2) Beaches and Dunes; and 3) Estuaries, Saltmarshes, and Tidal Flats.

Freshwater Wetlands (Figure 8): Lakes, ponds, impoundments, rivers, and marshes. The defining characteristic is the presence of fresh water, which stimulates the growth of such plants as cattails, sedges, and bulrushes. Freshwater wetlands are among the most endangered of habitats in our region, perhaps because they tend to limit human economic activities, and so are dammed, dredged, drained, channeled, and filled out of existence. Those that remain serve mainly as conduits for waste.

Grassland (Figure 9): Most of what now occurs of this habitat in our region has been created by human activity, such as the grasslands now covering reclaimed strip mines in Pennsylvania's Clarion County, the broad pasture lands of Virginia's Piedmont, or Delaware's hayfields, although natural

Figure 8. Freshwater tidal marsh at John Heinz National Wildlife Refuge, Tinicum, Philadelphia.

Figure 9. Grassland at Chincoteague National Wildlife Refuge, Virginia.

coastal cordgrass prairies still occur along the immediate coast, especially in places like Brigantine and Barnegat marshes in New Jersey. The primeval forests of the eastern U.S. are legendary, where supposedly at the time of Captain John Smith, "a squirrel could travel from the coast to the Mississippi, and never touch the ground." Nevertheless, the fact that there were extensive areas of grassland has been well documented. Wayland (1989), for instance, reports that much of the broad Shenandoah Valley was grassland at the time of the arrival of the first European settlers, perhaps maintained by Indians with fire. At least one bird was native to grasslands, the Heath Hen, an eastern subspecies of the Greater Prairie-Chicken (*Tympanuchus cupido*), now extinct.

Broadleaf Deciduous and Mixed Forest (Figure 10): Five of Küchler's (1975) forest communities are lumped within this designation including the Beech (*Fagus*)-Maple (*Acer*) Forest as found near Sharon and New Castle in western Pennsylvania; the Mixed Mesophytic Forest of maple, beech, oak, tulip poplar, and horse chestnut found in much of the Appalachian Plateau of West Virginia; Appalachian Oak Forest (and formerly American Chestnut), which covers the lower slopes of the Appalachian Highlands of Virginia, Maryland, Pennsylvania, and New Jersey; Northeastern Oak-Pine Forest, otherwise known as the "Pine Barrens" of central and southern New Jersey; and Oak-Hickory-Pine Forest found throughout most of the Piedmont. The reason for lumping these different plant communities together is that they are quite similar in terms of the animal communities they support - at least in their current stage of recovery from nineteenth century clearing.

Northern Mixed Hardwoods (Figure 11): Maple, birch, beech, hemlock, and white pine forest, often referred to as "Transitional Forest" (from C. Hart Merriam's classification), as in transition from the primarily deciduous forests of the temperate

Figure 10. Broadleaf deciduous and mixed forest near Compton Gap, Shenandoah National Park, Virginia.

Figure 11. Northern mixed hardwoods near Kinzua Dam, northwestern Pennsylvania.

portions of the continent to the mainly coniferous forests of the boreal regions.

Highland Coniferous Forest (Figure 12): This forest type is referred to as Northeastern Spruce-Fir Forest by Küchler (1975), and actually represents an outlier of the vast Holarctic boreal forests that cover much of Canada, Siberia, and northern Europe. During the Wisconsin glaciation (22,000–10,000 years before the present), this habitat covered vast areas, and extended hundreds of kilometers southward. Now, it remains only in small isolated patches at high elevations, e.g., above 1200 m (4000 ft) in the southern mountains of our region, lower in more northern latitudes.

Southern Floodplain Forest (Figure 13): Like the previous forest type, southern floodplain forest is an outlier in our region (of Dice's "Austroriparian Biotic Province" and Merriam's Lower Austral Zone), representing the northernmost extent of the great southeastern bottomlands that once choked every river with tangled swamps and bayous along the Coastal Plain from southeastern Virginia all the way to east Texas, but now persist only as remnants. The upland longleaf pine/palmetto savanna of the Austroriparian, principal habitat of the endangered Red-cockaded Woodpecker, is now even more rare than the bottomland forest in our region.

Agriculture and Residential (Figure 14): Unfortunately, many of the beautiful and unique habitats of the Mid-Atlantic States have been converted to these universal types: a corn field looks pretty much the same in coastal Maryland as it does in Iowa, or Ecuador for that matter, and plowed dirt is plowed dirt, no matter where you find it. These habitats share common species in many parts of the world - such as Norway Rats, Rock Pigeons, and House Mice. Still, some species native to the region's aboriginal grasslands and woodlands, e.g., the Meadow Vole, Northern Leopard Frog, and Smooth Greensnake, can still be found in "improved" pastures, orchards, suburban lawns and gardens, and similarly altered environments.

Figure 12. Highland coniferous forest on Spruce Knob, Spruce Knob Rocks National Recreation Area, West Virginia.

Figure 13. Southern floodplain forest at Great Dismal Swamp National Wildlife Refuge, Virginia.

Figure 14. Agricultural and residential habitat near Cearfoss, Maryland.

Wildlife and Conservation Issues for the Mid-Atlantic Region

There is considerable debate concerning the effects of humans on North American biota prior to European colonization, e.g., the role of *Homo sapiens* in large mammal extinctions during the late Pleistocene and early Holocene, 10,000 years ago (Martin and Klein 1984, MacPhee 1999). In contrast, basic facts associated with anthropogenic influence on the biota since Europeans first established permanent settlements on the continent are well documented, including the following: 1) forest cover in the eastern United States declined from >90% in 1600 to about 30% in 1900, recovering to almost 60% at present (Powell and Rappole 1986), although almost none is old growth forest; 2) widespread extirpation of species resulting from hunting by the early 1900s, with recovery subsequent to protection for some species, e.g., Diamond-backed Terrapin, White-tailed Deer, and Wild Turkey; 3) extensive poisoning of river systems by pollutants, e.g., DDT and acid rain, with subsequent ripple effects on species up through the food chain; 4) drastic reduction, regional extirpation, or elimination of species resulting from intentional or unintentional introductions of species from the Old World (e.g., Chestnut Blight Fungus and Dutch Elm Fungus); 5) shifts in breeding distribution of 150 species of migratory birds resulting from global climate change (Matthews et al. 2004).

Extinction of species like the Carolina Parakeet and Passenger Pigeon is emblematic of the profound environmental changes that have taken place resulting from human influence. Among the data summarized in this book are the species-by-species conservation assessments performed by state Natural Heritage programs and departments of natural resources. All the states in the Mid-Atlantic follow The Nature Conservancy's suggested rankings of degree of threat to wild populations (S1 = Critically Imperiled, S2 = Imperiled, S3=Vulnerable, etc.). Of the 546 species of birds, mammals, reptiles, and amphibians in the Mid-Atlantic, 287 are considered to be imperiled in one or more of our states. Furthermore, many are considered as already extirpated from the region (e.g., Gray Wolf, Martin, Whooping Crane, Heath Hen, Eskimo Curlew) (See Appendix). These data show that the problem of vertebrate species decline and disappearance in the Mid-Atlantic is much more profound than the short list of species known to be extinct might indicate. Analysis of the known or suspected reasons for the decline or disappearance of these species is informative (Table 1). These are discussed below. For some species, the reasons are obvious. Trumpeter Swans, Elk, and Whooping Cranes by human hunting for food; American Martens and Fishers by trapping for sale of pelts; wolves and mountain lions by predator removal programs, often subsidized by government bounties. However, for many species, the reasons are not obvious. For instance, 162 species of migratory birds that occur in the Mid-Atlantic region have been listed as S1, S2, or S3 in one or more states, are extinct or extirpated, or have been found to have undergone significant declines in the region or throughout their range, based on National Breeding Bird surveys (DeGraaf and Rappole 1995). For some of these birds, regional loss of breeding habitat seems like the obvious cause. However, for migratory species with complex life cycles, significant parts of which may be spent thousands of kilometers from the Mid-Atlantic in areas with their own conservation problems, it is very hard to know whether observed declines are the result of breeding ground threats, wintering ground threats, or threats to critical habitats somewhere in between (Robbins et al. 1989, DeGraaf and Rappole 1995, Rappole 1995).

Habitat Change During One or More Phases of the Life Cycle for Migratory Species: Thirty-one species of migratory birds are listed as S1, S2, or S3 in the region for which the reasons for decline are not obvious (Table 1). Such declines are not restricted to the Mid-Atlantic, but have been recorded for many migratory birds across their entire continental ranges (DeGraaf and Rappole 1995). The complex life cycles of migrants, in which populations are dependent on different habitats in widely separated geographic areas at different times of the year, make identification of the specific portion of the life cycle responsible for the decline difficult.

Many researchers have suggested that fragmentation of breeding habitat is responsible (Askins et al. 1990), and, in fact, there is good evidence from several studies to indicate that size of a piece of habitat is positively correlated with numbers of breeding pairs ("Area Effect") (Robbins et al. 1989). However, as Rappole and McDonald (1994, 1998) point out, loss of transient or wintering habitat could provide a sound explanation for this observation. In addition, amount of habitat for many forest-related migrants has actually increased in the Mid-Atlantic region; yet populations of many such species continue to show declines (Rappole 1995).

Loss of Old Field and Grassland Habitat: There were few areas of extensive grassland in eastern North America prior to European colonization (Küchler 1975). This situation changed rapidly during the 1800s as human populations expanded rapidly westward, replacing forest with farm fields and various stages of second growth (Powell and Rappole 1986). Many species of flora and fauna responded quickly to this change in habitat, expanding their ranges into the Mid-Atlantic where they had not previously been recorded, e.g., Bachman's Sparrow, a bird originally of southeastern pine savanna (Dunning 1993), and Bewick's Wren, whose precolonization distribution presumably was restricted to the Great Plains and western North America (Kennedy and White 1997). By the 1920s, various economic and social forces began to cause small farms in the Mid-Atlantic and elsewhere in eastern North America to be abandoned, and by the late 1900s, large areas of grassland and old field habitat had reverted to forest or been converted to intensive agriculture or urban use, resulting in the increasing rarity or disappearance of many members of open habitat and early seral stage communities (Askins 2000) (Table 1).

Disturbance and Loss of Coastal Breeding Habitat: Coastal habitats of the Mid-Atlantic in general, and wetlands in particular, are under intense pressure. The waterways, marshes, beaches, dunes, and other habitats unique to the immediate coast are being developed for residential housing and recreation at an extremely rapid rate (Carter 1988). In addition, those habitats not actually developed are increasingly subjected to intense disturbance, either directly by people (e.g., recreational beach traffic) or by increased populations of predators associated with increased human use, including dogs, cats, and introduced species, e.g., feral pigs (Rappole 1982). Coastal wetlands are also subject to pollution from agricultural runoff, industrial waste, poultry farms, and human sewage from the immediate vicinity, or brought to the coast by the region's major rivers (Carter 1988). One of the more ironic sources of human coastal disturbance is that provided by those conducting beach surveys for nesting by endangered sea turtles. These surveys are conducted in spring and summer at night along several of the region's barrier-island beaches, usually by vehicles driven along the shore above high-tide line, destroying preferred nesting sites for beach-nesting shorebirds (e.g., Wilson's Plover, American Oystercatcher, Least Tern) (Rappole 1982).

Hunting: Destruction of wildlife populations for personal or market consumption is not currently a serious threat to North American vertebrate fauna, with the possible exception of the Black Scoter. However, hunting and trapping caused complete or near-complete extirpation of many Mid-Atlantic species during the 1800s and early 1900s. Eastern populations of some species decimated by hunting or trapping are recovering or have recovered (e.g., Snowy Egret, White-tailed Deer, American Beaver, Canada Goose); some, however, are extinct (e.g., Passenger Pigeon, Carolina Parakeet), regionally extirpated (e.g., Elk, Bison, Timber Wolf, Cougar, Whooping Crane, Trumpeter Swan), or have never fully recovered (e.g., Short-billed Dowitcher, Hudsonian Godwit, American Avocet) (Table 1).

Highland Forest and Bog Disappearance: In many parts of the Mid-Atlantic, boreal forest and bogs remain as scattered remnants of the most recent Ice Age, persisting only at higher elevations. Where these exist, they provide relatively isolated islands of habitat, home to unique assemblages of animals and plants that are quite vulnerable to disturbance. Several threats affect these habitats, including logging, acid rain, mining, spruce budworm, and wooly adelgid infestation (Little 1995), causing local disappearance of boreal wildlife communities. To these threats must now be added the influence of rapid global warming, which is likely to change the climate sufficiently in the next century to cause many remaining boreal sites in the Mid-Atlantic to be replaced by more temperate habitats (Matthews et al. 2004).

Conversion of Flood Plain Forest: Like boreal forest, flood plain forest is restricted in distribution by its very nature; although perhaps more like a peninsula or isthmus than an island, these are the aboriginal habitats lining the region's waterways (Küchler 1975). Often such sites present the richest soils for crops and timber products, as well as the most pleasant sites for recreational homes. As a result this habitat is threatened by logging and clearing for agriculture, and by residential development, in much of the Mid-Atlantic, threatening the natural communities associated with it as well (Noss et al. 1995) (Table 1).

Roosting/Breeding Cave Destruction and Disturbance: Like migratory birds, several bat species are dependent on widely separated areas and habitats at different times of the year. These localities are often caves where large portions of the known regional population of a species can be found in a relatively few sites. In addition to the normal kinds of disturbance threatening Mid-Atlantic habitats, e.g., pollution and development, bat caves are extremely vulnerable to simple human trespass. Spelunkers, visiting and exploring caves, can cause massive bat die-offs as a result of disturbing breeding mothers, forcing abandonment of newborn offspring (Tuttle 1979), or starvation of bats awakened from hibernation at winter roosts, causing depletion of precious fat stores (Brady et al. 1982).

Wetland Destruction or Pollution: Wetland communities are especially vulnerable, not only because they are highly prized for both residential and agricultural development, but also because the habitats that replace them are completely unsuitable for most community members. In addition, of course, wetlands can be made uninhabitable for wildlife by pollution as well, often occurring far from the site where the pollutants collect. As a result, wetland drainage and despoilation are among the most serious environmental problems in the

United States in general, and the Mid-Atlantic in particular (Noss et al. 1995). As is clear from Table 1, more species are endangered by threats to this habitat type (81) than from any other single cause.

Control of Human, Crop, Livestock, or Fisheries Predators: Vertebrates that pose direct potential threats to human life, or to resources valued by humans, have been subjected to intense and extremely effective control measures in the past. Kercheval (1925) for instance, reported that 256 wolf heads were turned in for bounty in Augusta County, Virginia in 1750. Thus, Gray Wolves, Red Wolves, and Mountain Lions have long been extirpated from the Mid-Atlantic, while Timber Rattlesnake populations have been decimated. In addition, such predators on livestock and fisheries as Golden Eagle, Double-crested Cormorant, and Common Merganser continue to be persecuted in parts of their range.

Pet Trade Collecting: Some rare amphibians and reptiles are especially prized by collectors, who will not hesitate to pay large amounts for specimens.

Wood Turtles, for instance, bring $35–$200 per animal, and illegal collecting has resulted in complete extirpation of the species from several waterways (Hammerson 2003).

Landscape change is a part of natural history. Huge changes occurred during the last advance of the Wisconsin ice sheet into Pennsylvania 22,000 years ago and its subsequent retreat a few thousand years later (Sevon and Fleeger 1999). Vast regions of habitat stretching thousands of kilometers were completely obliterated or pushed southward hundreds of kilometers. Nevertheless, such changes are not quite the same as those resulting from the human effects discussed above. For one thing, anthropogenic changes can occur over a matter of decades rather than millennia; for another, they can be quite insidious and difficult to document, if, nonetheless, pervasive; and lastly, and perhaps most importantly, they are not inevitable. Once we know what effects we are having on our environment, we have a choice.

TABLE 1. Terrestrial vertebrate species identified as extirpated or having declined in the Mid-Atlantic[1], along with hypothesized reason.

Hypothesized Reason for Decline[2]	Declining Species
Habitat change during some phase of the life cycle for migratory species.	Mississippi Kite, Broad-winged Hawk, Red Knot, Yellow-billed Cuckoo, Black-billed Cuckoo, Common Nighthawk, Chuck-will's-widow, Whip-poor-will, Chimney Swift, Olive-sided Flycatcher, Eastern Wood-Pewee, Acadian Flycatcher, Purple Martin, Barn Swallow, Wood Thrush, Veery, Swainson's Thrush, Bicknell's Thrush, Bachman's Warbler, Black-throated Blue Warbler, Blackburnian Warbler, Kirtland's Warbler, Cerulean Warbler, Swainson's Warbler, Ovenbird, Canada Warbler, Wilson's Warbler, Scarlet Tanager, Rose-breasted Grosbeak, Bobolink
Loss of old field and grassland habitat through reversion to forest or conversion to intensive agriculture.	Eastern Fence Lizard, Six-lined Racerunner, Short-headed Gartersnake, Northern Bobwhite, Black Vulture, Upland Sandpiper, Barn Owl, Short-eared Owl, Fish Crow, Bewick's Wren, Horned Lark, Loggerhead Shrike, Blue-winged Warbler, Golden-winged Warbler, Chestnut-sided Warbler, Prairie Warbler, Yellow-breasted Chat, Dickcissel, Eastern Towhee, Vesper Sparrow, Grasshopper Sparrow, Bachman's Sparrow, Field Sparrow, Savannah Sparrow, Henslow's Sparrow, Blue Grosbeak, Indigo Bunting, Eastern Meadowlark, Orchard Oriole, American Goldfinch, Least Shrew, Eastern Harvest Mouse, Golden Mouse, Southern Bog Lemming, Meadow Jumping Mouse, Least Weasel
Disturbance and predation at coastal breeding sites and development and destruction of coastal habitats.	Brown Pelican, Tricolored Heron, Great Egret, Snowy Egret, Little Blue Heron, Cattle Egret, Yellow-crowned Night-Heron, White Ibis, Glossy Ibis, Black Rail, Clapper Rail, Wilson's Plover, Piping Plover, American Oystercatcher, Black-necked Stilt, Willet, Herring Gull, Great Black-backed Gull, Royal Tern, Roseate Tern, Caspian Tern, Gull-billed Tern, Sandwich Tern, Forster's Tern, Common Tern, Least Tern, Black Skimmer, Saltmarsh Sharp-tailed Sparrow, Boat-tailed Grackle, Delmarva Fox Squirrel, Marsh Rice Rat
Hunting	Black Scoter, Trumpeter Swan, Tundra Swan, Whooping Crane, Passenger Pigeon, Carolina Parakeet, Buff-breasted Sandpiper, Eskimo Curlew, Short-billed Dowitcher, American Marten, Elk, Bison, American Avocet, Harlequin Duck, American Black Duck, Hudsonian Godwit

TABLE 1 continued

Hypothesized Reason for Decline[2]	Declining Species
Disappearance and pollution of mature highland forest, stream, and bog habitats resulting from timbering, acid rain, spruce budworm, mining, wooly adelgid infestation, and other factors	Allegheny Mountain Dusky Salamander, Black-bellied Salamander, Blue Ridge Dusky Salamander, Black Mountain Salamander, Blue Ridge Two-lined Salamander, Black Mountain Salamander, Cheat Mountain Salamander, Shenandoah Salamander, Cumberland Plateau Salamander, Bog Turtle, Peaks of Otter Salamander, Cow Knob Salamander, Wehrle's Salamander, Yonahlossee Salamander, Weller's Salamander, Green Salamander, Mountain Chorus Frog, Coal Skink, Pinesnake, Sharp-shinned Hawk, Northern Goshawk, Long-eared Owl, Northern Saw-whet Owl, Yellow-bellied Sapsucker, Alder Flycatcher, Yellow-bellied Flycatcher, Blue-headed Vireo, Common Raven, Red-breasted Nuthatch, Brown Creeper, Winter Wren, Golden-crowned Kinglet, Hermit Thrush, Nashville Warbler, Magnolia Warbler, Northern Waterthrush, Mourning Warbler, Red Crossbill, Purple Finch, Pine Siskin, American Water Shrew, Appalachian Cottontail, Snowshoe Hare, Northern Flying Squirrel, Appalachian Woodrat, Rock Vole, North American Porcupine
Conversion of mature flood plain forest, swamp, and savanna to commercial timber, agriculture, residential, or other development	Eastern Spadefoot, Common Five-lined Skink, Broad-headed Skink, Little Brown Skink, Eastern Glass Lizard, Eastern Worm Snake, Scarletsnake, Cornsnake, Eastern Hog-nosed Snake, Common Kingsnake, DeKay's Brownsnake, Red-bellied Snake, Southeastern Crowned Snake, Milksnake, Barred Owl, Red-cockaded Woodpecker, Pileated Woodpecker, Ivory-billed Woodpecker, Willow Flycatcher, Yellow-throated Vireo, Warbling Vireo, White-breasted Nuthatch, Brown-headed Nuthatch, Northern Parula, Black-throated Green Warbler, Yellow-throated Warbler, Black-and-white Warbler, American Redstart, Worm-eating Warbler, Louisiana Waterthrush, Kentucky Warbler, Hooded Warbler, Summer Tanager, Cotton Mouse
Small populations located in a few roosting caves vulnerable to disturbance	Gray Myotis, Small-footed Myotis, Northern Long-eared Myotis, Indiana Myotis, Townsend's Big-eared Bat
Freshwater wetland habitat destruction or pollution	Common Mudpuppy, Mabee's Salamander, Mole Salamander, Tiger Salamander, Mud Salamander, Red Salamander, Green Tree Frog, Oak Toad, Hellbender, Lesser Siren, Greater Siren, Dwarf Waterdog, Jefferson Salamander, Spotted Salamander, Marbled Salamander, Long-tailed Salamander, Cave Salamander, Four-toed Salamander, Shovel-nosed Salamander, Many-lined Salamander, Fowler's Toad, Northern Cricket Frog, Gray Tree Frog, Cope's Tree Frog, Pine Barrens Tree Frog, Barking Tree Frog, Little Grass Frog, Southeastern Chorus Frog, Northern Leopard Frog, Southern Leopard Frog, Carpenter Frog, Eastern Mud Turtle, Loggerhead Musk Turtle, River Cooter, Florida Cooter, Northern Red-bellied Turtle, Pond Slider, Chicken Turtle, Spotted Turtle, Northern Map Turtle, Spiny Softshell, Smooth Softshell, Kirtland's Snake, Plain-bellied Watersnake, Rough Greensnake, Rainbow Snake, Queen Snake, Eastern Ribbonsnake, Smooth Earthsnake, Glossy Crayfish Snake, Cottonmouth, American Black Duck, Gadwall, Blue-winged Teal, Hooded Merganser, Pied-billed Grebe, American Bittern, Great Blue Heron, Black-crowned Night-Heron, Least Bittern, Wood Stork, Northern Harrier, Bald Eagle, Osprey, Peregrine Falcon, King Rail, Virginia Rail, Sora, Common Moorhen, American Coot, Spotted Sandpiper, Wilson's Snipe, Black Tern, Bank Swallow, Cliff Swallow, Sedge Wren, Marsh Wren, Prothonotary Warbler, Common Yellowthroat, Swamp Sparrow, River Otter
Control measures aimed at species presumed to cause damage to fisheries, livestock, or crops, or to threaten humans directly	Red Wolf, Gray Wolf, Bobcat, Mountain Lion, Golden Eagle, Double-crested Cormorant, Red-breasted Merganser, Common Merganser, Timber Rattlesnake
Collecting for the pet trade	Pigmy Salamander, Weller's Salamander, Spotted Turtle, Wood Turtle, Bog Turtle

[1] "Decline" means that each species listed has been classified as S1 (critically imperiled), S2 (imperiled), or S3 (vulnerable) by one or more states for species that have breeding populations in the Mid-Atlantic. For migratory species, declines have also been identified through reference to National Breeding Bird Survey data. For extinct or extirpated species, status has been assessed through the appropriate literature.

[2] Reasons for declines for many species are not known, or are attributed to more than one factor.

Terrestrial Vertebrate Biota of the Mid-Atlantic

The Mid-Atlantic region has 546 species that occur as regular members of the terrestrial vertebrate biota. For each of these species, an illustration (photo or drawing), map, and species account is provided. The illustration is of a typical, adult individual. The maps were prepared based on detailed, published information on each species. They are as accurate as available data could make them. For amphibians, reptiles, and mammals, distribution is shown county-by-county for the entire region. Readers should consider the following caveats in consulting these maps:

1) Not all states have publications that provide county-by-county distribution. In these cases, regional or species-specific sources were consulted to obtain the most accurate representation possible.

2) Some counties may be shown on the map of a particular species as included within the range when, in fact, there are large parts of the county from which the species is absent. I have attempted to correct egregious examples of this problem by drawing range lines to include only those parts where the species is known to be found. However, some errors of this type are inevitable.

3) Published records of a species do not necessarily include every county in which the species is found. This problem is very common, and is generally handled by including counties with similar habitat in the range of a species despite the lack of specific documentation. In general, I have followed that procedure here.

4) Bird maps are not shown county-by-county because, with the exception of nesting records, these kinds of data are not readily available. Also, because of the extreme mobility of birds, an individual sighting has far less meaning than a specimen record for a member of one of the other terrestrial vertebrate groups.

The maps and illustrations are presented in association with each individual species account. These accounts form the core of the work.

Organization of Species Accounts

Each account begins with the species' English name followed by the scientific name in italics. Nomenclature follows Crother (2000) for reptiles and amphibians except as explained in the text; for birds—the American Ornithologists' Union Checklist, Seventh Edition (1998) as amended by supplements (American Ornithologists' Union 2000, 2001, 2002, 2003, 2004, 2005); for mammals—Baker et al. (2003) is used. The order in which species are listed in the text follows "taxonomic order," i.e., an organization of species based on the evolutionary relationships between members of each Class as understood by experts in the field. The same experts are followed for taxonomic order as those used for nomenclature except in the case of amphibians and reptiles, where I follow the order used by Conant (1984) for the most part, because Crother (2000) lists genera and species of these Classes in alphabetical rather than taxonomic order. Average size of the species is given in metric units (with English units following in parentheses): L = total length, tip of nose or bill to tip of tail; W (birds only) = wingspan. This heading is followed by several subheadings defined as follows:

Description: A description of the average adult. For birds, the basic description is for an adult male in breeding plumage. Other age or sex category descriptions are provided where necessary. For birds, these descriptions often involve reference to special nomenclatural terms shown in Figure 15 (also referenced in the Glossary, Appendix 2). A photo or illustration is provided for all species treated.

Habitat and Distribution: The abundance (for birds), distribution, and habitat use by the species in the Mid-Atlantic region are provided in this section. Seasonal occurrence is given for species that migrate, aestivate, or hibernate. Normally only regular (common, uncommon, rare) occurrences of the species

Figure 15. Bird parts.

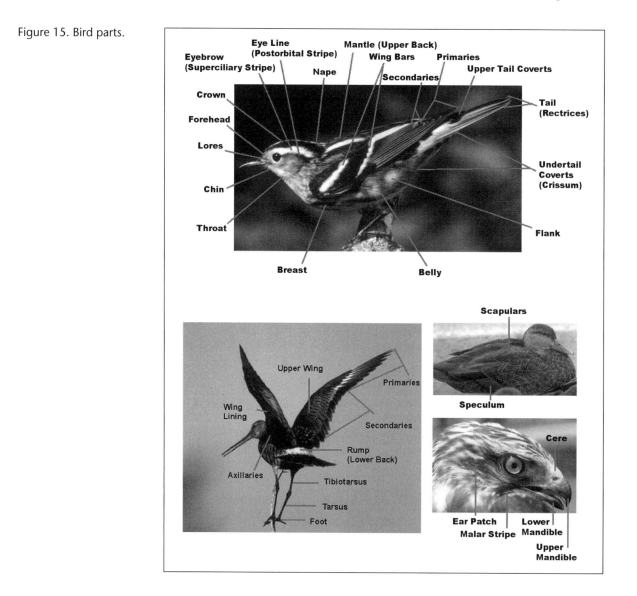

within the region are reported. All bird species that have been recorded as breeding in the region are marked with an asterisk (*) following a statement of their residency status. The meanings of the abundance categories used for birds are as follows: Common—Ubiquitous in specified habitat; high probability of finding several individuals (>5) in a day; Uncommon—Present in specified habitat; high probability of finding a few individuals (<5) in a day; Rare—Scarce in specified habitat, with only a few records per season; low probability of finding the species; Casual—A few records per decade; Accidental—Not expected to recur.

Habits: Comments are provided in this section on noteworthy aspects of ecology and life history.

Diet - Main prey items for the species in the Mid-Atlantic region.

Reproduction: For species that breed regularly in the Mid-Atlantic region, information is provided in this section on mating system, timing of mating, description of nest or den location, average clutch or litter size, age at departure from nest or den, duration of parental care, and number of broods or litters per season. Data provided are based on averages or summaries, and when not well documented, a "?" is inserted. Life cycle comments are also included in this section for amphibians.

Conservation Status: This section is included only for those species for which a particular threat has been identified within the Mid-Atlantic region or, in some cases, for species that cause conservation problems for other species. In addition to brief discussion of the threat or threats known, official threat categories are given using those provided by The Nature Conservancy (TNC), through their Explorer NatureServe web site, U.S. Fish and Wildlife Service, and each of the six states included within the region as defined: Delaware, Maryland, New Jersey, Pennsylvania, Virginia, and West Virginia.

State TNC rankings as determined by the Natural Heritage programs within each state, for those states that have such programs, are as follows (adapted from Roble 2003):

S1 Extremely rare and critically imperiled with 5 or fewer occurrences or very few remaining individuals in the designated state; or because of some factor(s) making it especially vulnerable to extinction.

S2 Very rare and imperiled with 6 to 20 occurrences or few remaining individuals in the designated state; or because of some factor(s) making it especially vulnerable to extinction.

S3 Rare to uncommon in the designated state with between 20 and 100 occurrences in the designated state; may be somewhat vulnerable to extirpation.

SH Formerly part of the designated state's fauna with some expectation that it may be redis-covered; generally applies to species that have not been verified in the designated state for an extended period (> 15 years) and for which some inventory has been attempted recently.

SX Believed to be extirpated in the designated state with virtually no likelihood of redis-covery.

Federal Status—as designated by the U. S. Fish and Wildlife Service (USFWS):

"Endangered". Threatened with extinction throughout all or a significant portion of its range.

"Threatened". Likely to become endangered in the foreseeable future.

"Extirpated". Once found in the country or state, but no populations are currently known.

Range: Total range (World) is provided in abbreviated form.

Key References: One or more references that provide critical life history information on the species are cited here.

In addition to these subheadings within the Species Accounts, a photo or drawing is provided for each species, as well as a map of the species' distribution in the Mid-Atlantic region.

Definitions for specialist terminology as well as scientific names for species mentioned in the text are given in the Glossary.

Class Amphibia—Amphibians

Amphibians are the most ancient group of terrestrial vertebrates, originating 380–400 million years ago during the late Devonian Period of the Paleozoic Era, although the fossil record documenting their evolution from fish is poor. The first amphibians known were members of the Sub-Class Labyrynthodontia, all members of which are now extinct. There are three modern orders of amphibians: salamanders (Caudata) with about 400 modern species; frogs (Anura) with about 3,800 modern species; and the worm-like, tropical caecilians (Apoda or Gymonophiona) with 170 species (Zug et al. 1993). Amphibians are characterized by glandular, permeable skin (no scales, feathers, or hair), which allows ready exchange of water and gases with their environment, but restricts most species of the group to habitats that either are extremely humid, or allow ready access to wetlands. The term "amphibian" means literally an animal that leads a double life, and, as their name implies, many amphibians have a life cycle that is divided into an aquatic egg and larval stage and a terrestrial adult stage, although more than half of amphibian species have evolved various strategies to reduce or eliminate the aquatic phase. Other characteristics of the group include lungs, 4 limbs with digits and claws (most salamanders and frogs have 4 toes on the forefeet and 5 toes on the hind feet); a 3-chambered heart (1 ventricle, 2 atria), external and internal nares, a double-loop blood circulation system (lungs and the rest of the body), and a shell-less egg. Amphibians respire using gills (aquatic forms) as well as through their skin or pumping air through their mouths (buccal respiration) into their lungs. There are 46 species of salamanders and 26 species of frogs and toads found regularly in the Mid-Atlantic region.

ORDER CAUDATA; FAMILY CRYPTOBRANCHIDAE

Hellbender *Cryptobranchus alleganiensis*

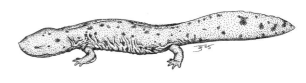

DESCRIPTION: L = 30–74 cm (12–29 in); a large, greenish to brown salamander mottled with brown; head is bluntly rounded with small, lidless eyes; short legs; a fleshy skin fold runs along the side of the body; vertically flattened, keeled tail.

HABITAT AND DISTRIBUTION: Denizen of clean, cool, shallow, fast-flowing streams and rivers with rocky bottoms in PA, WV, and western MD and VA.

HABITS: Aquatic in all life cycle stages, using its powerful arms to forage under rocks and stones on stream bottoms.

DIET: Crayfish, minnows, aquatic invertebrates.

REPRODUCTION AND LIFE CYCLE: Breeding occurs August–October or later; males dig nests under large rocks on stream bottoms; females lay 150–400 eggs, 5–7 mm (0.2 in) in diameter, in the nest, which are subsequently fertilized by the male; males guard the nest (from fish and other hellbenders that might eat the eggs) until young hatch in winter, 2 months or so after laying; newly hatched young have external gills, which they retain until attaining a length of 120 mm (5 in) or so; adults show gill slits throughout their lives.

CONSERVATION STATUS: S1—MD; S2—WV, VA; S3—PA; preservation of unpolluted streams and rivers with high oxygen content is principal concern.

RANGE: Eastern United States in the Allegheny and Ozark mountains from southern NY to northern SC, GA, AL, and MS and west to MO.

KEY REFERENCES: Mitchell (1991). Petranka (1998).

FAMILY SIRENIDAE

Lesser Siren *Siren intermedia*

DESCRIPTION: L = 13–38 cm (5–15 in); a long, thin, eel-like salamander, black, dark brown, or olive in color, with external gills, laterally compressed tail, and tiny forelegs; lacks hind legs; hatchlings (1–2 cm) (0.4–0.8 in) show red band on snout and side of head.

HABITAT AND DISTRIBUTION: Blackwater cypress swamps and ponds of VA Coastal Plain.

HABITS: Sirens communicate using click sounds made by snapping of jaws; aquatic; entire life cycle spent under water; able to store fat and aestivate for months or  years encapsulated in mucus-lined cocoons in dried mud of ephemeral pools.

DIET: Forages nocturnally in benthic mud for crayfish, annelids, molluscs, and aquatic insects.

REPRODUCTION AND LIFE CYCLE: Fertilization believed to be external; female lays eggs in spring in a hole in the pool bottom; hatching occurs two months later.

CONSERVATION STATUS: S2—VA; wetland habitat for this species is limited in the Mid-Atlantic region, and threatened by drainage and pollution.

RANGE: Coastal Plain of the eastern United States and northeastern Mexico from VA to Tamaulipas; also north along the Mississippi Valley to IL.

KEY REFERENCES: Virginia Department of Game and Inland Fisheries (2004).

Greater Siren *Siren lacertina*

DESCRIPTION: L= 0.5–0.9 m (20–35 in); a very long, thin, gray, eel-like salamander with external gills, laterally compressed tail, and tiny forelegs; lacks hind legs; hatchlings (2–3 cm) (0.8–1.2 in) show a light red or yellow stripe along the side of the body.

HABITAT AND DISTRIBUTION: Ponds, swamps, lakes, streams, and flooded ditches of VA Coastal Plain.

HABITS: Sirens communicate using click sounds made by snapping of jaws; aquatic; entire life cycle spent under water; able to store fat and aestivate for months or years encapsulated in mucus-lined cocoons in dried mud of ephemeral pools.

DIET: Forages nocturnally in benthic mud for molluscs, crayfish, aquatic insects, and fish.

REPRODUCTION AND LIFE CYCLE: Fertilization believed to be external; female lays eggs in March in several, small, dispersed clumps on water body bottom, with hatching two months later.

CONSERVATION STATUS: S3—VA; small population size threatened by wetland drainage and pollution.

RANGE: Coastal Plain of the eastern United States from VA to AL.

KEY REFERENCES: Martof et al. (1980).

FAMILY SALAMANDRIDAE

Eastern Newt *Notophthalmus viridescens*

DESCRIPTION: L = 6–11 cm (2.4–4.4 in); subadult (eft) is orange-red with scarlet, brown-rimmed spots down each side of the back; adult is dark brown or olive with scarlet spots above; yellow below; laterally compressed tail; adult breeding male has a higher tail fin than the female and develops dark pads on the legs.

HABITAT AND DISTRIBUTION: Larvae and adults live in pools, lakes, ponds, and quiet streams throughout the Mid-Atlantic; efts live on land in litter of moist forests.

HABITS: Poisonous to eat for predators, as signaled by the bright aposematic coloration of the slow-moving, rough-skinned eft, which is significantly more poisonous than adults; efts hibernate, but adults may remain active through the winter, even under ice.

DIET: Aquatic newts feed on a wide variety of vertebrate and invertebrate prey, including insect eggs, larvae, and adults, frog and salamander eggs, molluscs, annelids, crustacea, and leeches; efts eat soil invertebrates.

REPRODUCTION AND LIFE CYCLE: Mating occurs in spring (Apr–Jun) and fall (Aug–Oct); females lay several hundred eggs individually on submerged pond debris; hatching takes place a month or so after laying depending on water temperature; larvae have external gills, and develop aquatically until summer or fall, when they emerge as efts with no external gills to spend 2–8 years on land, returning to water as adults for the remainder of their life cycle; some populations are entirely

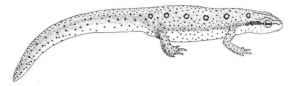

aquatic throughout their lives with no terrestrial eft stage.

RANGE: Eastern North America from southern Canada to northeastern Mexico.

KEY REFERENCES: DeGraaf and Yamasaki (2001).

FAMILY PROTEIDAE

Common Mudpuppy *Necturus maculosus*

DESCRIPTION: L = 20–50 cm (8–20 in); compact, flattened, grayish, brownish, or greenish body with dark spots; reddish filamentous external gills; 4 legs with 4 toes each; laterally compressed, keeled tail; young are brownish, often with pale lateral stripes.

HABITAT AND DISTRIBUTION: Slow-moving areas of streams, lake shores, rivers, canals, and drainage ditches at scattered localities in the western portion (Appalachian Plateau) of the region.

HABITS: Aquatic throughout life cycle; mostly nocturnal; remains active through the winter.

DIET: Forages on stream bottom for aquatic insects, crustacea, annelids, frog eggs, tadpoles, small fish, and other aquatic vertebrates and invertebrates.

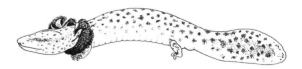

REPRODUCTION AND LIFE CYCLE: Mating occurs in the fall; 60–100 eggs are laid the following spring (May) attached singly on stalks, often under a rock or sunken log in shallow water (15–20 cm) (6–8 in); female guards eggs until they hatch, 40–60 days post-laying; sexual maturity is reached in 5–6 years.

CONSERVATION STATUS: S2—VA; S1—MD; wetland habitat for this species is threatened by drainage and pollution.

RANGE: Eastern North America west of the Piedmont from southern Canada to east TX.

KEY REFERENCES: Pauley (2005a).

Dwarf Waterdog *Necturus punctatus*

DESCRIPTION: L = 10–20 cm (4–8 in); slender, sausage-shaped body, dark brownish above and paler below; orangish filamentous external gills; 4 legs with 4 toes each; laterally compressed, keeled tail; young are uniformly brownish.

HABITAT AND DISTRIBUTION: Quiet streams with muddy or sandy bottoms in southeastern VA (e.g., Chowan River).

DIET: Small aquatic inverte-brates and vertebrates.

REPRODUCTION AND LIFE CYCLE: Entirely aquatic throughout life cycle; reaches maturity at 5 years; nests are not described, but probably located under submerged rocks or logs.

CONSERVATION STATUS: S2—VA; wetland habitat for this species is threatened by drainage and pollution.

RANGE: Coastal Plain of VA, NC, SC, and GA.

KEY REFERENCES: Martof et al. (1980).

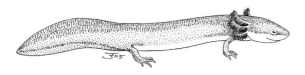

FAMILY AMPHIUMIDAE

Two-toed Amphiuma *Amphiuma means*

DESCRIPTION: L = 0.5–1.1 m (20–44 in); large, eel-like body with pointed snout and long tail; dark brown above, paler below; single pair of gill slits; tiny, useless-looking legs with two toes on each foot.

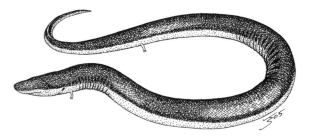

HABITAT AND DISTRIBUTION: Swamps, sloughs, drainage ditches, ponds, streams, wet forest litter of the southeastern VA Coastal Plain.

HABITS: Almost entirely aquatic, although it will slither out of the water on occasion, especially on wet nights; mostly nocturnal; bites savagely when captured.

DIET: Forages by waiting concealed in an underwater hole or crevice to seize passing crustacea, fish, insects, annelids, snakes, frogs, and other aquatic animals.

REPRODUCTION AND LIFE CYCLE: Poorly known; females lay eggs in long, attached strings in winter under debris in wet areas, and guard the nest until eggs hatch at about 5 months; aquatic larvae are 5–6 cm in length.

RANGE: Coastal Plain and neighboring Piedmont of the southeastern United States from VA to east TX, and up the Mississippi Valley to IL.

KEY REFERENCES: Petranka (1998).

FAMILY AMBYSTOMIDAE

Jefferson Salamander *Ambystoma jeffersonianum*

DESCRIPTION: L = 10–20 cm (4–8 in); medium-sized salamander; dark brown or gray body, paler below, with small bluish spots on the sides (fading with age); 4 long toes on the front legs, 5 on the back; vertically compressed tail.

HABITAT AND DISTRIBUTION: Litter, rotted logs, and debris of moist, mixed and deciduous forest and swamps in the Ridge and Valley and Appalachian Plateau provinces of the Mid-Atlantic to southern VA; breeds in forest pools, ponds, and swamps.

HABITS: Hibernates in forest duff in winter (Nov–Feb); nocturnal; secretes mucus when attacked.

DIET: Forages in litter for small invertebrates, e.g., insects, millipedes, and spiders.

REPRODUCTION AND LIFE CYCLE: Adults migrate to breeding pools on rainy nights in spring (Feb–Apr); males deposit spermatophores on underwater leaves, logs, twigs, or rocks, which the female nips off and picks up with her vent (internal fertilization); female lays eggs in several cylindrical clumps of 30 or so eggs each attached to vegetation just below the water surface; the eggs hatch in 13–45 days; larvae remain aquatic for 56–125 days when they become terrestrial; breeding age is reached at about 18 months.

CONSERVATION STATUS: S3—NJ, MD; wetland breed-

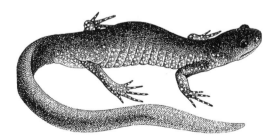

ing habitat for this species is threatened by drainage and pollution.

RANGE: Northeastern North America from Labrador and central Manitoba south to IL, IN, KY, and VA.

KEY REFERENCES: Martof et al. (1980), DeGraaf and Yamasaki (2001).

Blue-spotted Salamander *Ambystoma laterale*

DESCRIPTION: L = 10–13 cm (4–5 in); medium-sized salamander; black or bluish-black body with bright white or blue spots, mostly on the sides and legs; head is narrow in comparison with *A. jeffersonianum*.

HABITAT AND DISTRIBUTION: Litter, rotted logs, and debris of moist, mixed and deciduous forest and swamps with sandy or loamy soils; found only in northern NJ in the Mid-Atlantic region; breeds in forest pools, ponds, and swamps.

HABITS: Hibernates in forest duff in winter (Nov–Feb), but can be active on warm days; nocturnal; secretes mucus when attacked.

DIET: Forages in litter for small invertebrates, e.g., insects, millipedes, and spiders.

REPRODUCTION AND LIFE CYCLE: Adults migrate to breeding pools on rainy nights in spring (Mar–Apr); males deposit spermatophores on underwater leaves, logs, twigs, or rocks, which the female nips off and picks up with her vent (internal fertilization); female lays eggs singly or in small loose clumps attached to rocks, sticks, or other bottom debris; eggs hatch in May–Jun;

larvae remain aquatic for 2–3 months.

CONSERVATION STATUS: S1—NJ; wetland breeding habitat for this species is threatened by drainage and pollution.

RANGE: Northeastern North America from Labrador and eastern Manitoba south to IA, IN, IL, and NJ.

KEY REFERENCES: Petranka (1998), Hulse et al. (2001).

Mabee's Salamander *Ambystoma mabeei*

DESCRIPTION: L = 8–12 cm (3–5 in); dark brown mole salamander with whitish spots and flecks on back and sides; small head; long, slender toes (4 on front legs, 5 on back); laterally compressed tail.

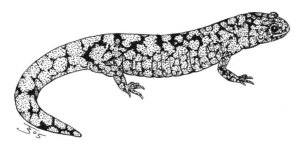

HABITAT AND DISTRIBUTION: Adults live in forest litter, holes, and burrows in mixed bottomland forests of Southampton, Gloucester, Isle of Wight, Suffolk, and York counties in the Coastal Plain of southeastern VA; they breed in vernal forest pools and ephemeral sinkhole ponds.

HABITS: Poorly known.

DIET: ?

REPRODUCTION AND LIFE CYCLE: Adults migrate to breeding ponds during rains in fall and winter; females lay light brown eggs singly or in small clumps (2–6 eggs) attached to leaves or twigs of emergent vegetation; eggs hatch in 9–14 days; larvae meta-

morphose and leave ponds in late spring at 5–6 cm in size.

CONSERVATION STATUS: S1—VA; chief threats are the small number of known breeding sites and wetland drainage resulting from urbanization.

RANGE: Coastal Plain of VA, NC, SC.

KEY REFERENCES: Mitchell (1991).

Spotted Salamander *Ambystoma maculatum*

DESCRIPTION: L = 15–25 cm (6–10 in); a short, thick-bodied mole salamander; black above, paler gray below; two irregular rows of large yellow or orange spots down the back.

HABITAT AND DISTRIBUTION: Litter of moist mixed forest, usually within 200 m (700 ft) of slow streams, ponds, or marshes nearly throughout the region; breeds in forest pools and ponds from which fish are absent.

HABITS: Fossorial, burrowing in leaf litter and humus of forest floor; nocturnal; hibernates in winter.

DIET: Terrestrial adults feed on insects, snails, worms, slugs, spiders, and other soil and litter invertebrates; larvae feed on zooplankton.

REPRODUCTION AND LIFE CYCLE: Adults migrate to breeding ponds on warm (> 10º C) rainy nights in spring (Mar–Apr); females lay eggs in 2–3 mucus-encapsulated masses with 100 or so eggs in each, which are attached to emergent vegetation several cm below the water surface; eggs hatch in 30–50 days; larvae transform to terrestrial habit 60–110 days post-hatching.

CONSERVATION STATUS: S2—DE; S3—NJ; limited wetland breeding habitat for this species is threatened by drainage and pollution.

RANGE: Eastern North America from southern Canada to the northern Gulf coast.

KEY REFERENCES: DeGraaf and Yamasaki (2001).

Marbled Salamander *Ambystoma opacum*

DESCRIPTION: L = 9–13 cm (3.6–5.2 in); a thick-bodied mole salamander; black with white cross bars and blotches, brighter in male than female; blunt rounded tail.

HABITAT AND DISTRIBUTION: Nonbreeding adults frequent drier habitats than other mole salamanders, including under debris, litter, logs, rocks, and other detritus in upland mixed forests and woodlands with sandy or gravelly soils; found nearly throughout the region except northern and central PA and most of northeastern WV and western MD; scarcer in highlands; breeding adults frequent moist, sandy areas bordering swamps, ponds, and streams.

HABITS: Nocturnal; secretes mucus when attacked.

DIET: Adults feed on soil invertebrates, e.g., insects, crustacea, molluscs, and annelids; aquatic larvae feed on zooplankton when small, and on other aquatic invertebrates as they grow.

REPRODUCTION AND LIFE CYCLE: Adults migrate to breeding sites in fall (Sep–Oct); unlike most other salamanders, the breeding sites are not wetlands,

rather they are depressions, hollows, or dry beds of temporary ponds likely to fill with water after rains; 50–200 eggs are laid singly under debris in these depressions, and are often guarded by the female until the eggs are submerged; once submerged, eggs hatch within a few hours or days; larvae hibernate over winter and metamorphose to terrestrial habit in spring (Apr–May); sexual maturity is reached at 15–18 months.

CONSERVATION STATUS: S3—PA, DE, NJ; limited wetland breeding habitat for this species is threatened by drainage and pollution.

RANGE: Eastern United States from NY and IL south to the northern Gulf coast.

KEY REFERENCES: Martof et al. (1980), DeGraaf and Yamasaki (2001).

Mole Salamander *Ambystoma talpoideum*

DESCRIPTION: L = 8–12 cm (3–5 in); thick, compact body with large, flattened head and laterally compressed tail; grayish-brown with bluish white dots; legs appear too large for the body.

HABITAT AND DISTRIBUTION: Soil, humus, and litter of mixed and deciduous forests, swamps, bottomlands, and pine flatwoods of Charlotte and Campbell counties, VA; breeds in woodland ponds and ephemeral pools.

HABITS: Fossorial as terrestrial adults; some larvae become terrestrial due to drying up of ponds prior to completion of metamorphosis (paedomorphosis), retaining external gills even as sexually mature adults; others remain permanently aquatic; larvae are nocturnal, remaining hidden in benthic pond vegetation during the day, and moving up the water column to feed at night.

DIET: Adult diet is unknown but presumably includes soil invertebrates; larvae feed on zooplankton, especially copepods and cladocera.

REPRODUCTION AND LIFE CYCLE: Adults migrate to breeding ponds in late fall, leaving to become terrestrial again by mid-May; females lay eggs singly or in small clumps attached to pond vegetation; larvae hatch in spring and metamorphose, apparently facultatively as pools dry up, in late summer.

CONSERVATION STATUS: S1—VA; there is only one known breeding pond in VA, so the species appears vulnerable to extirpation.

RANGE: Coastal Plain of the southeastern United States from SC to northern FL and LA; also scattered localities in VA, NC, TN, IL, KY, MO, and OK.

KEY REFERENCES: Mitchell (1991).

Tiger Salamander *Ambystoma tigrinum*

DESCRIPTION: L = 18–25 cm (7–10 in); large mole salamander, dark brown above with yellow or yellowish-brown spots; almost entirely yellowish below and on laterally compressed tail with some dark mottling.

HABITAT AND DISTRIBUTION: Loose, sandy or loamy soil in pine savanna, open woodlands, and old fields; breeds in vernal pools, flooded sinkholes, gravel pits, and farm ponds that are unpolluted and without fish; Augusta, Mathews, and York counties, VA; Charles, Dorchester, and Somerset counties, MD; Atlantic and Cumberland counties, NJ; also recorded for DE.

HABITS: Adults are fossorial, burrowing in litter, soil, and manure piles; larvae are nocturnal.

DIET: Terrestrial adults feed on soil invertebrates; larvae feed on zooplankton, other aquatic invertebrates, and vertebrates (e.g., frog and salamander eggs and larvae).

REPRODUCTION AND LIFE CYCLE: Adults migrate to breeding ponds in late winter (Jan–Mar); eggs are laid in clumps attached to submerged vegetation, logs, or sticks and hatch in about a month; larvae transform to terrestrial adults in May–Jul.

CONSERVATION STATUS: S1—VA, DE; S2—NJ, MD; SX—PA; main threats are to breeding ponds from drainage, fish-stocking, and alteration of ground water levels resulting from agricultural and urban use.

RANGE: Much of North America from central Canada to northern Mexico, though absent from highlands.

KEY REFERENCES: Mitchell (1991).

FAMILY PLETHODONTIDAE

Green Salamander *Aneides aeneus*

DESCRIPTION: L = 8–14 cm (3–5.6 in); flattened body is grayish or brownish covered dorsally with yellowish-green blotches; long legs; toe tips are expanded and squared off.

HABITAT AND DISTRIBUTION: Damp crevices in rocky outcrops, talus slopes and scree, crags, and cliffs; under bark of trees and rotted stumps and logs; highlands of WV, western VA, and MD, and southwestern PA.

HABITS: Mostly nocturnal; hibernates (Nov–Mar).

DIET: Terrestrial invertebrates, e.g., insects, spiders, snails, and slugs.

REPRODUCTION AND LIFE CYCLE: Breeds in damp crevices in late spring, summer, or early fall;

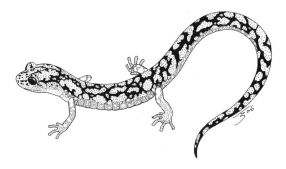

females lay 10–30 eggs in clusters on crevice walls and guard them until they hatch in 80–90 days; young are like miniature adults (lack gills, not aquatic).

CONSERVATION STATUS: S1—PA; S2—MD; S3—WV, VA; destruction and pollution of highland old growth forest threaten small, isolated populations.

RANGE: Highlands of AL, GA, SC, TN, NC, VA, MD, KY, and WV; also southern OH and IN.

KEY REFERENCES: Martof et al. (1980), Petranka (1998).

Southern Dusky Salamander
Desmognathus auriculatus

DESCRIPTION: L = 8–16 cm (3–6 in); rounded (not flattened) body; laterally compressed, keeled tail; short, blunt snout; dark brown body with lateral white spots; back legs noticeably longer than front.

HABITAT AND DISTRIBUTION: Terrestrial adults are found in leaf litter and other detritus of bottomland forests, swamps, bayous, and quiet stream borders; southeastern VA; larvae are found in

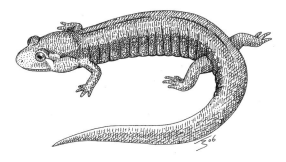

blackwater swamps, pools, stagnant streams, and other bottomland wetlands.

HABITS: Nocturnal; hibernates/aestivates.

DIET: Adults feed on spiders, insects, worms, and other soil invertebrates; larvae feed on zooplankton and other aquatic invertebrates.

REPRODUCTION AND LIFE CYCLE: Breeding takes place on land in Sep–Oct; 9–20 eggs are laid in debris, moss, or under logs on forest floor immediately adjacent to water body, and guarded until hatching; newly hatched larvae enter water and remain in aquatic phase until late spring when they transform and begin terrestrial existence as adults.

RANGE: Coastal Plain of the southeastern United States from VA to east TX.

KEY REFERENCES: Martof et al. (1980), Petranka (1998).

Northern Dusky Salamander
Desmognathus fuscus

DESCRIPTION: L = 6–14 cm (2.4–5.6 in); a relatively stout-bodied salamander; hind legs larger and heavier than front; variable in coloration—various shades of gray or brown with a distinct dark lateral line of spots along back, fading in older specimens, which may be uniformly dark; whitish line extends from the eye to the jaw; tail is keeled and basal third of tail is flattened laterally

HABITAT AND DISTRIBUTION: Under stones, rocks, and woody debris along springs, brooks, woodland streams, or moist woodlands; found nearly throughout the region except southeastern VA, the southern Delmarva Peninsula, and southern NJ.

HABITS: Nocturnal; larvae are aquatic; juveniles and adults are terrestrial, foraging in moist or wet woodlands or stream-border detritus.

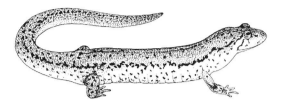

DIET: Feeds mostly on insects, although other invertebrates are also taken.

REPRODUCTION AND LIFE CYCLE: Breeds in spring and fall along streams, pools, and pond borders; eggs are laid in stalked clusters of 8–28 in depressions under moss, leaves, or rocks, or logs near water (< 0.5 m); the female guards the eggs until they hatch in 5–10 weeks; larvae move into nearby water body; transformation occurs in 7–12 months; juveniles become sexually mature adults at 2–5 years of age.

RANGE: Eastern United States from ME to SC exclusive of the Coastal Plain; west to MO.

KEY REFERENCES: Petranka (1998), DeGraaf and Yamasaki (2001).

Shovel-nosed Salamander *Desmognathus marmoratus*

DESCRIPTION: L = 9–14 cm (3.6–5.6 in); a large, stocky salamander; body color is variable, but often dark brown above and pale below with either 2 rows of dorsal blotches or irregular light markings; snout is flattened and wedge-shaped; tail is keeled.

HABITAT AND DISTRIBUTION: Mountain rills, runs, and creeks with stony bottoms, most often in riffles; known from 4 tributary streams of Laurel Creek on Whitetop Mountain; Washington, Smythe, and Grayson counties, VA.

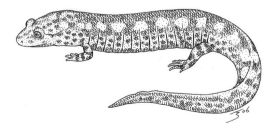

HABITS: Nocturnal; aquatic in all life stages.

DIET: Feeds mainly on aquatic insects.

REPRODUCTION AND LIFE CYCLE: Breeds in summer (Jun–Jul); female attaches eggs to underside of rocks in streambed and guards them until hatching occurs in 2–3 months; metamorphosis in 1–3 years.

CONSERVATION STATUS: S2—VA; potential destruction of limited stream habitat is the principal threat.

RANGE: GA, SC, NC, TN, and VA.

KEY REFERENCES: Mitchell (1991).

Seal Salamander *Desmognathus monticola*

DESCRIPTION: L = 8–15 cm (3–6 in); long body; front legs smaller than back; buff, gray, or light brown in color with dark blotches on the back and tiny white spots on the sides and belly; distal half of long tail is laterally compressed and keeled.

HABITAT AND DISTRIBUTION: Adults frequent boggy edges and borders of brooks, streams, and creeks, where they can be found under rocks, logs, and litter of hemlock, mixed, and other montane forests during the day; larvae are aquatic; highland regions of VA, WV, MD, and PA.

HABITS: Nocturnal.

DIET: Adults feed on soil and litter invertebrates and small vertebrates (e.g., other salamanders); larvae feed on zooplankton and other aquatic invertebrates.

REPRODUCTION AND LIFE CYCLE: Breeds in early summer; females lay clusters of 15–40 eggs in depressions under logs or rocks on stream banks or in streams, and guard the nest until the young hatch in late summer; larvae enter nearby water body where they remain for 9–10 months; after

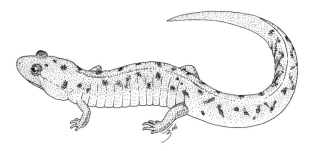

transformation, juveniles require 2–7 years to reach sexual maturity.

RANGE: Appalachians and neighboring Piedmont from western PA to GA; also parts of western GA and central and southern AL.

KEY REFERENCES: Martof et al. (1980), Petranka (1998).

Allegheny Mountain Dusky Salamander *Desmognathus ochrophaeus*

DESCRIPTION: L = 7–11 cm (3–4.4 in); long body with long, tapered tail; front legs smaller than back; body color is variable from dark brown to beige or gray with lighter splotches or spots dorsally and tiny spots laterally and on the belly; northern VA specimens often show a light dorsal band.

HABITAT AND DISTRIBUTION: Adults are found in wet litter, crevices, or seepage areas, often under logs or other debris, in spruce-fir and mixed montane forest; aquatic larvae use forest streams and brooks; highlands of VA, WV, MD, PA, and NJ.

HABITS: More active at night than during the day; hibernates in winter.

DIET: Adults feed on soil and litter invertebrates, including fly and beetle eggs and larvae and

worms; aquatic larvae eat zooplankton and other small aquatic invertebrates.

REPRODUCTION AND LIFE CYCLE: Breeding occurs in spring or summer; female lays eggs in clusters of 5–40 in a wet crevice or mossy seepage, which she guards until hatching in 6 weeks; hatchlings move into nearby water body where they live as aquatic larvae until transformation at 2–8 months.

CONSERVATION STATUS: SH—NJ; limited wetland larval habitat for this species is threatened by drainage and pollution.

RANGE: Highland and boreal regions of eastern North America from Quebec to TN; except Ontario and New England.

KEY REFERENCES: Martof et al. (1980), Petranka (1998).

Blue Ridge Dusky Salamander *Desmognathus orestes*

DESCRIPTION: L = 7–11 cm (3–4.4 in); long body with long, tapered tail; front legs smaller than back; body color is variable from light brown to beige with lighter splotches or spots dorsally and tiny spots laterally and on the belly. This species was formerly considered conspecific with the Allegheny Mountain Dusky Salamander (*Desmognathus ochrophaeus*) until separated based on allozyme analysis (Tilley and Mahoney 1996). It is very similar in appearance to *D. ochrophaeus*, but tends to have a wavy, irregular dorsal stripe, as opposed to the relatively straight and distinct dorsal stripe of *D. ochrophaeus*.

HABITAT AND DISTRIBUTION: Adults can be found under rotted logs, rocks, and forest debris in montane coniferous and mixed forest; breeding sites are wet crevices, rocks, logs, or mossy seepages adjacent to

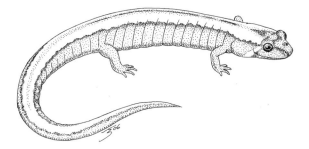

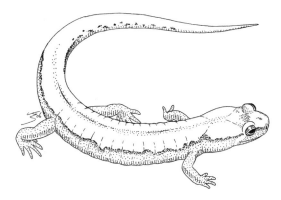

wetlands in the Blue Ridge Province of southwestern VA; aquatic larvae live in bogs, springs, and forest pools.

HABITS: Adults are more active at night, but will forage on rainy days; hibernates in winter.

DIET: Adults feed on soil and litter invertebrates; aquatic larvae eat zooplankton and other aquatic invertebrates.

REPRODUCTION AND LIFE CYCLE: Presumably similar to *D. ochrophaeus*.

CONSERVATION STATUS: S3—VA; limited wetland larval habitat for this species is threatened by drainage and pollution.

RANGE: Blue Ridge Mountain area of western VA, NC, and TN.

KEY REFERENCES: Petranka (1998), NatureServe (2005).

Black-bellied Salamander
Desmognathus quadramaculatus

DESCRIPTION: L = 10–21 cm (4–8 in); a heavy-bodied salamander with laterally compressed, keeled tail; large, bulging eyes; body is dark dorsally with

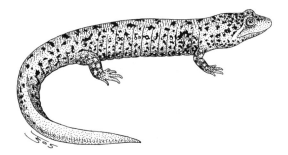

greenish-yellow splotches; tiny light spots in a lateral row from armpit to groin; belly black, sometimes splotched with yellowish-white; throat paler.

HABITAT AND DISTRIBUTION: Found in or along mountain cataracts and rocky streams in the Blue Ridge and Ridge and Valley provinces of VA and WV > 1,000 m (3,300 ft).

HABITS: Adults are mainly nocturnal and aquatic foragers, but occasionally bask exposed on wet rocks in spray zone of rapids and waterfalls.

DIET: Aquatic invertebrates.

REPRODUCTION AND LIFE CYCLE: Breeds in summer (Jun–Jul); female deposits 20–60 eggs singly or in a cluster under a stream-bed rock or tree root, which she guards until hatching in Aug–Sep; aquatic larvae metamorphose in 2–4 years; sexual maturity is reached at 7–10 years.

CONSERVATION STATUS: S3—WV; limited, wetland larval habitat for this species is threatened by drainage and pollution.

RANGE: Mountains of GA, NC, SC, TN, VA, and WV.

KEY REFERENCES: Martof et al. (1980), NatureServe (2005).

Black Mountain Salamander
Desmognathus welteri

DESCRIPTION: L = 8–17 cm (3–6.7 in); a large salamander; front legs smaller than back; brown or tan in color, usually with dark spots or blotches on the back; vent yellowish; black-tipped toes; formerly considered a subspecies of *Desmognathus fuscus*.

HABITAT AND DISTRIBUTION: Adults live under rocks and logs or in crevices in the splash zone of mountain cataracts; larvae and juveniles inhabit brooks, seeps, springs, puddles; McDowell, Mercer,

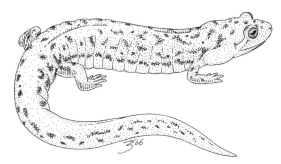

Summers, and Wyoming counties in WV; Scott, Lee Wise, Dickinson, Buchanan, Russell, Smyth, and Washington counties in VA.

HABITS: Nocturnal.

DIET: Aquatic insects, worms, crustacea.

REPRODUCTION AND LIFE CYCLE: Breeds Mar–Nov; female lays a clutch of 20–30 eggs, and curls herself around them to brood until they hatch; prolonged larval stage of 2 years before transformation.

CONSERVATION STATUS: S2—WV; S3—VA; limited wetland habitat for this species is threatened by drainage and pollution.

RANGE: KY, TN, WV, and VA.

KEY REFERENCES: Martof et al. (1980), NatureServe (2005).

Pigmy Salamander *Desmognathus wrighti*

DESCRIPTION: L = 4–5 cm (1.6–2 in); a small salamander; chestnut, bronze, or reddish-brown in color with a light dorsal stripe and dark chevrons down the back; light stripe from eye to mouth.

HABITAT AND DISTRIBUTION: Lives under leaf litter, bark, moss, logs, stumps, and rocks in highland spruce-fir and mixed forests; moves to burrows and crevices of springs and seeps during the

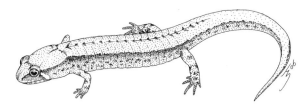

winter; sometimes climbs up tree trunks on rainy nights; Mt. Rogers, Whitetop, and Pine mountains of Grayson, Washington, and Smith counties, VA.

HABITS: Nocturnal; often fossorial; hibernates in winter.

DIET: Small arthropods, e.g., mites and gnats.

REPRODUCTION AND LIFE CYCLE: Breeds in fall, winter, or spring; male uses his mouth to grasp and hold the female during mating, which can last several hours; female lays eggs in underground seeps, and remains with them until hatching; no larval stage; young are born similar to adults with no gills.

CONSERVATION STATUS: S2—VA; threats derive from destruction of spruce-fir forest from acid rain and spruce budworm, and from herp collectors.

RANGE: NC, TN, VA.

KEY REFERENCES: Mitchell (1991).

Two-lined Salamander *Eurycea bislineata*

DESCRIPTION: L = 6–12 cm (2.4–4.8 in); yellowish, bronze or tan body with two dark lines down the side of the back and a series of dark spots down the center; some authors recognize populations from southern VA as a separate species, the Blue Ridge Two-lined Salamander (*E. b. wilderae*). The treatment here follows Petranka (1998), which includes this form along with the Northern Two-lined Salamander (*E. b. bislineata*) as subspecies of the Two-lined Salamander.

HABITAT AND DISTRIBUTION: Under rocks, logs, and other debris along or in streams, springs, pools, and brooks of mixed and deciduous woodlands nearly throughout the region.

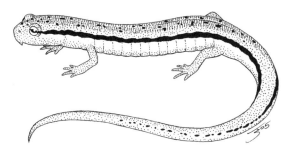

HABITS: Adults are nocturnal, and mainly terrestrial, based on diet; shrews evidently find them distasteful (poisonous to eat?); larvae forage on stream bottoms.

DIET: Adults feed on small arthropods and worms; larvae feed on aquatic arthropods, incorporating fish, frog, and salamander eggs and young in their diet as they grow and approach transformation.

REPRODUCTION AND LIFE CYCLE: Courtship occurs in the fall; eggs are laid in winter and spring under a rock or log, often in running water with several females often using the same site; at least one female remains with the eggs until hatching in 4–10 weeks; aquatic larvae require 2–3 years before transformation to subadult stage; sexual maturity occurs in second fall after metamorphosis.

CONSERVATION STATUS: S2—VA; limited wetland habitat for this species is threatened by drainage and pollution.

RANGE: Northeastern North America from southern Quebec to northern VA, west to OH.

KEY REFERENCES: DeGraaf and Yamasaki (2001).

Three-lined Salamander *Eurycea guttolineata*

DESCRIPTION: L = 9–20 cm (3.5–7.8 in); long-tailed salamander; yellowish or tan with a dark median stripe and two broad, dark stripes along the

sides; belly mottled with gray; formerly considered conspecific with the Long-tailed Salamander, *Eurycea longicauda*.

HABITAT AND DISTRIBUTION: Under rocks and logs bordering ponds, streams, ditches, and vernal pools of floodplain forests in the Piedmont and Coastal Plain of VA.

HABITS: Nocturnal; hibernates.

DIET: Adults feed on small soil and litter invertebrates, e.g., insects, snails, and worms; aquatic larvae presumably feed on zooplankton and other

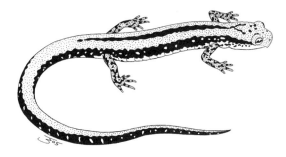

aquatic invertebrates.

REPRODUCTION AND LIFE CYCLE: Breeding occurs in the fall in forest seeps, springs, and ponds; eggs evidently are laid in holes, cavities, or burrows, since few have been found; newly hatched aquatic larvae occur in March in pools and ponds, transform to terrestrial subadults in 4–5 months, and reach sexual maturity the following year.

RANGE: Southeastern United States from VA and KY to FL and LA.

KEY REFERENCES: NatureServe (2005).

Long-tailed Salamander *Eurycea longicauda*

DESCRIPTION: L = 9–20 cm (3.5–7.9 in); long, slender salamander with tail nearly 2/3 length of body; yellowish or reddish body with a median line of dark spots and lateral lines of spots and blotches forming vertical bars on the tail.

HABITAT AND DISTRIBUTION: Under litter, logs, rocks, stones, and shale of stream banks, springs, and

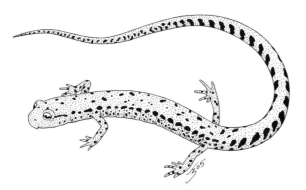

limestone caves in woodlands of the Blue Ridge, Ridge and Valley, and Appalachian Plateau provinces of the region, although possibly absent from NJ; aquatic larvae are found in forest ponds, streams, sink holes, mines, and springs with cool, slow-moving water.

HABITS: Nocturnal; hibernates.

DIET: Adults feed on soil and litter invertebrates; aquatic larvae feed on zooplankton and small aquatic invertebrates.

REPRODUCTION AND LIFE CYCLE: Breeds in ponds and streams in winter; eggs are laid in burrows, cavities, or under rocks in streams and pools or attached to rocks or boards suspended above water in caves and mines; aquatic larvae hatch in winter; transformation to subadult stage occurs in early summer, and sexual maturity is reached the following summer.

CONSERVATION STATUS: S1—DE; limited wetland habitat for this species is threatened by drainage and pollution.

RANGE: Eastern United States from NY south to FL and west to KS and OK.

KEY REFERENCES: Martof et al. (1980), Virginia Department of Game and Inland Fisheries (2004), NatureServe (2005).

Cave Salamander *Eurycea lucifuga*

DESCRIPTION: L = 12–18 cm (4.7–7.1 in); a slender salamander, flattened below, with a long, prehensile tail; broad head with bulging eyes; orangish, red, or yellowish in color with numerous black spots, sometimes in rows, on the back.

HABITAT AND DISTRIBUTION: Adults are found on walls and ledges in or near the cave mouths of limestone karst regions in the Ridge and Valley Province of eastern WV and western VA; also under rocks and logs along stony brooks and streams in wet periods; aquatic larvae are found in cave pools.

HABITS: Active day or night in caves, but chiefly nocturnal on forays outside caves.

DIET: Adults feed on small soil and litter invertebrates, e.g., insects, spiders, isopods, mites, earthworms, and snails; aquatic larvae feed on zooplankton and small aquatic invertebrates.

REPRODUCTION AND LIFE CYCLE: Breeds in fall and winter; female attaches 50–90 eggs to the underside of a rock or ledge over, in, or near a cave pool or rocky stream; young hatch in winter and metamorphose 12–18 months later; sexual maturity is achieved the following summer.

CONSERVATION STATUS: S3—WV; limited wetland habitat for this species is threatened by drainage and pollution.

RANGE: Limestone karst regions of the eastern United States from IN south to AL and VA west to OK.

KEY REFERENCES: NatureServe (2005).

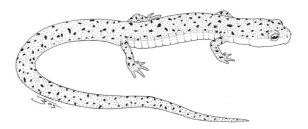

Spring Salamander *Gyrinophilus porphyriticus*

DESCRIPTION: L = 12–22 cm (4.7–8.7 in); a rather large, stocky salamander, rounded on the sides and flattened below; body is brown, orangish, or yellowish with indistinct dark spotting on the back and sides; subadults may have purplish ground color; parallel light and dark stripes from eye to nostril; truncate snout; top of tail is keeled.

HABITAT AND DISTRIBUTION: Adults are found under stones, logs, rocks, and other debris bordering mountain streams, fens, and springs in coniferous and mixed forest at scattered highland localities throughout the region.

HABITS: Nocturnal; shrews find them distasteful (poisonous to eat?).

DIET: Adults feed on insects, worms, centipedes, crayfish, isopods, other salamanders, and a variety of other small soil invertebrates and vertebrates; aquatic larvae feed on zooplankton and small aquatic invertebrates.

REPRODUCTION AND LIFE CYCLE: Breeds in fall and winter; female lays 20–60 eggs in spring and summer in a benthic burrow or cavity or attached under rocks in streams, springs, bogs, fens, or pools; female guards nest until the young hatch in the fall; aquatic larvae transform as much as 4 years post-hatching, and reach sexual maturity the following year.

RANGE: Eastern North America from Ontario and Quebec to GA and MS.

KEY REFERENCES: DeGraaf and Yamasaki (2001), NatureServe (2005).

Four-toed Salamander *Hemidactylium scutatum*

DESCRIPTION: L = 5–10 cm (2.0–3.9 in); head and body are brownish above, heavily mottled with dark spots; white belly with black spotting; head is flattened; 4 toes on front *and* back feet (5 on back feet in most salamanders); tail is constricted at the base.

HABITAT AND DISTRIBUTION: Found in vegetation of forested wetlands bordering fens, bogs, swamps, seeps, and ponds, especially sphagnum, probably throughout the region (few records from western WV).

HABITS: Hibernates.

DIET: Adults feed on moss-inhabiting invertebrates, e.g., insects, spiders, slugs, worms, centipedes, and snails; aquatic larvae feed on zooplankton.

REPRODUCTION AND LIFE CYCLE: Breeds in fall or early spring; female lays 30–50 eggs under damp pool-side vegetation, often in communal nests in sphagnum moss; one or more females remain with the eggs until hatching in 2–3 months; aquatic larvae enter pools where they remain until transformation in 6–18 weeks; subadults reach sexual maturity 1.5–2.5 years later.

CONSERVATION STATUS: S1—DE; S3—NJ; draining and pollution of limited bogland habitat are the principal threats.

RANGE: Eastern North America.

KEY REFERENCES: DeGraaf and Yamasaki (2001), NatureServe (2005).

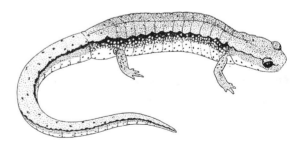

Eastern Red-backed Salamander *Plethodon cinereus*

DESCRIPTION: L = 6–13 cm (2.4–5.1 in); a small, slender, somewhat flattened salamander; two color phases: 1) broad reddish, orange, or gray stripe down the back bordered by dark lateral pigment, or 2) uniformly gray or black..

HABITAT AND DISTRIBUTION: Rotten logs, leaf litter, rocks, and other detritus in deciduous, coniferous, or mixed woodlands nearly throughout the region except western WV and parts of southern VA.

HABITS: Terrestrial in all stages of the life cycle; nocturnal; hibernates.

DIET: Forest soil and litter invertebrates, e.g., insects, spiders, molluscs, and mites.

REPRODUCTION AND LIFE CYCLE: Breeds in fall or spring; female lays a small clutch (1–15 eggs), which she guards (sometimes accompanied by male) until hatching in 6–9 weeks; young pass through metamorphosis in the egg and emerge as small, terrestrial subadults; sexual maturity reached in 2–3 years.

RANGE: Eastern North America from central Canada south to TN and NC and west to MN.

KEY REFERENCES: Martof et al. (1980), DeGraaf and Yamasaki (2001), NatureServe (2005).

Slimy Salamander *Plethodon glutinosus*

DESCRIPTION: L = 12–21 cm (4.7–8.3 in); a large salamander with tubular body and flattened head; long legs; black with silvery, white, or golden spots above; slate below. This species has been split into several species by some taxonomists (Highton et al. 1989) on the basis of allele frequencies unsubstantiated by morphological, ecological, life history, or reproductive information. Treatment of the group as a single, polytypic species, as suggested by Petranka (1998), is followed here (includes Northern Slimy Salamander *Plethodon glutinosus*, White-spotted Slimy Salamander *Plethodon cylindraceus*, and Atlantic Coast Slimy Salamander *Plethodon chlorobryonis* from our region).

HABITAT AND DISTRIBUTION: Duff, litter, rotten logs, manure piles, shale piles, and rock crevices, especially along stream banks and ravines in deciduous, mixed, or, occasionally, coniferous forest or second growth throughout most of the region except for southern NJ, MD, and the Delmarva Peninsula.

HABITS: Produces copious, sticky secretions when attacked; active both day and night, though more often at night; entirely terrestrial throughout life cycle; hibernates underground in winter.

DIET: Feeds on soil and litter invertebrates, e.g., insects (mainly ants and beetles in one study), slugs, spiders, snails, and centipedes.

REPRODUCTION AND LIFE CYCLE: Breeding appears to occur in fall and spring in the north; Martof et al. (1980) state that breeding occurs in late summer or fall in the south along the Coastal Plain, but only every other year in spring in southern mountains; eggs are laid in burrows, rocky crevices, cave walls, and rotted logs, and protected by the female (she curls around them) until hatching 2–3 months later; metamorphosis occurs in the egg, so young are born as terrestrial subadults, reaching sexual maturity at 4–5 years of age.

RANGE: Eastern United States.

KEY REFERENCES: Martof et al. (1980), DeGraaf and Yamasaki (2001), Virginia Department of Game and Inland Fisheries (2004), NatureServe (2005).

Valley and Ridge Salamander
Plethodon hoffmani

DESCRIPTION: L = 8–14 cm (3.1–5.5 in); a small, slim, short-legged, long-tailed salamander; black body (sometimes brown or tan) with white or brassy flecks. This species is part of the Ravine Salamander (*Plethodon richmondi*) superspecies complex, and was considered conspecific with that taxon prior to 1972; some taxonomists recognize the *P. hoffmani* from the mountains of northwestern VA and neighboring eastern WV as a separate species, *P. virginia*, the Shenandoah Mountain Salamander.

HABITAT AND DISTRIBUTION: Lives under rotted logs, litter, rocks, and in crevices in mature hardwood

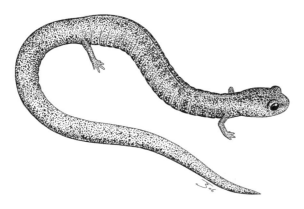

forests of the Ridge and Valley Province of VA, WV, MD, and PA.

HABITS: Nocturnal; hibernates; life history is not well-studied.

DIET: Forest litter invertebrates.

REPRODUCTION AND LIFE CYCLE: Breeding probably occurs in the spring; female lays a clutch of 3–8 eggs, which she protects until hatching in 2–3 months; young metamorphose in the egg, and hatch as subadults, reaching sexual maturity at 2–3 years.

RANGE: Ridge and Valley Province of VA, WV, MD, and PA.

KEY REFERENCES: Martof et al. (1980), NatureServe (2005).

Peaks of Otter Salamander
Plethodon hubrichti

DESCRIPTION: L = 8–12 cm (3.1–4.7 in); a small, thin salamander with relatively long legs and bulging eyes; black body with gold or greenish blotches and spots forming an irregular dorsal band; gray or black below with some mottling. Recognition as a species separate from the closely related Cheat Mountain Salamander (*Plethodon nettingi*) is open to question (Petranka 1998).

HABITAT AND DISTRIBUTION: Hides under litter, logs, and rocks; climbs into ferns and other understory vegetation while foraging in mature cove hardwood forest, predominantly on north-facing slopes in the mountains of Bedford, Botetourot, and Rockbridge counties of VA.

HABITS: Nocturnal.

DIET: Forest litter invertebrates?

REPRODUCTION AND LIFE CYCLE: Poorly known; eggs laid in May or June; clutch size 10?, presumably laid under rocks or decaying logs; metamorphosis occurs during the egg stage, and young hatch as subadults.

CONSERVATION STATUS: S2—VA; restricted range and poor understanding of life history.

RANGE: Bedford, Botetourot, and Rockbridge counties of VA.

KEY REFERENCES: Mitchell (1991).

Cumberland Plateau Salamander
Plethodon kentucki

DESCRIPTION: L = 10–17 cm (3.9–6.7 in); a large black salamander similar to the Slimy Salamander, *P. glutinosus* (with which it was formerly considered conspecific) but somewhat smaller, has a lighter throat, and smaller white spots on the back.

HABITAT AND DISTRIBUTION: Under forest litter, rocks, and logs in highland mixed hardwood forest; southern WV and southwestern VA.

HABITS: Nocturnal; terrestrial in all life stages.

DIET: Soil and litter invertebrates, e.g., mites, spiders, insects, and worms.

REPRODUCTION AND LIFE CYCLE: Breeding evidently occurs in winter or spring; eggs are laid in underground brood chambers in Jul, and are guarded by the female until hatching in Oct, and perhaps several weeks post-hatching; clutch size is small (9–12); sexual maturity is reached at 3–5 years.

CONSERVATION STATUS: S3—WV, VA; limited range in mature highland forest makes species vulnerable.

RANGE: Southern WV, southwestern VA, eastern KY, and extreme northeastern TN.

KEY REFERENCES: Mitchell (1991), Pague et al. (2005).

Cheat Mountain Salamander
Plethodon nettingi

DESCRIPTION: L = 8–12 cm (3.1–4.7 in); a small, slender, somewhat flattened salamander; dark brown, black, or gray body with brassy or white flecks on the back. Recognition as a species separate from the closely related Peaks of Otter Salamander (*Plethodon hubrichti*) and Shenandoah Salamander (*P. shenandoah*) is open to question (Petranka 1998).

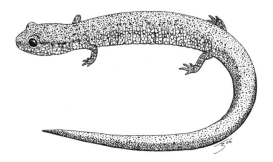

HABITAT AND DISTRIBUTION: Under rocks, logs, and leaf litter of spruce-fir and mixed forest in the highlands of Grant, Pendleton, Pocahontas, Randolph, and Tucker counties, WV.

HABITS: Nocturnal; terrestrial throughout life cycle.

DIET: Soil and litter invertebrates, e.g., ants, flies, beetles, mites, and springtails.

REPRODUCTION AND LIFE CYCLE: Poorly known; breeds apparently in spring; female lays a small clutch of eggs under a rock or rotted log, and guards them until hatching in Aug or Sep; young metamorphose in the egg, and hatch as terrestrial subadults.

CONSERVATION STATUS: S2—WV; "Threatened"—USFWS; small, endemic population is vulnerable to habitat loss and possible competition from closely related species.

RANGE: Grant, Pendleton, Pocahontas, Randolph, and Tucker counties, WV.

KEY REFERENCES: Pauley (2005a), Hammerson and Qureshi (2005).

Cow Knob Salamander *Plethodon punctatus*

DESCRIPTION: L = 10–16 cm (3.9–6.3 in); a small, slim, long-tailed salamander; thin legs and feet with webbed toes; dark brown to black body color with yellowish dots on the back; pale throat. This species is closely related to the Slimy Salamander *Plethodon glutinosus.*

HABITAT AND DISTRIBUTION: Under forest litter, rocks, and logs in highland hemlock and mixed hardwood forests, often with rocky outcrops, in Augusta, Rockingham, and Shenandoah counties, VA, and Pendleton, Hardy, and Hampshire counties, WV.

HABITS: Nocturnal; terrestrial in all life stages.

DIET: Soil and litter invertebrates, e.g., mites, spiders, insects (especially Hymenoptera and Collembola), and worms.

REPRODUCTION AND LIFE CYCLE: Breeding evidently occurs in winter or spring; clutch size is small (13 eggs in two gravid females examined); young metamorphose in the egg and hatch as terrestrial subadults, reaching sexual maturity in 3 years.

CONSERVATION STATUS: S1—WV; S2-VA; limited range makes species vulnerable.

RANGE: Part of Ridge and Valley Province along the VA, WV border.

KEY REFERENCES: Mitchell (1991), Pague et al. (2005).

Northern Gray-cheeked Salamander *Plethodon montanus*

DESCRIPTION: L = 8–12 cm (3.1–4.7 in); a small, slim salamander with long tail and protuberant eyes; variable in coloration, but a common phase is entirely black above, slate below with lighter gray cheeks and pale throat. This taxon is part of a superspecies complex formerly called Jordan's Salamander, a close relative of the Slimy Salamander *Plethodon glutinosus,* with which it hybridizes in some areas. Currently the group of species previously recognized as Jordan's Salamander has been divided into seven species: *P. jordani, P. metcalfi, P. amplus, P. meridianus, P. shermani, P. cheoah,* and the only member of the group found in the Mid-Atlantic region, *P. montanus.*

HABITAT AND DISTRIBUTION: Under roots, stones, logs, and litter of highland mixed forest in the Blue Ridge and Ridge and Valley provinces of southwestern VA.

HABITS: Nocturnal; terrestrial in all life stages.

DIET: No specific information, but likely feeds on soil and litter invertebrates.

REPRODUCTION AND LIFE CYCLE: Poorly known; breeds in the fall; young metamorphose in the egg and hatch as terrestrial subadults.

RANGE: Portions of the Valley and Ridge and Blue Ridge provinces of TN, NC, and VA.

KEY REFERENCES: Mitchell (1991), Petranka (1998).

Ravine Salamander *Plethodon richmondi*

DESCRIPTION: L = 8–14 cm (3.1–5.5 in); a slender salamander with a long tail and small front and rear legs; body uniformly dark with variable amounts of silver or bronze flecks; some taxonomists split this species into two species: the Northern Ravine Salamander (*P. electromorphus*) and the Southern Ravine Salamander (*P. richmondi*). This species was considered conspecific with *P. hoffmani* (the Valley and Ridge Salamander) prior to 1972.

HABITAT AND DISTRIBUTION: Under logs, rocks, and forest litter of slopes in moist highland deciduous and mixed forests of VA and WV.

HABITS: Nocturnal; aestivates during dry summers; hibernates in winter; terrestrial in all life stages.

DIET: Soil and litter invertebrates, e.g., snails, insects, worms, and spiders.

REPRODUCTION AND LIFE CYCLE: Breeds in spring; female lays a small clutch of eggs under a log or

rock or in a burrow; young metamorphose in the egg and hatch late summer or fall as terrestrial subadults; 2 years are required for the female to form eggs prior to breeding.

RANGE: PA, VA, WV, KY, TN, NC.

KEY REFERENCES: Pauley (2005a), Martof et al. (1980), Virginia Department of Game and Inland Fisheries (2004), Hammerson (2005a).

Shenandoah Salamander *Plethodon shenandoah*

DESCRIPTION: L = 8–11 cm (3.1–4.3 in); a skinny salamander; two color phases: 1) dark body with central yellow or reddish dorsal stripe; 2) uniformly dark body with brass flecks. This taxon is part of a superspecies complex that includes *P. hubrichti* and *P. nettingi*. Whether these groups deserve recognition as separate species has not been resolved (Petranka 1998).

HABITAT AND DISTRIBUTION: Under rocks, leaf litter, and logs in pockets of soil on north or northwest-facing talus slopes of The Pinnacle, Stony Man Mountain, and Hawksbill Mountain, Page and Madison counties, VA.

HABITS: Nocturnal; terrestrial in all life stages.

DIET: Soil and litter invertebrates, e.g., insect larvae and mites.

REPRODUCTION AND LIFE CYCLE: Poorly known; female lays clutch of 4–18 eggs, presumably in an underground nest; young undergo metamorphosis in the egg, and hatch in late summer or fall as terrestrial subadults, reaching sexual maturity in 2–3 years; females lay eggs every other year.

CONSERVATION STATUS: S1—VA; "Endangered"—USFWS; natural forest succession seems to allow the related *P. cinereus* to invade habitats occupied by *P. shenandoah*, outcompeting and/or interbreeding with the latter taxon.

RANGE: Blue Ridge Mountains of northern VA.

KEY REFERENCES: Mitchell (1991), Petranka (1998), Virginia Department of Game and Inland Fisheries (2004).

Wehrle's Salamander *Plethodon wehrlei*

DESCRIPTION: L = 10–17 cm (3.9–6.7 in); a large salamander with prominent eyes and webbed hind toes (more obvious than in other large *Plethodon* species); dark body variably patterned with white dots dorsally and white or yellowish blotches laterally; paler below with white blotches on the throat.

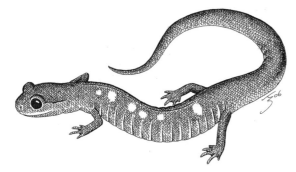

HABITAT AND DISTRIBUTION: Burrows under rocks, logs, and litter, often at cave entrances and rock crevices on hillsides of upland spruce-fir and mixed deciduous forests of WV, western VA, southwestern and central PA, and western MD.

HABITS: Fossorial; nocturnal; terrestrial in all life stages.

DIET: Soil and litter invertebrates, e.g., spiders, mites, ants, and beetles.

REPRODUCTION AND LIFE CYCLE: Breeds in Mar and Apr; female lays a small clutch of eggs in a nest in a damp, rotten log, moss, burrow, or cave crevice in May, which hatch 3 months later; metamorphosis occurs at time of hatching so young emerge as terrestrial subadults; females produce eggs only every other year.

CONSERVATION STATUS: S2—MD; limited distribution.

RANGE: Appalachian Plateau Province from southwestern NY to TN.

KEY REFERENCES: Martof et al. (1980), Petranka (1998), Virginia Department of Game and Inland Fisheries (2004), Hammerson (2005a).

Weller's Salamander *Plethodon welleri*

DESCRIPTION: L = 6–8 cm (2.4–3.1 in); a small salamander; toes are partially webbed; dark above with gold flecks or blotches; paler below.

HABITAT AND DISTRIBUTION: Under logs, rocks, and litter in spruce-fir forest of Whitetop, Mount Rogers, and Pine Mountain in Grayson, Smyth, and Washington counties of VA.

HABITS: Nocturnal; terrestrial in all life stages.

DIET: Soil and litter invertebrates, e.g., spiders, mites, beetles, and insect larvae.

REPRODUCTION AND LIFE CYCLE: Breeds in fall or spring; female lays a clutch of 4–13 eggs under

moss or a rotten log in spring; young hatch in late summer as terrestrial subadults (metamorphosis occurs in the egg).

CONSERVATION STATUS: S2—VA; threats include restriction to spruce-fir forest threatened by acid rain, spruce budworm, and logging, and also illegal collecting by herp collectors.

RANGE: Mountains of the Blue Ridge Province of eastern TN, western NC, and southwestern VA.

KEY REFERENCES: Mitchell (1991), Virginia Department of Game and Inland Fisheries (2004), Clausen Hammerson (2005).

Yonahlossee Salamander *Plethodon yonahlossee*

DESCRIPTION: L = 11–22 cm (4.3–8.7 in); a large salamander; brick-red back; light gray sides; dark gray head and tail; belly and throat paler and variously blotched.

HABITAT AND DISTRIBUTION: Deciduous forest, 440–1700 m (1400–5600 ft) elevation, under logs, rocks, and litter of rocky hillsides and ravines, especially those covered with mosses and ferns, in the southern portion of the Blue Ridge Province of VA.

HABITS: Crepuscular or nocturnal; fossorial; terrestrial in all life stages.

DIET: Soil and litter invertebrates, including centipedes, millipedes, mites, spiders, worms, and insects.

REPRODUCTION AND LIFE CYCLE: Poorly known; presumably breeds in spring; female lays eggs in a burrow; young undergo metamorphosis in the egg, and hatch in the late summer or fall, 2–3 months after laying, as terrestrial subadults; sexual maturity is reached at 3 years.

CONSERVATION STATUS: S3—VA; relatively restricted range and habitat make the species vulnerable.

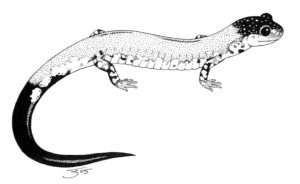

RANGE: Blue Ridge Province of southern VA, western NC, and eastern TN.

KEY REFERENCES: Martof et al. (1980), Virginia Department of Game and Inland Fisheries (2004), Hammerson (2005a).

Mud Salamander *Pseudotriton montanus*

DESCRIPTION: L = 7–20 cm (2.8–7.9 in); thick-bodied salamander with relatively short tail; red or pink body with dark spots; brown eye; montane and Coastal Plain populations of this taxon are sometimes recognized as separate species: the Midland Mud Salamander (*Pseudotriton diasticus*) and the Eastern Mud Salamander (*Pseudotriton montanus*), respectively.

HABITAT AND DISTRIBUTION: Under logs or stones or in burrows in muddy margins of swamps, beaver ponds, quiet streams, and springs in the Coastal Plain and adjacent Piedmont of VA, MD, DE, and NJ; the montane form occurs in the highlands of southwestern VA and southern WV.

HABITS: Fossorial.

DIET: Poorly known, but includes worms, arthropods, and occasionally, smaller salamanders.

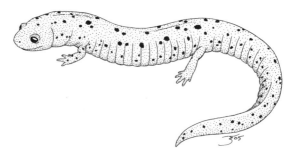

REPRODUCTION AND LIFE CYCLE: Breeds in the fall; female lays 60–200 eggs in winter, attaching them to submerged vegetation in quiet pools, ponds, and springs; aquatic larvae hatch in late winter and transform to subadult status in 14–32 months; reaches sexual maturity in 3–4 years.

CONSERVATION STATUS: S1—DE, PA, WV; limited range and habitat in some portions of the region.

RANGE: Eastern United States from OH and PA to LA and FL.

KEY REFERENCES: Martof et al. (1980), Pauley (2004), Hammerson (2005a).

Red Salamander *Pseudotriton ruber*

DESCRIPTION: L = 8–18 cm (3.1–7.1 in); a bulky salamander; red body with dark spots, like the Mud Salamander, but with short, broad snout and yellow (not brown) eye.

HABITAT AND DISTRIBUTION: Under logs, leaf litter, and rocks in or near cold, clear, gravel or rocky-bottomed streams, creeks, and springs nearly throughout the region.

HABITS: Nocturnal.

DIET: Insects, worms, small salamanders; larvae feed on zooplankton and aquatic invertebrates.

REPRODUCTION AND LIFE CYCLE: Breeds in summer and fall; female lays 50–100 eggs in the fall, attaching them to the underside of rocks in pools; young hatch in winter; larvae transform 18–33 months after hatching; sexual maturity is reached at 4–5 years.

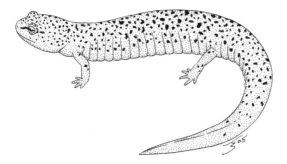

CONSERVATION STATUS: S3—DE, WV; threats include deforestation; stream pollution from acid mine drainage.

RANGE: Eastern North America from southern Canada to LA and FL.

KEY REFERENCES: Martof et al. (1980), Petranka (1998), Pauley (2005a).

Many-lined Salamander *Stereochilus marginatus*

DESCRIPTION: L = 6–11 cm (2.4–4.3 in): a small salamander with flattened head and relatively short, keeled tail; brown dorsal color, paler below; dark line through eye running the length of the body as well as several additional parallel dark lines along the sides, some incomplete or composed of dots.

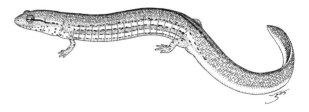

HABITAT AND DISTRIBUTION: Sluggish streams, blackwater swamps, and cypress bays of southeastern VA.

HABITS: Mostly aquatic in all life stages; benthic.

DIET: Feeds mostly on or near the bottom of aquatic habitats on invertebrates, including crustacea, isopods, insect larvae, and amphipods.

REPRODUCTION AND LIFE CYCLE: Breeds in fall; female lays 60–100 eggs in submerged vegetation, e.g., log, moss, or leaf litter, and remains guarding the clutch until hatching in spring; aquatic larvae transform in 13–28 months; sexual maturity is reached in 3–4 years.

CONSERVATION STATUS: S3—VA; threats include limited distribution in wetlands subject to development pressure.

RANGE: Coastal Plain of the southern Atlantic states from VA to FL.

KEY REFERENCES: Martof et al. (1980), Petranka (1998), Virginia Department of Game and Inland Fisheries (2004).

ORDER ANURA; FAMILY PELOBATIDAE

Eastern Spadefoot *Scaphiopus holbrookii*

DESCRIPTION: L = 4–7 cm (1.6–2.8 in); a small, brown-backed toad with light yellow lines running down the back, yellower in males than females; whitish below and on the face; protuberant eyes with gold-flecked iris and vertical pupils; skin can produce an irritating secretion; dark, horny nail or tubercule ("spade") on the inner toe of the hind feet is used for digging.

HABITAT AND DISTRIBUTION: Adults forage on the ground in open fields and woodlands with loose, sandy soils; mostly found in the Coastal Plain and Piedmont, but also in scattered localities elsewhere in the region.

HABITS: Adults are terrestrial, fossorial, and nocturnal; aestivate; hibernate.

DIET: Adults feed on soil, litter, and understory insects, worms, molluscs, and other invertebrates; tadpoles feed on zooplankton and other small aquatic invertebrates and vertebrates.

REPRODUCTION AND LIFE CYCLE: Spadefoots breed after heavy spring or summer rains, migrating to temporary pools and puddles for courtship and

external fertilization; females lay 1000–2500 eggs in sticky strings on pool bottoms; eggs hatch in 2–14 days; tadpoles transform in 2–8 weeks; sexual maturity is reached at 15–19 months.

CONSERVATION STATUS: S1—PA, WV; preferred floodplain habitats with good soils are under pressure from agriculture and urbanization.

RANGE: Eastern United States from NY and IL south to LA and FL.

KEY REFERENCES: DeGraaf and Yamasaki (2001), Pauley (2005b), Hammerson and Dirrigl (2005).

FAMILY BUFONIDAE

American Toad *Bufo americanus*

DESCRIPTION: L = 5–11 cm (2.0–4.3 in); a bulky-bodied toad with two parotid glands separate from cranial ridges behind eye; dorsal color is variable—brown, gray, olive, or even reddish above, often with a light dorsal stripe; one or two warts in each dark, dorsal spot; lighter below, often with spots on the chest and belly; females tend to be more brightly patterned than males; also males are smaller than females, and have dark throats and horny tubercules on the first and second toes of the front feet.

HABITAT AND DISTRIBUTION: Open woodlands and fields with loose, moist soils nearly throughout the region, although scarcer along the Coastal Plain.

HABITS: Terrestrial as adults; crepuscular; fossorial; hibernates in winter.

DIET: Adults feed on a variety of forest and field litter invertebrates, e.g., beetles, ants, isopods, millipedes, spiders, and slugs.

REPRODUCTION AND LIFE CYCLE: Breeds in early spring and summer, migrating to breeding pools

and ponds at night during heavy rains; female lays several thousand eggs in long strings threaded through underwater vegetation; eggs hatch in 6–12 days as aquatic larvae; transformation to terrestrial phase occurs in 5–10 weeks, depending on food supply and temperature; sexual maturity is reached in 2–4 years.

RANGE: Eastern North America.

KEY REFERENCES: DeGraaf and Yamasaki (2001), Virginia Department of Game and Inland Fisheries (2004), Hammerson (2005a).

Oak Toad *Bufo quercicus*

DESCRIPTION: L = 2–3 cm (0.8–1.2 in); a small toad covered uniformly with small tubercules on the back; reddish tubercules on soles of feet; dorsal color of gray, brown, or black, with dark blotches; median yellow stripe down the back; deflated vocal sack of male looks like a triangular bib.

HABITAT AND DISTRIBUTION: Adults frequent grassy

undergrowth of pine and oak savannas with sandy soil in southeastern VA.

HABITS: Diurnal; fossorial; hibernate; aestivate.

DIET: Adults feed on litter invertebrates, e.g., ants, beetles, and spiders; aquatic larvae are herbivorous, feeding on algae and aquatic plant parts.

REPRODUCTION AND LIFE CYCLE: Breeds in spring and summer, migrating to temporary pools, ponds, ditches, or puddles at night during heavy rains; females lay several hundred eggs in many small strands; eggs hatch into aquatic larvae, which metamorphose to terrestrial subadults in 2 months.

CONSERVATION STATUS:

S1—VA; limited range where wetland breeding sites are vulnerable to drainage or pollution and adult pine-oak habitat to logging.

RANGE: Coastal Plain of the southeastern United States from VA to LA.

KEY REFERENCES: Mitchell (1991).

Southern Toad *Bufo terrestris*

DESCRIPTION: L = 4–10 cm (1.6–3.9 in); cranial ridges form prominent knobs on the back of the head; dorsal color can be brown, gray, black, or reddish, with a light median stripe; grayish below with spots on the chest; female is larger than the male, and often lighter in color.

HABITAT AND DISTRIBUTION: Adults frequent fields and woodlands with loose, sandy soils; southeastern VA.

HABITS: Adults are terrestrial, foraging in the evening and night; fossorial, digging burrows for hiding during inactive periods; hibernates; aestivates.

DIET: Adults presumably feed on litter invertebrates, though diet is not well-documented; aquatic larvae are herbivorous, feeding on aquatic plant parts and algae.

REPRODUCTION AND LIFE CYCLE: Breeds in the spring, migrating to breeding pools, ditches, and ponds during rains; females lay several thousand eggs in long strings, which hatch into tadpoles in a few days; aquatic larvae metamorphose to terrestrial subadults in 1–2 months.

RANGE: Coastal Plain of southeastern United States from VA to LA.

KEY REFERENCES: Martof et al. (1980), Hammerson (2005a).

Fowler's Toad *Bufo fowleri*

DESCRIPTION: L = 5–8 cm (2.0–3.1 in); a medium-sized toad, similar to the American Toad but with parotid gland touching cranial ridge, 3–4 warts per dark blotch on the back, and no spots on chest and belly; dorsal color is usually brown or gray with dark blotches and a light, median stripe; white chest and belly, often with a single dark spot; males are smaller than females, and have a black throat; considered to be an eastern

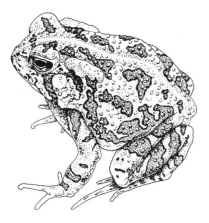

race of Woodhouse's Toad (*Bufo woodhousii*) by some taxonomists.

HABITAT AND DISTRIBUTION: Adults frequent riparian areas, lowlands, lake shores, beaches, and marsh borders with sandy soils nearly throughout the region except northern PA.

HABITS: Normally crepuscular, but sometimes active in daylight hours; fossorial, digging burrows for hiding during inactive periods; hibernates; aestivates.

DIET: Adults eat litter invertebrates, e.g., ants, beetles, snails, slugs, worms, and spiders; aquatic larvae are herbivorous.

REPRODUCTION AND LIFE CYCLE: Breeds in late spring or summer, migrating to breeding ponds, pools, and ditches on rainy nights; females lay several thousand eggs in long, double-row strings; eggs hatch in about 7 days; aquatic larvae transform at 1–2 months; sexual maturity is reached at 1–4 years.

CONSERVATION STATUS: S3—PA; limited wetland habitat for this species is threatened by drainage and pollution.

RANGE: Much of eastern North America from southern Canada south to TX and FL.

KEY REFERENCES: DeGraaf and Yamasaki (2001), Martof et al. (1980), Hammerson (2005a).

FAMILY HYLIDAE

Northern Cricket Frog *Acris crepitans*

DESCRIPTION: L = 1.5–2.5 cm (0.6–1.0 in); a small frog with numerous tiny skin tubercles and long back legs with webs between the toes; variable in dorsal color—green, brown, or orange, with a triangular dark or light patch between the eyes and a median line or "Y"-shaped patch on the back; light below; similar to the Southern Cricket Frog (*A. gryllus*), which has longer legs (heel beyond snout when extended forward) and sharp-edged

thigh stripe (irregular in *A. crepitans*); *Acris* tadpoles have a black tail tip.

HABITAT AND DISTRIBUTION: Adults frequent marshy margins of lakes, ponds, streams, and swamps in the eastern portion of the region, scarce or absent in the mountains.

HABITS: Adults are terrestrial; active day and night; great leapers, able to jump a distance 40 times their own length; hibernate; aestivate.

DIET: Adults eat small invertebrates captured in or near water; aquatic larvae are herbivorous.

REPRODUCTION AND LIFE CYCLE: Breeds in spring or summer; females attach eggs to submerged vegetation or benthic substrate; tadpoles metamorphose to adults in late summer; sexual maturity is reached in the first year.

CONSERVATION STATUS: S2—WV; S3—PA; declines have occurred in the northwestern part of the range for reasons not understood—seemingly related to the worldwide declines occurring in many anurans.

RANGE: Much of eastern North America from southern Canada south to FL and TX and west to NM, CO, and SD.

KEY REFERENCES: Martof et al. (1980), Hammerson (2005a).

Southern Cricket Frog *Acris gryllus*

DESCRIPTION: L = 1.5–2.5 cm (0.6–1.0 in); a small frog with numerous tiny skin tubercules and long back legs with webs between the toes; variable in dorsal color—green, brown, or orange, with a triangular dark or light patch between the eyes and a median line or"Y"-shaped patch on the back; light below; similar to the Northern Cricket Frog (*A. crepitans*), which has shorter legs (heel fails to reach snout when extended forward) and irregular thigh stripe (sharp-edged in *A. gryllus)*; *Acris* tadpoles have a black tail tip. Populations in our region are sometimes recognized as the "Coastal Plain Cricket Frog."

HABITAT AND DISTRIBUTION: Marshy borders of swamps, bayous, ponds, lakes, ditches, and pools in the Coastal Plain of southeastern VA.

HABITS: Adults are terrestrial; active throughout the year, day and night; great leapers (even better than *A. crepitans*), able  to jump a distance 40 times their own length.

DIET: Adults eat small insects, spiders, and other invertebrates; tadpoles are herbivorous.

REPRODUCTION AND LIFE CYCLE: Breeds Feb–Oct, depending on temperature and rainfall; females attach eggs to submerged vegetation or benthic substrate, which hatch in a few days; tadpoles metamorphose in 2–3 months.

RANGE: Coastal Plain of the southeastern United States from VA to LA.

KEY REFERENCES: Martof et al. (1980), Virginia Department of Game and Inland Fisheries (2004), Hammerson (2005a).

Pine Barrens Treefrog *Hyla andersonii*

DESCRIPTION: L = 3–6 cm (1.2–2.4 in); smooth-skinned with a broad, flattened head; large eye; green above; stripes of yellow and purple run from the snout along the eye and side to the thigh; lighter below; orange on inner surfaces of legs; maroon-colored feet.

HABITAT AND DISTRIBUTION: Shrubs and understory vegetation of woodlands adjacent to swamps and acidic bogs of the NJ pine barrens, especially areas with sphagnum moss.

HABITS: Adults are arboreal or terrestrial; crepuscular; nocturnal; hibernate; aestivate; hide in tree cavities or under bark during the day.

DIET: Adults feed mainly on flies, crickets, and other insects; larvae feed on benthic detritus, algae, and other plant material.

REPRODUCTION AND LIFE CYCLE: Breeds May–Jul, migrating to breeding pools at night; females lay about 500 eggs singly, attached to benthic vegetation of pools,

puddles, ponds and ditches; eggs hatch in 3 days; tadpoles transform in 1.5–2 months.

CONSERVATION STATUS: S3—NJ; limited wetland habitat for this species is threatened by drainage and pollution.

RANGE: Pine barrens of NJ; also localities in NC, SC, the western FL panhandle, southern AL.

KEY REFERENCES: Martof et al. (1980), Hammerson (2005a).

Gray Treefrog *Hyla versicolor*

DESCRIPTION: L = 3–6 cm (1.2–2.4 in); relatively large for a treefrog; gray (sometimes brown or green) with dark reticulations on the back and sides, and usually with a whitish patch below the eye; orange thighs; numerous tubercles cover the body; toes have enlarged pads; indistinguishable from the closely related Cope's Gray Treefrog (*H. chrysoscelis*) except by chromosome count, erythrocyte size, or voice: *H. versicolor* has a slower, melodious trill with about 25 notes per second, lasting 1–3 seconds and ending abruptly, while *H. chrysoscelis* has a faster, buzzy trill with about 45 notes per second.

HABITAT AND DISTRIBUTION: Adults forage in woodland trees, shrubs, and understory nearly throughout the region except western WV, eastern VA, and the southern Delmarva Peninsula.

HABITS: Adults are arboreal or terrestrial; crepuscular; nocturnal; hibernate; aestivate; hide in tree cavities or under bark during the day.

DIET: Adults feed mainly on insects; tadpoles are herbivorous.

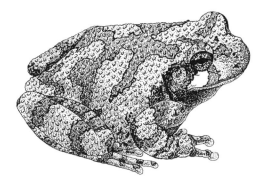

REPRODUCTION AND LIFE CYCLE: Breeds in spring and summer, migrating to breeding pools at night; females lay scattered clumps of 10–40 eggs on the surface of pools, puddles, ponds and ditches; eggs hatch in 4–5 days; tadpoles transform in 1.5–2 months.

CONSERVATION STATUS: Gray Treefrogs may be undergoing declines for reasons related to worldwide declines in anurans.

RANGE: Most of eastern North America.

KEY REFERENCES: Martof et al. (1980), Virginia Department of Game and Inland Fisheries (2004), Hammerson (2005a), Pauley (2005b).

Cope's Gray Treefrog *Hyla chrysoscelis*

DESCRIPTION: L = 3–6 cm (1.2–2.4 in); relatively large for a treefrog; gray (sometimes brown or green) with dark reticulations on the back and sides, and usually with a whitish patch below the eye; orange thighs; numerous tubercules cover the body; toes have enlarged pads; indistinguishable from the closely related Gray Treefrog (*H. versicolor*) except by voice: *H. versicolor* has a slower, melodious trill with about 25 notes per second, lasting 1–3 seconds and ending abruptly, while *H. chrysoscelis* has a faster, buzzy trill with about 45 notes per second.

HABITAT AND DISTRIBUTION: Adults forage in woodland trees, shrubs, and understory in many parts

of the region although not recorded from PA, and patchily distributed in WV, VA, MD, and perhaps other states.

HABITS: Adults are arboreal or terrestrial; crepuscular; nocturnal; hibernate; aestivate; hide in tree cavities or under bark during the day.

DIET: Adults feed mainly on insects; tadpoles are herbivorous.

REPRODUCTION AND LIFE CYCLE: Breeds in spring and summer, migrating to breeding pools at night; females lay scattered clumps of 10–40 eggs on the surface of pools, puddles, ponds and ditches; eggs hatch in 4–5 days; tadpoles transform in 1.5–2 months.

CONSERVATION STATUS: S2—NJ, DE; Cope's Gray Treefrog may be undergoing declines for reasons related to worldwide declines in anurans.

RANGE: Most of eastern North America.

KEY REFERENCES: Martof et al. (1980), Virginia Department of Game and Inland Fisheries (2004), Hammerson (2005a), Pauley (2005b).

Green Treefrog *Hyla cinerea*

DESCRIPTION: L = 3–6 cm (1.2–2.4 in); a relatively large, slim treefrog; usually green above and white below; often with a white stripe running along the side from the upper lip to the thigh (may be absent in northern populations); pads on toes.

HABITAT AND DISTRIBUTION: Adults forage in trees and shrubs bordering swamps, ponds, and lakes, or on floating or emergent vegetation in the Coastal Plain of VA, MD, and DE.

HABITS: Crepuscular; nocturnal; inactive in cold periods; hides on green vegetation during the day; hibernates; aestivates.

DIET: Adults feed mainly on flying insects; tadpoles are herbivorous.

REPRODUCTION AND LIFE CYCLE: Breeds in spring and summer; females lay eggs on water surface amidst floating vegetation, which hatch in a few days; tadpoles transform in about 2 months.

CONSERVATION STATUS: S3—DE; limited range in the state makes the DE population vulnerable.

RANGE: Eastern United States from MD and IL south to TX and FL; introduced into Puerto Rico.

KEY REFERENCES: Martof et al. (1980), Virginia Department of Game and Inland Fisheries (2004), Hammerson (2005a).

Pine Woods Treefrog *Hyla femoralis*

DESCRIPTION: L = 2–4 cm (0.8–1.6 in); a small, slim treefrog, reddish-brown or gray in dorsal color with dark blotches; whitish below; similar to the Squirrel Treefrog, but with distinctive orange or yellow spots on the thighs; pads on toes.

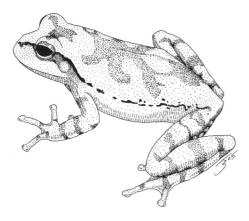

HABITAT AND DISTRIBUTION: Adults forage throughout woodland vegetation from the floor to the canopy of pine flatwoods and southern floodplain forest in the Coastal Plain of southeastern VA; hides under logs or forest litter when inactive.

HABITS: Arboreal; nocturnal; hibernates; aestivates.

DIET: Adults feed on beetles, grasshoppers, crickets, wasps, spiders, caddisflies, ants, and a variety of other small invertebrates; tadpoles are herbivorous.

REPRODUCTION AND LIFE CYCLE: Migrates to breeding sites in cypress bays, swamps, bogs, ponds, and pools on rainy nights in spring and summer; females lay clumps of 100 or so eggs on vegetation on or near the water surface; larvae transform in 2–3 months.

RANGE: Coastal Plain of the southeastern United States from VA to LA.

KEY REFERENCES: Martof et al. (1980), Conant (1984), Virginia Department of Game and Inland Fisheries (2004), Hammerson (2005a).

Barking Treefrog *Hyla gratiosa*

DESCRIPTION: L = 5–7 cm (2.0–2.8 in); a large treefrog with warty skin; green, gray, or brown dorsal color, often with large, round spots; individuals can change color; light and dark lateral lines from upper jaw to thigh, and on legs; whitish or yellowish green below; large toe pads; males have a yellowish gular pouch. The English name derives from sharp, doglike barks given by males.

HABITAT AND DISTRIBUTION: Forages in trees, shrubs, and litter of pine savanna, cypress bays, swamps, farmland, and bottomlands, generally with sandy soil, at scattered sites along the

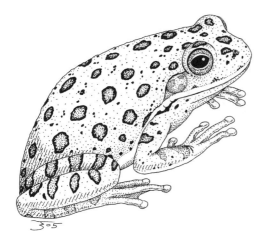

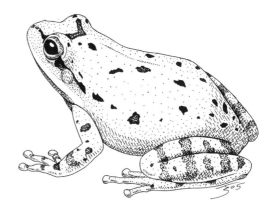

Coastal Plain of DE, MD, and VA; introduced in NJ, but probably extinct.

HABITS: Arboreal; fossorial; hibernates; aestivates; hides in burrows during inactive periods.

DIET: Adults feed on arboreal insects; also take a variety of small invertebrates, e.g., crickets, from ground litter and soil; tadpoles are herbivorous.

REPRODUCTION AND LIFE CYCLE: Breeds in spring and summer, migrating to shallow water of ponds, pools, or swamps; female lays 2000 or so eggs singly on benthos, which hatch in a few days; tadpoles transform at 1.5–2.5 months.

CONSERVATION STATUS: S1—VA, DE, MD; principal threat appears to be logging of native pine forest and savanna with drainage of wetlands.

RANGE: Mainly in the Coastal Plain of the southeastern United States from DE to LA; also KY and TN.

KEY REFERENCES: Mitchell (1991), U.S. Geological Survey (2005).

Squirrel Treefrog *Hyla squirella*

DESCRIPTION: L = 2–4 cm (0.8–1.6 in); a small treefrog that can change its dorsal color from brown to green; light lateral line; whitish below; large toe pads; distinguished from similar *Hyla* by 1) lack of a light spot below the eye, 2) lack of an "X"-shaped mark on the back, and 3) lack of yellow or

orange spots on white thigh; males have gular vocal pouch.

HABITAT AND DISTRIBUTION: Adults forage in vegetation of open woodlands, parklands, and residential gardens in southeastern VA; hides under bark, in tree cavities, or in thick vegetation when inactive.

HABITS: Nocturnal; arboreal; active most of the year.

DIET: Adults feed mainly on arboreal insects; tadpoles are herbivorous.

REPRODUCTION AND LIFE CYCLE: Breeds in spring and summer, migrating to shallow ponds, pools, or sloughs on rainy nights; female lays 800–1000 eggs, singly or in small clumps, on the pond bottom; tadpoles metamorphose in 6–7 weeks.

RANGE: Coastal Plain of the southeastern United States from VA to TX; also OK.

KEY REFERENCES: U.S. Geological Survey (2005), Hammerson (2005a).

Spring Peeper *Pseudacris crucifer*

DESCRIPTION: L = 2–3.5 cm (0.8–1.4 in); a small frog; tan, brown, or gray above with a dark "X" (crucifer) on the back and a chevron between the eyes; pads on toes; males are smaller and darker than females, and have dark throats; formerly considered to be a true treefrog (genus *Hyla*), placed for the present with the chorus frogs (*Pseudacris*); well known

for its "peeping" song given from pools at dusk in spring and early summer.

HABITAT AND DISTRIBUTION: Adults forage on the ground, in woodland understory, or emergent vegetation throughout the region, hiding underground or under litter, logs, or rocks when inactive; found throughout the region.

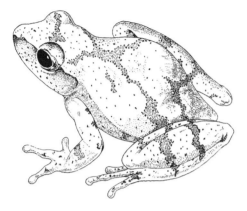

HABITS: Mainly crepuscular and nocturnal, but sometimes active in the day; hibernates; aestivates.

DIET: Adults feed primarily on insects, e.g., ants, springtails, flies, beetles, and Homoptera; tadpoles are herbivorous.

REPRODUCTION AND LIFE CYCLE: Adults migrate to wetlands to breed in early spring; female deposits about 1000 eggs singly on benthic vegetation; eggs hatch in 6–12 days, and tadpoles transform in 3–4 months.

RANGE: Eastern North America from Canada south to TX and FL.

KEY REFERENCES: DeGraaf and Yamasaki (2001), Virginia Department of Game and Inland Fisheries (2004), Hammerson (2005a).

Little Grass Frog *Pseudacris ocularis*

DESCRIPTION: L = 1–2 cm (0.4–0.8 in); a tiny frog, tan, brown, gray, or reddish on the back with a dark, triangle-shaped blotch between the eyes and a dark line from the nostril, through the eye, and down the side; whitish below; males have a dark throat and large gular vocal pouch.

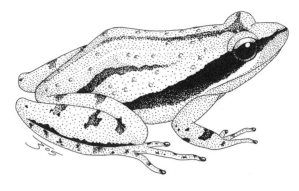

HABITAT AND DISTRIBUTION: Wet grasslands and marshy borders of cypress bays, ponds, bogs, and swamps in southeastern VA.

HABITS: Nocturnal; active throughout the year.

DIET: Adults feed mainly on insects; larvae are herbivorous.

REPRODUCTION AND LIFE CYCLE: Breeds in spring and summer, migrating to pools, puddles, ponds, and ditches on rainy nights; females lay a few hundred eggs in several small clumps attached to aquatic vegetation or the pond bottom, which hatch in 4–5 days; tadpoles transform in 1.5–2.5 months.

CONSERVATION STATUS: S3—VA; limited range in floodplain forest is threatened by logging, drainage, and development.

RANGE: Coastal Plain of the southeastern United States from VA to AL.

KEY REFERENCES: Virginia Department of Game and Inland Fisheries (2004), U.S. Geological Survey (2005), Hammerson (2005a).

Mountain Chorus Frog *Pseudacris brachyphona*

DESCRIPTION: L = 2.5–4 cm (1.0–1.6 in); a small frog, gray or brown on the back with a dark line from nostril, through eye, and down each side; white upper lip; dark inverted triangle-shaped blotch between the eyes, and other blotches on back and legs; expanded toe disks; males have dark throats.

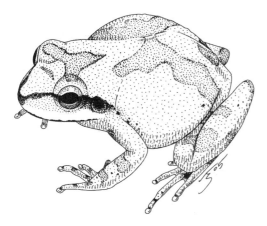

HABITAT AND DISTRIBUTION: Adults forage in forest litter of woodland slopes in highlands of VA, WV, MD, and PA; hides under logs, rocks, and duff during inactive periods.

HABITS: Nocturnal; mostly terrestrial, but occasionally climbs into understory; aestivates; hibernates.

DIET: Adults feed on litter invertebrates; tadpoles are herbivorous.

REPRODUCTION AND LIFE CYCLE: Breeds in spring, migrating to vernal pools, puddles, and ponds on rainy nights; females lay 300–1500 eggs in clumps of 10–50 attached to submerged vegetation, which hatch in 3–4 days; larvae metamorphose in 7–9 weeks.

CONSERVATION STATUS: S2—MD; limited distribution vulnerable to habitat alteration.

RANGE: Highlands of the eastern United States from PA and OH south to GA and MS.

KEY REFERENCES: Hammerson (2005a), Martof et al. (1980), Virginia Department of Game and Inland Fisheries (2004).

Brimley's Chorus Frog *Pseudacris brimleyi*

DESCRIPTION: L=2.5–3 cm (1.0–1.2 in); a small frog tan or brown above with 3 irregular, parallel lines down the middle of the back and a broad black stripe down the side from snout to thigh; yellow below; black stripe on leg; chest often spotted; expanded toe disks.

HABITAT AND DISTRIBUTION: Forages on the ground and in understory of floodplain forests, swamps, and marshes in southeastern VA..

HABITS: Nocturnal.

DIET: Adults climb in low vegetation, feeding on insects and other small invertebrates; larvae are herbivorous.

REPRODUCTION AND LIFE CYCLE: Breeds in ditches, pools, or shallow ponds in late winter or early spring; females lay small batches of eggs on submerged vegetation; tadpoles transform in 1.5–2 months.

RANGE: Coastal Plain of VA, SC, NC, and GA.

KEY REFERENCES: Hammerson (2005a), Martof et al. (1980), Virginia Department of Game and Inland Fisheries (2004).

Western Chorus Frog *Pseudacris triseriata*

DESCRIPTION: L = 2–4 cm (0.8–1.6 in); a small frog; tan, brown, or gray above with 3 irregular stripes (or lines of blotches) down the back; broad stripe from nostril to thigh along the side; blotch between the eyes; white upper lip; yellowish below, sometimes

with mottling on the chest; toes have small pads; some taxonomists (e.g., Platz 1989) have suggested that *P. triseriata* may be composed of three separate species currently recognized as subspecies of the taxon: Western Chorus Frog (*P. t. triseriata*), Southeastern or Upland Chorus Frog (*P. t. feriarum*), and New Jersey Chorus Frog (*P. t. kalmi*).

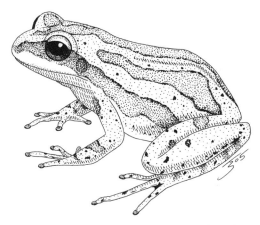

HABITAT AND DISTRIBUTION: Ground and understory of moist woodlands and riparian areas; marshy borders of swamps, swales, bogs, and ponds; hides under forest litter, logs, or in burrows during inactive periods; most of MD, DE, NJ; parts of VA, PA, and eastern WV.

HABITS: Active day or night, but mainly nocturnal for breeding; hibernates; aestivates.

DIET: Adults feed mainly on insects and other small invertebrates; larvae are herbivorous.

REPRODUCTION AND LIFE CYCLE: Adults migrate to breeding pools in late winter or early spring; females lay several hundred eggs in clumps of 40–60 attached to submerged vegetation, which hatch in 3–4 days; tadpoles transform in 2–3 months; sexual maturity is reached in 1–3 years.

CONSERVATION STATUS: S2—WV, PA; limited wetland habitat for larvae of this species potentially threatened by drainage and pollution.

RANGE: Eastern United States from PA and IN south to FL and west to OK and TX.

KEY REFERENCES: Pauley (2005b), Hammerson (2005a), Martof et al. (1980), Virginia Department of Game and Inland Fisheries (2004).

FAMILY RANIDAE

American Bullfrog *Rana catesbeiana*

DESCRIPTION: L = 9–20 cm (3.5–7.9 in); a large frog; brown or olive above with dark, horizontal bars on the legs; white below with some dark mottling; males have yellow throat; loud, deep, "chug-a-rum" call is a familiar night sound of many large inland waterbodies on the continent.

HABITAT AND DISTRIBUTION: Shorelines and shallows of ponds, swamps, marshes, lakes, and rivers throughout the region; hides in benthic material during inactive periods.

HABITS: Aquatic throughout life cycle; active day or night, but most calling is crepuscular or nocturnal; hibernates; aestivates; tadpoles are distasteful (poisonous to eat?) for many fish.

DIET: Adults feed on a wide variety of invertebrate and small vertebrate prey; tadpoles are primarily herbivorous, but will take small invertebrates or vertebrates, especially as they approach metamorphosis.

REPRODUCTION AND LIFE CYCLE: Breeds in spring and summer; females lay thousands of eggs in large, floating masses attached to emergent vegetation, which hatch in a few to several days; tadpoles usually overwinter before transforming to adults; sexual maturity is reached in 1–5 years.

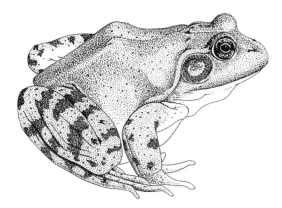

CONSERVATION STATUS: Bullfrogs can be voracious predators on fish fry and amphibian larvae, causing serious depletions and extirpations of some species, especially in areas where bullfrogs are not native, as in the western United States.

RANGE: North America from Canada to southern Mexico; Hawaii; Greater Antilles; distribution in pre-colonial times probably restricted to eastern North America.

KEY REFERENCES: DeGraaf and Yamasaki (2001), Hammerson (2005a), Virginia Department of Game and Inland Fisheries (2004).

Green Frog *Rana clamitans*

DESCRIPTION: L = 5–9 cm (2.0–3.5 in); a relatively large frog, green or brown above with prominent ridge along the side of the back; white below; males have a yellow throat and large tympanum, foreleg, and thumb.

HABITAT AND DISTRIBUTION: Shoreline and shallows of ponds, lakes, creeks, and pools throughout the region; hides in benthic material in water or under logs or litter during inactive periods.

HABITS: Active day or night; mostly aquatic in all life stages (except adult dispersal); hibernates; aestivates.

DIET: Adults feed on insects (often beetles), crustacea, spiders, salamanders, fish, worms, and other small aquatic and terrestrial invertebrates as dictated by abundance; tadpoles are herbivorous.

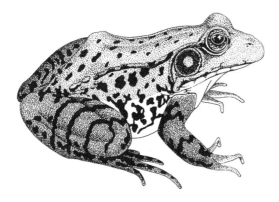

REPRODUCTION AND LIFE CYCLE: Breeds in spring and summer; females lay 3000–5000 eggs in floating masses attached to vegetation, which hatch in 3–6 days; tadpoles transform in late summer or fall or overwinter and transform the following spring.

RANGE: Eastern North America from southern Canada to northern FL and east TX.

KEY REFERENCES: DeGraaf and Yamasaki (2001), Hammerson (2005a), Virginia Department of Game and Inland Fisheries (2004).

Pickerel Frog *Rana palustris*

DESCRIPTION: L = 4–9 cm (1.6–3.5 in); a medium-sized, slim frog; green, tan, brown, or gray with irregular, square-shaped blotches in parallel lines down the back and sides (rounded blotches in the similar Leopard Frog); dorsolateral fold running down each side of the back; white below although chin is sometimes mottled; inside of thigh is orange or yellow (white in Leopard Frog).

HABITAT AND DISTRIBUTION: Shoreline, borders, and edges of wetlands, lakes, ponds, swales, and bogs at scattered localities throughout the region; also uses wet meadows and open woodlands; hides in benthic mud or under stones or logs on land during inactive periods.

HABITS: Adults are both terrestrial and aquatic; active day or night; hibernates; skin secretions may be noxious to predators.

DIET: Adults feed mostly on terrestrial arthropods; tadpoles are herbivorous.

REPRODUCTION AND LIFE CYCLE: Breeds in spring; female lays 2000–3000 eggs in globular clumps attached to submerged vegetation of ponds, bogs, and swales, which hatch in 10–20 days; tadpoles transform in 2–3 months.

RANGE: Eastern North America from southern Canada south to northern GA and eastern TX.

KEY REFERENCES: DeGraaf and Yamasaki (2001).

Northern Leopard Frog *Rana pipiens*

DESCRIPTION: L = 4–9 cm (1.6–3.5 in); a medium-sized, slim frog; green, tan, or brown with round or oval blotches on the back and sides (irregular, square blotches in lines in the similar Pickerel Frog); dorsal ridge running down each side of the back; dark blotches on the snout and nose (lacking in the Southern Leopard Frog); white below (inside of thigh is orange or yellow in Pickerel Frog).

HABITAT AND DISTRIBUTION: Wet meadows, open woodlands, grasslands, marshes, wetlands at scattered localities in WV and PA.

HABITS: Terrestrial; aquatic; diurnal; hibernates.

DIET: Adults feed mainly on insects, including crickets, caterpillars, wasps, beetles, and ants; tadpoles are herbivorous.

REPRODUCTION AND LIFE CYCLE: Breeds in spring; females lay several thousand eggs in oblong globular clumps attached to submerged vegetation, which hatch in 2–3 weeks; young transform in 2–4 months.

CONSERVATION STATUS: S2—WV; S3—PA; wetland habitat threatened by drainage and pollution.

RANGE: Most of North America except southeastern and south-central United States, where replaced by the Southern Leopard Frog (southeast) and the Rio Grande Leopard Frog (south-central).

KEY REFERENCES: DeGraaf and Yamasaki (2001).

Southern Leopard Frog *Rana sphenocephala*

DESCRIPTION: L = 5–9 cm (2.0–3.5 in); a medium-sized, slim frog; green, tan, or brown round dark blotches on the back and sides; dorsal ridge running down each side of the back; white below; white spot in center of tympanum (dark in the similar Northern Leopard Frog); lacks dark blotches on nose and snout area (present in Northern Leopard Frog).

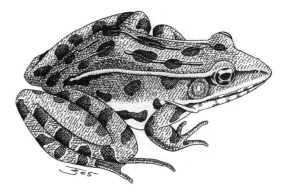

HABITAT AND DISTRIBUTION: Wet meadows, open woodlands, grasslands, marshes, wetlands at scattered localities nearly throughout the Coastal Plain in VA, MD, DE, PA, and NJ.

HABITS: Terrestrial; aquatic; diurnal; hibernates.

DIET: Adults feed mainly on insects, including crickets, caterpillars, wasps, beetles, and ants; tadpoles are herbivorous.

REPRODUCTION AND LIFE CYCLE: Breeds in spring; females lay several clumps of eggs attached to submerged vegetation just below the water's surface, which hatch in 1–2 weeks; young transform in 2–3 months.

CONSERVATION STATUS: S1—PA; wetland habitat threatened by drainage and pollution.

RANGE: Coastal Plain of the eastern United States from NJ to eastern TX; also north in central region to IL, IN, and MO.

KEY REFERENCES: Martof et al. (1980).

Wood Frog *Rana sylvatica*

DESCRIPTION: L = 4–8 cm (1.6–3.1 in); medium-sized frog; brown or tan with dark mask; lateral ridge on each side of back; dark bars on legs; white below with a dark spot on each side of chest; females are larger and more brightly patterned than males.

HABITAT AND DISTRIBUTION: Woodlands and fields nearly throughout the region except Piedmont and Coastal Plain of VA; hides under logs, rocks, and forest litter during inactive periods.

HABITS: Adults are terrestrial during nonbreeding periods; diurnal; hibernates.

DIET: Adults feed mainly on insects, e.g., beetles, wasps, and flies; tadpoles are herbivorous.

REPRODUCTION AND LIFE CYCLE: Adults migrate to vernal pools, puddles, ponds, swales, or quiet streams in early spring to breed; females lay eggs in globular masses attached to submerged vegetation; eggs hatch in 10–30 days, and tadpoles transform in 1–3 months; sexually mature at 2–3 years.

RANGE: Northern North America, south in the United States in mountains of northern GA; also scattered localities in the mid-western and western United States.

KEY REFERENCES: Martof et al. (1980), DeGraaf and Yamasaki (2001).

Carpenter Frog *Rana virgatipes*

DESCRIPTION: L = 4–7 cm (1.6–2.8 in); medium-sized frog, olive, brown or black above with 4 light lateral lines—2 down each side of the back, and one down each side from snout to thigh; no lateral ridges; paler below with dark mottling.

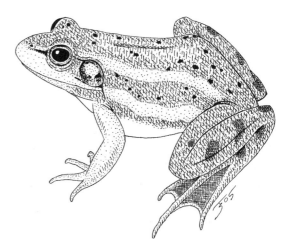

HABITAT AND DISTRIBUTION: Bogs and swamps, especially sphagnum moss bogs in pine savanna; parts of the Coastal Plain of DE, MD, and VA.

HABITS: Mostly aquatic throughout life cycle.

DIET: Adults feed on terrestrial and aquatic invertebrates; tadpoles are herbivorous.

REPRODUCTION AND LIFE CYCLE: Breeds in spring and summer; females lay 200–600 eggs in a clump attached to submerged vegetation; tadpoles may take more than a year to transform.

CONSERVATION STATUS: S1—DE; S2—MD; S3—VA; chief threats are drainage and pollution of limited bog habitats.

RANGE: Coastal Plain of the eastern United States from DE to northeastern FL.

KEY REFERENCES: Martof et al. (1980), Mitchell (1991).

FAMILY MICROHYLIDAE

Eastern Narrow-mouthed Toad
Gastrophryne carolinensis

DESCRIPTION: L = 2–4 cm (0.8–1.6 in); a small, plump, smooth-skinned frog with a small head, pointed snout, and a fold of skin across the head just behind the eyes; no webs between the toes; dark brown, gray, or chestnut down the center of the back, lighter along the sides; heavily mottled below; males have dark throats; individuals can change their color.

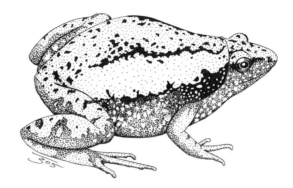

HABITAT AND DISTRIBUTION: Adults frequent a variety of terrestrial habitats under rocks, logs, and litter in moist, loose soil during nonbreeding periods; mostly along the Coastal Plain of MD and VA.

HABITS: Terrestrial; fossorial; aestivates; hibernates.

DIET: Adults feed mostly on soil and litter invertebrates, especially ants; tadpoles are herbivorous.

REPRODUCTION AND LIFE CYCLE: Adults migrate to puddles, ponds, swamps, and quiet streams on rainy nights in spring and summer; females lay clumps of eggs on the water's surface; tadpoles transform in 1–2 months.

CONSERVATION STATUS: S1—MD; limited range in MD.

RANGE: Southeastern United States from MD south to FL and west to eastern OK and TX.

KEY REFERENCES: Martof et al. (1980).

Class Reptilia—Reptiles

Reptiles appear in the fossil record during the late Carboniferous Period of the Paleozoic Era, about 300 million years ago. Like amphibians, most have 4 limbs with digits tipped with claws (except snakes and some lizards), a 3-chambered heart (except crocodilians, which have 4), double-loop circulation, and lungs. Unlike amphibians, they have keratinized scales, impermeable to air or water, and eggs with shells that are also impermeable to air or water. These 2 differences have allowed reptiles considerably more access to terrestrial environments than has been available to amphibians, although, like amphibians, reptiles are limited by their inability to thermoregulate (i.e., they are "cold blooded"). There are 4 modern orders of reptiles: turtles (Testudines)—330 species; snakes and lizards (Squamata)—5600 species; crocodiles and alligators (Crocodilia)—25 species; and the Sphenodon or Tuatara (Rhynchocephalia)—1 species. There are 20 species of turtles, 33 snakes, and 9 lizards found regularly in the Mid-Atlantic region.

ORDER CHELONIA; FAMILY CHELYDRIDAE

Sex of most turtles in our region (except the soft-shelled turtles [*Apalone*]) is determined by incubation temperature.

Snapping Turtle *Chelydra serpentina*

DESCRIPTION: L = 20–30 cm (7.9–19.7 in); massive, leathery head, a dark, oval, flattened carapace with 3 rows of horny plates, and a long, spiny tail; legs are thick and powerful with prominent claws; plastron is relatively small and cross-shaped (+); in males the cloaca is located beyond the edge of the shell, while in females it is in front of the edge of the shell.

HABITAT AND DISTRIBUTION: Common inhabitant of still or slowly moving bodies of fresh or brackish water, particularly those with muddy bottoms, abundant aquatic vegetation, and sunken debris, throughout the region.

HABITS: Active day and night, alternating bouts of active hunting with periods of quiescence, including floating on the water's surface or

sit-and-wait foraging while partially buried in benthic muck; active foraging involves swimming along the bottom, poking and probing, using mostly olfactory cues evidently to locate prey. Northern and highland populations hibernate in detritus Nov–Feb. Mostly aquatic, but individuals are occasionally seen on land as they disperse to new ponds or search for appropriate nesting sites (females). In the water, they are relatively docile, as they have few aquatic predators that threaten them. However, those disturbed on land are extremely aggressive.

DIET: Omnivorous, feeding on fish, amphibians, aquatic invertebrates, birds, reptiles, small mammals, algae, and aquatic plants including duckweed (*Lenna* sp.).

REPRODUCTION: Breeds Mar–Nov; egg deposition Apr–Sep with a peak in late May and early June. Females leave the water to locate appropriate nest sites, usually located in sand or similar loose, dry soil. Once an appropriate site is found, they dig a shallow, urn-shaped hole, lay 20–30 eggs, and then refill the hole and return to the water. Studies have shown that females may move as much as 8 km (5 mi) over land to locate a nesting site, although < 1 km is usual, and are able to demonstrate site fidelity to both home pond and nest locality. Hatching occurs 55–125 days later, depending on ambient temperature. Once hatched, the young burrow out of the hole, and disperse in an attempt to locate a body of water.

CONSERVATION STATUS: Humans are the principal predators on adults, and can extirpate local populations. Considered a game species throughout the region; NJ and DE have open season on adults for most of the year except during the spring egglaying; take is regulated in PA. There are few data on acceptable harvest levels for maintaining healthy populations.

RANGE: North America east of the Rockies from southern Canada south through Mexico and Central America; also in northwestern South America; introduced in CA and elsewhere in the far West.

KEY REFERENCES: DeGraaf and Rudis (1983), Mitchell (1994), Hulse et al. (2001).

FAMILY KINOSTERNIDAE

Eastern Mud Turtle *Kinosternon subrubrum*

DESCRIPTION: L = 8–12 cm (3.1–4.7 in); a small turtle with oval, domed carapace, sloped anteriorly, and a plastron with 2 hinges allowing the animal to close itself up completely within its shell; carapace varies from horn color to dark brown; plastron is lighter; differs from the similar Common Musk Turtle (*Sternotherus odoratus*) in having a more domed carapace, a larger plastron, and lacking the 2 prominent yellow stripes on the head of that species.

HABITAT AND DISTRIBUTION: Quiet waters of swamps, ponds, lakes, sloughs, slow streams, and ditches; tolerates brackish water of coastal marshes; Coastal Plain and adjacent Piedmont of VA, MD, DE, PA (at least formerly), and NJ.

HABITS: Both aquatic and terrestrial; mainly crepuscular, although it can be active day or night; aestivates and hibernates buried in loose soil on land.

DIET: Mud turtles forage actively by stalking along the bottom to capture a variety of invertebrates (e.g., crustacea, insects, and molluscs), small vertebrates, plants, and carrion; they also forage at times on land.

REPRODUCTION: Breeds in spring; mating takes place under water; female lays a clutch of 2–4 eggs on land under loose soil or debris; eggs hatch in 3–4 months; eggs or hatchlings may overwinter in the nest in northern populations.

CONSERVATION STATUS: SH—PA; limited range in an area of intense development (Philadelphia) makes the PA population of this species vulnerable, and, perhaps, extirpated in the state.

RANGE: Eastern United States from Long Island, NY south to FL and west to OK and east TX.

KEY REFERENCES: Frazier et al. (1991).

Striped Mud Turtle *Kinosternon baurii*

DESCRIPTION: L = 8–12 cm (3.1–4.7 in); a small turtle with oval, domed, dark brown carapace; the three broad carapace stripes for which the species is named are absent in VA; head usually has a yellow stripe between eye and nostril; plastron with 2 hinges allowing the animal to close itself up completely within its shell; *K. baurii* populations in VA were considered to be a variety of the Common Mud Turtle (*K. subrubrum*) before studies by Lamb and Lovich (1990).

HABITAT AND DISTRIBUTION: Quiet waters of cypress swamps, ponds, slow streams, and ditches; also uses upland habitats, e.g., wet grasslands and forests; tolerates brackish water of coastal marshes; Coastal Plain of VA.

HABITS: Both aquatic and terrestrial; active day or night; aestivates and hibernates buried in loose soil on land; aggressive when handled.

DIET: Forages for plant seeds, leaves, algae, insects, crayfish, molluscs, and carrion by crawling along stream and pond bottoms; also forages on land.

REPRODUCTION: Breeds in summer; female lays a clutch of 1–4 eggs in late summer or fall on land under loose soil or debris; eggs hatch in 3–4 months.

RANGE: Coastal Plain from VA to FL.

KEY REFERENCES: Mitchell (1994).

Loggerhead Musk Turtle *Sternotherus minor*

DESCRIPTION: L = 7–11 cm (2.8–4.3 in); a small turtle with a high-backed, keeled (in young), yellowish carapace patterned with dark blotches; lighter, unblotched plastron with a single anterior hinge; striking mottlings of yellow and black on the head, neck, and legs.

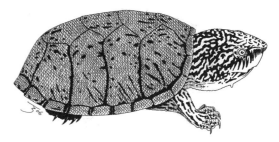

HABITAT AND DISTRIBUTION: Muddy bottoms of streams and rivers, often near submerged debris; Lee and Scott counties in the southwestern corner of VA.

HABITS: Aquatic except for dispersal and egglaying; active day or night.

DIET: Forages on stream or river bottom, mainly for molluscs, insects, crustacea, and fish.

REPRODUCTION: Breeds in fall and spring in FL, but timing not well known for VA; female lays 3–4 clutches of 2–3 eggs in a nest on land, often dug in loose soil at the base of a stump or log; hatching occurs in about 3 months; sexual maturity is reached in 3–6 years.

CONSERVATION STATUS: S2—VA; limited range in VA rivers makes this species vulnerable to pollution.

RANGE: Southwestern VA, eastern TN, western NC, GA, AL, northern FL, and southeastern LA.

KEY REFERENCES: Martof et al. (1980), Mitchell (1994), Hammerson (2005a).

Stinkpot *Sternotherus odoratus*

DESCRIPTION: L = 8–14 cm (3.1–5.5 in); a small turtle with dark, oblong, domed carapace, often covered with algae; small plastron with a single, weak, anterior hinge; head has a pointed snout, barbels on the chin and throat, and prominent yellow lines running from the snout above and below the eye.

HABITAT AND DISTRIBUTION: Shallow, muddy-bottomed, vegetation-choked, freshwater wetlands throughout much of the region, although mostly

absent from the Appalachian Plateau of PA and parts of WV.

HABITS: Aquatic except for dispersal and egglaying; sometimes basks on horizontal limbs above water; mainly crepuscular or nocturnal, but sometimes active during the day; hibernates; aestivates; aptly dubbed the "stinkpot" for exuding a disagreeable odor when attacked by predators.

DIET: Benthic forager for molluscs, insects, fish, tadpoles, and carrion.

REPRODUCTION: Breeds in spring and summer; female lays a clutch of 2–4 eggs in a nest dug in mud or sandy soil or under a rotted log, which hatch in 2–3 months; sometimes several females will use the same nest site.

RANGE: Eastern North America from Canada to Mexico.

KEY REFERENCES: Hulse et al. (2001), Hammerson (2005a).

FAMILY EMYDIDAE

River Cooter *Pseudemys concinna*

DESCRIPTION: L = 14–32 cm (5.5–12.6 in); a fairly large turtle with a slightly domed carapace, somewhat flattened and serrated along the back edge; carapace is patterned with dark and light concentric circles; head and legs are brown or greenish, broadly

striped with yellow; plastron yellow with dark markings.

HABITAT AND DISTRIBUTION: Rivers and lakes of VA west of the Coastal Plain, and southern WV.

HABITS: Aquatic except for dispersal and nesting; diurnal; hibernates; aestivates; basks on logs and rocks.

DIET: Forages along river bottoms on benthic vegetation (e.g., eel grass and elodea), invertebrates, vertebrates, and carrion.

REPRODUCTION: Breeds in spring or early summer; female lays one or more clutches of about 20 eggs in a nest excavated on land near the shoreline in loose soil; hatching occurs in 3–4 months; some hatchlings overwinter in the nest cavity.

CONSERVATION STATUS: S2—WV; limited range in WV makes this species vulnerable to habitat degradation.

RANGE: Eastern United States from VA south to FL and west to eastern KS and TX.

KEY REFERENCES: Mitchell (1994).

Florida Cooter *Pseudemys floridana*

DESCRIPTION: L = 22–40 cm (8.7–15.7 in); a fairly large turtle with slightly domed carapace serrated along the back edge; carapace is brown or greenish with yellow markings and vertical bars along the edge; plastron is mostly yellow; head and legs are

brown or greenish with yellow striping; considered a subspecies of *P. concinna* by Seidel (1994).

HABITAT AND DISTRIBUTION: Muddy-bottomed swamps and ponds with abundant aquatic vegetation and basking sites along the Coastal Plain of southeastern VA.

HABITS: Aquatic except for egglaying, dispersal, and basking; diurnal; hibernates.

DIET: Adults feed mainly on aquatic plants; juveniles eat invertebrates.

REPRODUCTION: Breeding occurs May–Aug; female digs nest in loose soil on land near the shoreline and lays a clutch of 5–18 eggs, which hatch in 3–4 months; hatchlings often overwinter in the nest; sexual maturity is reached in 3–6 years.

CONSERVATION STATUS: S3—VA; limited range makes this species vulnerable in VA.

RANGE: Coastal Plain of the southeastern United States from VA to La; also in IL.

KEY REFERENCES: Mitchell (1994).

Northern Red-bellied Cooter
Pseudemys rubriventris

DESCRIPTION: L = 25–40 cm (9.8–15.7 in); a fairly large turtle with an oblong, slightly domed carapace serrated along the back edge; carapace is brown or greenish with yellow markings and vertical orange or yellow bars along the edge; plastron is red or pink-

ish; head and legs are brown or greenish with yellow striping; the front edge of the upper jaw is notched with a pair of tooth-like cusps.

HABITAT AND DISTRIBUTION: Muddy-bottomed rivers, swamps, marshes, and lakes with abundant aquatic vegetation and basking sites near deep water; can use brackish wetlands; Coastal Plain and adjacent Piedmont of VA, PA, MD, DE, and NJ; also in WV.

HABITS: Main activity period is Mar–Oct; diurnal; aquatic except for dispersal, egglaying, and basking; hibernates in benthic mud.

DIET: Adults are mainly herbivorous, feeding on leaves, stems, and seeds of aquatic plants, e.g., water-lily and pickerelweed; juveniles are omnivorous.

REPRODUCTION: Breeds in spring and early summer; female digs a nest in loose soil, at times several hundred m from the shoreline, and lays a clutch of 8–20 eggs, which hatch in 2–3 months; hatchlings often remain in the nest to emerge in spring.

CONSERVATION STATUS: S2—PA, WV; limited range in PA and WV makes this species vulnerable; harvested for food in some parts of range, with unknown effects on populations.

RANGE: Eastern United States from NY and MA south to NC.

KEY REFERENCES: Mitchell (1994), Hulse et al. (2001), Hammerson (2005a).

Painted Turtle *Chrysemys picta*

DESCRIPTION: L = 11–18 cm (4.3–7.1 in); a fairly small turtle; the oval, slightly domed carapace is dark with yellowish lines along the seams of the broad, dorsal scutes and reddish marks along the lateral and posterior edges; yellowish plastron; head and legs are dark with reddish or yellowish lines; large yellowish spot behind the eye.

HABITAT AND DISTRIBUTION: Muddy-bottomed swamps, rivers, lakes, ponds, and marshes with abundant submerged vegetation and basking sites (emergent rocks, logs, or banks) nearly throughout the region except for parts of WV and western VA.

HABITS: Active Mar–Oct; diurnal; aquatic except for dispersal, egglaying, and basking; hibernates in benthic mud; incubation temperature determines sex (males < 27% C; females > 30% C).

DIET: Aquatic plants, invertebrates, and vertebrates.

REPRODUCTION: Mating occurs fall and spring; females lay one or more clutches of 2–8 eggs in spring and summer, digging a nest in sandy, well drained sites, well exposed to the sun; hatchlings generally winter in the nest in northern portions of the range, and emerge in spring; sexual maturity is reached in 5–8 years.

RANGE: Much of North America from Canada to northern Mexico.

KEY REFERENCES: Hulse et al. (2001).

Pond Slider *Trachemys scripta*

DESCRIPTION: L = 13–29 cm (5.1–11.4 in); a medium-sized turtle; the oval, moderately domed carapace is brown with broad yellow stripes across the back and smaller, vertical stripes along the sides; plastron is yellow with large black spots along the sides; head and legs are brown with fine yellow lines; large yellow patch behind the eye; some older males are almost entirely dark.

HABITAT AND DISTRIBUTION: Muddy-bottomed fresh or brackish water swamps, lakes, ponds, rivers, streams, and marshes with abundant aquatic vegetation and basking sites; most of the naturally occurring Mid-Atlantic sliders are *T. scripta scripta*, the

Yellowbelly Slider, which is found mainly in the Coastal Plain of southeastern VA; there is a native population of the closely related Red-eared Slider (*T. scripta elegans*) in WV as well as apparently introduced populations of this taxon at scattered localities nearly throughout the region, especially along rivers and streams in urban areas. These urban Red-eared Sliders presumably derive from escaped pets since this species is commonly sold as hatchlings in the pet trade.

HABITS: Active Apr–Oct; diurnal; aquatic except for dispersal, egglaying and basking; hibernates.

DIET: Omnivorous, feeding on aquatic plant parts, invertebrates, vertebrates, and carrion.

REPRODUCTION: Females lay one or more clutches of 6–15 eggs in a nest dug in loose soil in late spring or early summer; young hatch in 2–3 months; sexual maturity is reached in 5–7 years.

CONSERVATION STATUS: S1—WV; limited range in WV makes this species vulnerable.

RANGE: Western Hemisphere from the eastern United States south to Argentina; introduced populations (*T. s. elegans*) in Canada and several western states.

KEY REFERENCES: Mitchell (2004), Hammerson (2005a).

Spotted Turtle *Clemmys guttata*

DESCRIPTION: L = 9–13 cm (3.5–5.1 in); a small turtle; black, oblong, moderately domed carapace with yellow or orange spots; plastron yellow with large black blotches; head is dark with yellow spots; legs are lined with orange or yellow; male has brown eyes and pale chin; female has yellow eyes and a yellow chin.

HABITAT AND DISTRIBUTION: Quiet, shallow pools, ponds, bogs, streams, and marshes with mucky bottom and abundant aquatic vegetation generally in or near woodlands; can tolerate brackish wetlands; northwestern and southeastern PA, NJ, DE, MD, eastern VA, and one locality in the eastern panhandle of WV (Jefferson County).

HABITS: Principal annual activity period is relatively limited, running from Mar–Jul; diurnal; aquatic except for dispersal, egglaying, and basking; hibernates in wetland mud, sometimes in communal hibernacula; often uses muskrat burrows in stream banks during inactive periods.

DIET: Benthic forager, mainly on invertebrates, especially insects, snails, and worms.

REPRODUCTION: Mating takes place in spring; female lays 1–2 clutches of 3–5 eggs in summer in a nest dug in loose, well drained soil; eggs hatch in 1.5–3 months; sexual maturity is reached in 7–10 years.

CONSERVATION STATUS: S1—WV; S3—PA, DE; reasons for declines are not well understood but presumed to be related to wetlands drainage and collecting for the pet trade.

RANGE: Great Lakes region from NY to IL and along the Atlantic Coast states from ME to northern FL.

KEY REFERENCES: Van Dam and Hammerson (2005).

Wood Turtle *Clemmys insculpta*

DESCRIPTION: L = 14–23 cm (5.5–9.1 in); a medium-sized turtle; rough horn-colored or dark brown carapace with each scute raised and ridged; relatively long tail; plastron yellowish with dark blotches; head and legs are brown and knobby; ventral portions of forelegs and neck with bright orange markings; lacks bright orange head markings of the smaller Bog Turtle (*C. muhlenbergii*); placed in the genus *Glyptemys* by some authors.

HABITAT AND DISTRIBUTION: Wood Turtles frequent a wide range of terrestrial and aquatic habitats; terrestrial habitats include alder thickets, woodlands, meadows, and agricultural fields; wetlands used include rivers and streams; in the Mid-Atlantic, they are found in most of PA, the eastern panhandle of WV, northern NJ, the northern tip of VA, and western MD.

HABITS: The annual activity period for Wood Turtles runs from Mar–Oct; during the cooler periods of

spring and fall, they spend most of their time in the water; during summer, they spend most of the day on land, but at night often return to water or bury themselves in wet depressions on land; in winter, they hibernate in benthic material.

DIET: Feeds on plant leaves and fruits (e.g.,strawberries), mushrooms, and invertebrates.

REPRODUCTION: Mating is mainly in the fall; nesting occurs in the summer; female lays a clutch of 3–13 eggs in a nest dug in loose soil at an exposed site, e.g., a stream bank or sand bar; young hatch in 2–3 months; sexual maturity is reached at 18–22 years.

CONSERVATION STATUS: S2—VA, WV; S3-NJ, PA; illegal collecting for the pet trade is a major threat to wild populations; prolonged time required to reach sexual maturity makes it difficult for populations to recover when adults are harvested; other threats include habitat destruction and highway mortality.

RANGE: Northeastern United States, neighboring Canada, and the Great Lakes region.

KEY REFERENCES: Mitchell (1991), Hulse et al. (2001), Soule and Hammerson (2005).

Bog Turtle *Clemmys muhlenbergii*

DESCRIPTION: L = 8–10 cm (3.1–3.9 in); a small turtle with dark carapace, plastron, head, limbs, and tail; head with bright orange markings; inside of legs mottled with yellow; males have concave plastron and long, thick tail; females have flat plastron and thinner, shorter tail; placed in the genus *Glyptemys* by some authors.

HABITAT AND DISTRIBUTION: Sphagnum bogs, fens, spring seeps, marshes, and meadows with shallow, muddy bottoms and low grass, sedge, or moss as emergent vegetation; found in scat-

tered localities in VA, PA, NJ, DE, and MD.

HABITS: Main annual activity period is Mar–Jun, extending into Sep in some areas; diurnal; basks on emergent vegetation; aestivates/hibernates in benthic muck, often in association with muskrats, the muddy floors of whose burrows serve as hibernacula.

DIET: Forages in wetlands, mainly for insects, but worms, crayfish, tadpoles, seeds, and fruits are also taken.

REPRODUCTION: Mating occurs Mar–Jun; egglaying May–Jul; female digs nest in moss, grass, or sedge mats, and lay 3–5 eggs; eggs hatch in 1.5–2.5 months; sexual maturity is reached in 5–8 years.

CONSERVATION STATUS: S1—DE, VA, PA; S2—MD, NJ; "Threatened"—USFWS; limited range, specialized habitat requirements, habitat degradation, and illegal collecting pose threats to populations.

RANGE: Spotty distribution in the eastern United States from MA and NY south to GA.

KEY REFERENCES: Mitchell (1991), Hulse et al. (2001), Hammerson (2005a).

Chicken Turtle *Deirochelys reticularia*

DESCRIPTION: L = 10–25 cm (3.9–9.8 in); a small to medium-sized turtle; elongate, moderately domed carapace is dark with light lines along the scute sutures; plastron is yellow, sometimes with dark blotches on the bridge; head and long neck with yellow lines;

broad yellow band on the forelimb; streaked yellow and black on the hindlimb thighs.

HABITAT AND DISTRIBUTION: Freshwater ponds among wooded dunes; found only at Seashore State Park in the city of Virginia Beach, VA.

HABITS: Diurnal; aestivates/hibernates; basks on logs.

DIET: Feeds mostly on aquatic insects, tadpoles, and crayfish.

REPRODUCTION: Nesting may be bimodal, with eggs laid in spring and fall; female digs nest in exposed area of loose sand and lays a clutch of 5–15 eggs; fall clutches may overwinter in nest; sexual maturity is reached at 2–4 years in males, 6–8 years in females.

CONSERVATION STATUS: S1—VA; restricted range with chronic low water levels and high predation rates from raccoons and snapping turtles threaten the small, vulnerable population.

RANGE: Southeastern United States from VA south to FL and west to eastern OK and TX.

KEY REFERENCES: Mitchell (1991), Martof et al. (1980), Buhlman (1995), Virginia Department of Game and Inland Fisheries (2004), Hammerson (2005a).

Blanding's Turtle *Emydoidea blandingii*

DESCRIPTION: L = 18–22 cm (7.1–8.7 in); a medium-sized turtle; carapace is dark with small yellow spots; hinged plastron is yellow with dark blotches; head is dark with a yellow throat; has been placed in the genus

Clemmys or the genus *Emys* by some taxonomists.

HABITAT AND DISTRIBUTION: Beaver swamps, marshes, ponds, and quiet streams; found only in northwestern PA in the Mid-Atlantic region.

HABITS: Mostly aquatic; diurnal; hibernate in benthic mud Oct–Apr.

REPRODUCTION: Mating takes place shortly after emergence from hibernation in spring; female digs a nest in a grassy area with sandy loamy soil; clutch size 9–16; incubation 73–106 days; age at sexual maturity for females is 14–20 years.

CONSERVATION STATUS: S1—PA; wetland habitat for this species is threatened by drainage and pollution.

RANGE: Northeastern North America from Nova Scotia west to western Ontario south to NE, IL, IN, OH, and, PA.

KEY REFERENCES: Hulse et al. (2001).

Northern Map Turtle *Graptemys geographica*

DESCRIPTION: L = 10–25 cm (3.9–9.8 in); a medium-sized turtle; distinctively shaped carapace is ridged down the center with a flared, serrated posterior edge; carapace is brown with light reticulations; plastron is yellow; head, neck, and limbs have yellow lines; yellow spot just behind the eye.

HABITAT AND DISTRIBUTION: Lakes, ponds, and rivers; in the Mid-Atlantic region it is found in

tributaries of the upper Tennessee River drainage in southwestern VA, Lake Erie and the Susquehanna River drainage in PA, Delaware River drainage in PA, NJ, and MD, and the Monongahela and Cheat rivers in northern WV.

HABITS: Aquatic except for dispersal and egglaying; diurnal; basks on logs, rocks, and snags; hibernates in winter in benthic mud.

DIET: Benthic forager on molluscs, crayfish, insects, and dead fish.

REPRODUCTION: Principal nesting period is May–Jul; female lays one or two clutches of 3–20 eggs in a nest dug in loose soil, which hatch in 2–3 months; young of later clutches may overwinter in nest; temperature controls sex of hatchlings, with low nest temperatures producing males and higher temperatures producing females (as is true for many turtle species).

CONSERVATION STATUS: S1—MD; S2—WV, VA; S3-PA, NJ; restricted distribution vulnerable to pollution.

RANGE: Scattered river drainages across the eastern and central United States and southern Canada from Quebec west to MN and south to GA and OK.

KEY REFERENCES: Mitchell (1994), White and White (2002), Virginia Department of Game and Inland Fisheries (2004), Hammerson (2005a).

Diamond-backed Terrapin
Malaclemys terrapin

DESCRIPTION: L = 10–14 cm (3.9–5.5 in) (males); L = 15–23 cm (5.9–9.1 in) (females); a small to medium-sized turtle; dark carapace with each scute scored with raised, concentric ridges; head, neck and limbs gray with dark spots or completely dark; plastron lighter, sometimes mottled.

HABITAT AND DISTRIBUTION: Saltmarsh, tidal creeks, bays, estuaries; immediate coast from NJ to VA.

HABITS: Annual activity period Mar–Oct; diurnal; hibernates in benthic mud of estuaries; salt glands allow these turtles to survive in brackish environments.

DIET: Benthic foragers on molluscs, crustacea, annelids, and carrion.

REPRODUCTION: Nests in spring and summer; female lays one or more clutches per year of 4–15 eggs in sand of estuary banks and beaches above high tide line, which hatch in 2–4 months; sexual maturity is reached in 2–3 years for males, 4–5 years for females.

CONSERVATION STATUS: This species was once highly prized for making turtle soup, resulting in near extirpation across much of its range. Although still harvested in some areas, most populations have recovered. The principal threat at present appears to be drowning in crab traps and highway mortality.

RANGE: Immediate coast from MA to TX.

KEY REFERENCES: Hulse et al. (2001), White and White (2002).

Eastern Box Turtle *Terrapene carolina*

DESCRIPTION: L = 11–17 cm (4.3–6.7 in); a small turtle; carapace is oval, high domed, and brightly patterned in yellow and brown; plastron is yel-

lowish, brown, or black, and hinged to allow the turtle to with draw completely within its shell; head and limbs are brown and yellow; males have a concave plastron and sometimes have red eyes; females have yellow or brownish eyes and flat plastron.

HABITAT AND DISTRIBUTION: Woodlands, fields, and marshlands throughout most of the region except northern PA; often in shallow water during summer hot periods.

HABITS: Annual activity period is Apr–Oct; diurnal; mostly terrestrial; hibernates in shallow pits under forest litter; aestivates during dry periods in mud of pools and marshes.

DIET: Feeds on leaves, seeds, grass, fruits, fungi, invertebrates, and salamanders.

REPRODUCTION: Mates on land in spring after emerging from hibernation or any time during the summer (females can retain viable sperm for months or years after mating); nesting period is May–Jul; female lays a clutch of 1–8 eggs in a nest dug in loose soil, which hatch 1.5–3 months; young may overwinter in nest; sexual maturity is reached at 4–10 years.

RANGE: Eastern North America from ME, southern Canada, and MI south to FL and west to TX and IA.

KEY REFERENCES: DeGraaf and Yamasaki (2001), Hulse et al. (2001).

FAMILY TRIONYCHIDAE

Spiny Softshell *Apalone spinifera*

DESCRIPTION: L = 12–24 cm (4.7–9.5 in) (males); 18–43 cm (7.1–16.9 in) (females); a medium-sized to large turtle; carapace is nearly round, slightly domed and brownish in color with scattered, eye-like spots (ocelli); plastron white; head with a long, pointed snout; the Spiny Softshell differs from the similar Smooth Softshell (*A. mutica*) in having spiny protrusions along the front edge of the carapace.

HABITAT AND DISTRIBUTION: Rivers, lakes, and ponds, usually with a sandy or muddy bottom, of western PA, southwestern VA, and western WV; also along the lower Delaware River in PA and NJ (possibly introduced).

HABITS: Annual activity period is Apr–Oct; aquatic except for egglaying; diurnal; basks on banks and logs; aestivates and hibernates in benthic mud.

DIET: Swims along the bottom to capture aquatic insects, crayfish, and fish.

REPRODUCTION: Nests from May–Aug; females lay a clutch of 8–33 eggs in a nest dug along an exposed shoreline or sandbar; hatching occurs in late summer or fall, and some hatchlings remain in the nest until spring.

CONSERVATION STATUS: S1—MD; S2—VA, S3—PA; restricted wetland distribution vulnerable to drainage and pollution.

RANGE: Southeastern Canada, much of the United States east of the Rockies, and northern Mexico

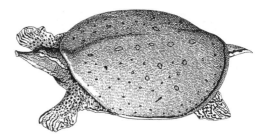

(Rio Grande and Colorado river drainages); absent from most of the northeast and peninsular FL; introduced in several western states.

KEY REFERENCES: Hulse et al. (2001), DeGraaf and Yamasaki (2001).

Smooth Softshell *Apalone mutica*

DESCRIPTION: L = 11–18 cm (4.3–7.1 in) (males); 17–35 cm (6.7–13.8 in) (females); a medium-sized to large turtle; carapace is nearly round, slightly domed with a central ridge, and brownish in color with scattered, eye-like spots (ocelli); plastron white; head with a long, pointed snout; distinguished from the similar Spiny Softshell (*A. spinifera*) by lack of spiny protrusions along the front edge of the carapace.

HABITAT AND DISTRIBUTION: Rivers, lakes, and ponds, usually with a sandy or muddy bottom; recorded from only three localities in the Mid-Atlantic-two in western PA, and Mason County, WV.

HABITS: Annual activity period is Apr–Oct; aquatic except for egglaying; diurnal; basks on banks and logs; aestivates and hibernates in benthic mud.

DIET: Swims along the bottom to capture aquatic insects, crayfish, and fish.

REPRODUCTION: Nests from May–Aug; females lay a clutch of 12–18 eggs in a nest dug along an exposed shoreline or sandbar; hatching occurs in late summer or fall, and some hatchlings remain in the nest until spring.

CONSERVATION STATUS: SX—PA; SH—WV; possibly extirpated in the Mid-Atlantic region; pollution of the few localities where the species was found may have been responsible.

RANGE: Central and eastern United States from western PA west to ND and south to FL and NM.

KEY REFERENCES: Hulse et al. (2001), Pauley (2005c), Hammerson (2005a).

ORDER SQUAMATA; FAMILY IGUANIDAE

Eastern Fence Lizard *Sceloporus undulatus*

DESCRIPTION: L = 10–18 cm (3.9–7.1 in); a rough, scaly lizard, brown or gray above with jagged dark and light horizontal markings down the back and tail; adult males have a bluish chin and greenish-blue blotches bordered with black on the belly.

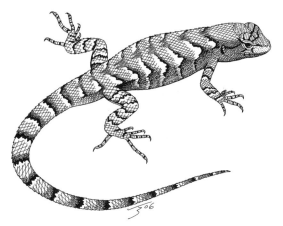

HABITAT AND DISTRIBUTION: A variety of open areas from forest clearings to fields, pastures, cliffs, and scree; preferred habitats have exposed sites for hunting and basking with nearby nooks and crannies for cover; southern half of PA and NJ south throughout the remainder of the Mid-Atlantic.

HABITS: Diurnal, sit-and-wait predator; solitary;

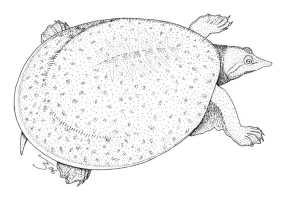

males are territorial; autotomizes tail when attacked by predators; hibernates from Sep–Apr.

DIET: Insects, spiders, millipedes, snails.

REPRODUCTION: Mating begins in Apr or May; females lay 1–2 clutches of 6–14 eggs in rotten logs, crevices, or holes in soil, which hatch in 6–8 weeks; males reach maturity in one year, females in 2–3 years.

CONSERVATION STATUS: S3— PA; limited range vulnerable to habitat change.

RANGE: United States and northern Mexico.

KEY REFERENCES: Mitchell (1994), Hulse et al. (2001).

FAMILY SCINCIDAE

Coal Skink *Eumeces anthracinus*

DESCRIPTION: L = 13–18 cm (5.1–7.1 in); a medium-sized lizard with smooth, shiny scales; brown and black above with 2 light stripes along each side (no central dorsal stripe); adult males have reddish along the side of the head and throat during the mating season; juveniles have bluish tails.

HABITAT AND DISTRIBUTION: Ground litter of moist forest, second growth, and fields; scattered localities in the highlands of PA, WV, and VA.

HABITS: Diurnal; readily autotomizes tail when attacked (up to 80% in some populations); aestivates; hibernates Oct–Mar.

DIET: Forages under logs, rocks, and forest duff for soil and litter invertebrates.

REPRODUCTION: Female lays a clutch of 4–11 eggs in a nest under a rock or log in May or Jun, which hatch in 4–6 weeks; female guards eggs.

CONSERVATION STATUS: S2—WV, VA; S3—PA; few localities and small population size make the species potentially vulnerable.

RANGE: Highlands of eastern United State from NY to GA; also southern and central United States from KY west to KS south to western FL and east TX.

KEY REFERENCES: Hulse et al. (2001), Hammerson (2005a).

Common Five-lined Skink *Eumeces fasciatus*

DESCRIPTION: L = 13–21 cm (5.1–8.3 in); a medium-sized shiny skink; dark brown above and pale below with 5 white or yellowish lines running from nose to tail (the central line splits at the back of the head and runs to either side of the snout); adult males have reddish head; juveniles have blue tails.

HABITAT AND DISTRIBUTION: Open, mesic woodlands

with ample logs, rocks, and litter; found throughout most of the Mid-Atlantic except northern PA.

HABITS: Diurnal; hibernates Oct–Mar; autotomizes tail when attacked.

DIET: Forages under rotten logs, rocks, and litter for soil invertebrates.

REPRODUCTION: Female lays 6–12 eggs in May or Jun, often in a nest burrowed in or under a rotten log; female guards and care for eggs, dampening nest as needed and shifting eggs daily until hatching in 4–6 weeks; sexual maturity is reached in 2 years.

CONSERVATION STATUS: S3—NJ; small population size and few localities make the species vulnerable.

RANGE: Eastern North America from southern Canada to Fl and east TX.

KEY REFERENCES: Mitchell (1994), Hulse et al. (2001), Hammerson (2005a).

Southeastern Five-lined Skink *Eumeces inexpectatus*

DESCRIPTION: L = 14–22 cm (5.5–8.7 in); a medium-sized shiny lizard, similar in appearance to the other lined skinks except that it lacks an enlarged row of scales under the tail; dark above and pale below with 5 white or yellowish lines running from nose to tail (the central line splits at the back of the head and runs to either side of the snout); adult males have reddish head; juveniles have blue tails.

HABITAT AND DISTRIBUTION: Open pine and mixed woodlands, fields, beaches, and disturbed areas with abundant logs, rocks, and similar ground cover; Piedmont and Coastal Plain of VA and neighboring MD.

HABITS: Diurnal; hibernates Nov–Mar; autotomizes tail when attacked.

DIET: Forages under forest duff for litter invertebrates.

REPRODUCTION: Female lays 6–11 eggs in May–Jun,

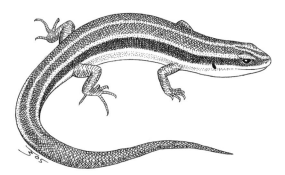

often in a nest burrowed under a board, rock, or rotten log; female attends eggs until hatching in 4–6 weeks; sexual maturity is reached in 2 years.

RANGE: Southeastern United States from MD and KY south to FL and LA.

KEY REFERENCES: Martof et al. (1980), Mitchell (1994).

Broad-headed Skink *Eumeces laticeps*

DESCRIPTION: L = 17–33 cm (6.7–13.0 in); a medium-sized to large shiny lizard; dark above and pale below with 5 white or yellowish lines running from nose to tail (the central line splits at the back of the head and runs to either side of the snout); adult males (depicted) have reddish head, and older individuals may be entirely coppery with no striping; juveniles have blue tails; long digits and claws for climbing.

HABITAT AND DISTRIBUTION: Open oak and mixed woodlands, fields, and disturbed areas with abundant logs, stumps, rocks, and similar ground cover; parts of the VA, PA, and MD Piedmont and Coastal Plain and a few highland records in VA; limited distribution in WV.

HABITS: Diurnal; semiarboreal, frequenting logs, trunks, and stumps; hibernates Sep–Mar; autotomizes tail when attacked.

DIET: Forages on tree trunks and stumps as well as under forest duff for invertebrates and small vertebrates, using olfaction and visual cues for detection of prey.

REPRODUCTION: Female lays 6–18 eggs in May–Jul, often in a nest burrowed under a board, rock, or rotten log; female attends eggs until hatching in 6–7 weeks; sexual maturity is reached in 2 years.

CONSERVATION STATUS: SH—DE; S1—PA; S2—WV; small population size and limited distribution make species vulnerable.

RANGE: Eastern United States from PA west to KS south to FL and east TX.

KEY REFERENCES: Martof et al. (1980), Mitchell (1994), Hammerson (2005a).

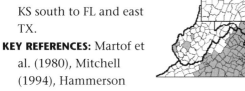

Little Brown Skink *Scincella lateralis*

DESCRIPTION: L = 8–13 cm (3.1–5.1 in); a small shiny skink with a brownish back and dark stripe along each side; pale below.

HABITAT AND DISTRIBUTION: Litter of deciduous and mixed woodlands and grassland; Piedmont and Coastal Plain of DE, MD, VA, NJ; scattered records in WV and highlands of VA.

HABITS: Mostly diurnal; hibernates Oct–Mar; autotomizes tail when attacked by predators.

DIET: Forages in forest litter for invertebrates.

REPRODUCTION: Female lays several small clutches (2–4 eggs) in rotting logs or stumps in Apr–Jun, which hatch in 4–8 weeks; sexual maturity is reached in first year.

CONSERVATION STATUS: S1—DE; S3—WV; few localities and low population size make this species vulnerable in some parts of its range.

RANGE: Eastern United States from NJ west to KS south to FL and east TX; mostly absent from Appalachians.

KEY REFERENCES: Martof et al. (1980), Mitchell (1994), Hammerson (2005a).

FAMILY TEIIDAE

Six-lined Racerunner *Cnemidophorus sexlineatus*

DESCRIPTION: L = 15–24 cm (5.9–9.5 in); a long, sandy-colored lizard with 3 thin, pale stripes running down each side; very long tail and digits; adult males are often bluish below while females are whitish; juveniles have bluish tails; recently placed in the genus *Aspidoscelis* by some authors.

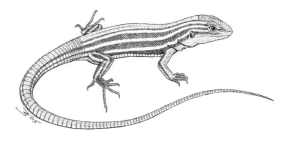

HABITAT AND DISTRIBUTION: Open areas including coastal dunes, fields, croplands, grasslands, and open woodlands; shelters in burrows under litter, logs, and rocks; Piedmont and Coastal Plain of VA and MD.

HABITS: Diurnal; hibernates Oct–Apr.

DIET: Forages in litter for invertebrates.

REPRODUCTION: Female lays one or more clutches of 1–6 eggs in burrows under litter, rotten logs, or rocks, which hatch in 6–8 weeks.

CONSERVATION STATUS: S1—WV; few localities and low population size.

RANGE: Eastern and central United States.

KEY REFERENCES: Mitchell (1994), Hammerson (2005a).

FAMILY ANGUIDAE

Slender Glass Lizard *Ophisaurus attenuatus*

DESCRIPTION: L = 56–118 cm (22–46 in); a long, legless, shiny lizard; snakelike but with movable eyelids, external ear openings, and a groove running along each side of its body; brown with several dark and light lines; differs from the Eastern Glass Lizard in having dark stripes or marks below the lateral groove.

HABITAT AND DISTRIBUTION: Pine savanna, grassland, old fields, second growth, and open woodlands with dense grassy undergrowth and litter; VA Piedmont and Coastal Plain.

HABITS: Diurnal; fossorial; hibernates Oct–Mar; autotomizes tail when attacked.

DIET: Soil and litter invertebrates and small vertebrates.

REPRODUCTION: Female lays a clutch of 3–19 eggs in a nest burrow dug in loose soil in May–Jun, and remains with the eggs until hatching 7–9 weeks later; sexual maturity is reached in 2 years.

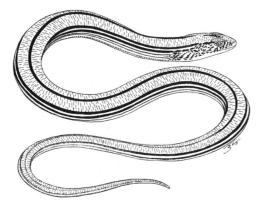

RANGE: Eastern and central United States from VA west to NE and south to FL and TX.

KEY REFERENCES: Mitchell (1994), Hammerson (2005a).

Eastern Glass Lizard *Ophisaurus ventralis*

DESCRIPTION: L = 46–107 cm (18–42 in); a legless, shiny lizard; snakelike but with movable eyelids, external ear openings, and a groove running along each side of its body; brown with several dark and light lines; differs from the Slender Glass Lizard in lacking dark stripes or marks below the lateral groove.

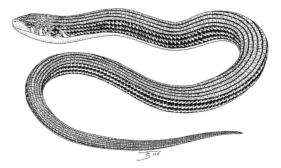

HABITAT AND DISTRIBUTION: Pine flatwoods, wet grasslands and open woodlands with grassy understory; there are only two known localities for the species in the Mid-Atlantic region: Back Bay National Wildlife Refuge and False Cape State Park, both in VA.

HABITS: Diurnal; fossorial; hibernates Oct–Mar; autotomizes tail when attacked.

DIET: Feeds mainly on grasshoppers, crickets, spiders, beetles, and snails.

REPRODUCTION: Female lays 7–15 eggs in May–Jun in a burrow or depression under a rotten log or rock, and guards nest until hatching in 7–9 weeks.

CONSERVATION STATUS: S1—VA; limited distribution vulnerable to habitat change.

RANGE: Southeastern United States along the Coastal Plain from VA to LA.

KEY REFERENCES: Mitchell (1991), Hammerson (2005a).

FAMILY COLUBRIDAE

Eastern Worm Snake *Carphophis amoenus*

DESCRIPTION: L = 19–32 cm (7.5–12.6 in); a small shiny snake; brown above, pinkish below; sharp spine-like tail tip.

HABITAT AND DISTRIBUTION: Moist woodlands with loose soils; found throughout most of the Mid-Atlantic region except western and northern PA, western MD, and parts of northern WV.

HABITS: Fossorial, burrowing in litter or sandy soil; active day or night; hibernates from Oct–Apr.

DIET: Earthworms are the main prey.

REPRODUCTION: Mating occurs in spring or fall; females lay a clutch of 2–6 eggs in burrows, moist litter piles, or under rotten logs in Jun–Jul, which hatch in 7–8 weeks; sexual maturity is reached in 2–3 years.

CONSERVATION STATUS: S3—PA, WV; limited distribution.

RANGE: Eastern and central United States from MA west to NE and south to GA and TX.

KEY REFERENCES: Hulse et al. (2001), Hammerson (2005a).

Scarletsnake *Cemophora coccinea*

DESCRIPTION: L = 36–66 cm (14–26 in); a beautiful, small to medium-sized snake, strikingly patterned with red, black, and white or yellowish bands above, whitish below; in the similarly patterned Scarlet Kingsnake, the red, black and white dorsal bands circle the body; Coral Snakes, which do not occur in the Mid-Atlantic region, have black snouts (not red as in Milksnake and Scarletsnake) and contiguous bands of red and yellow (not red and black).

HABITAT AND DISTRIBUTION: Pine and mixed woodlands with loose soil; NJ (Pine Barrens) and parts of the Piedmont and Coastal Plain of VA, MD, and DE; scattered records in VA highlands.

HABITS: Fossorial; nocturnal; hibernates from Oct–Apr.

DIET: Forages in sandy soil and litter for reptile eggs, insects, and mice.

REPRODUCTION: Female lays a clutch of 3–8 eggs under forest litter in Jun–Jul, which hatch in 10–12 weeks.

CONSERVATION STATUS: SH—DE; S3—MD; low numbers and secretive habits make the species vulnerable.

RANGE: Eastern and central United States from NJ west to OK and south to FL and TX.

KEY REFERENCES: Hulse et al. (2001), Hammerson (2005a).

Kirtland's Snake *Clonophis kirtlandii*

DESCRIPTION: L = 36–45 cm (14–18 in); a small to medium-sized, slender snake; brown or reddish-brown above with two dorsal rows of large, dark spots; red belly with a row of small spots along each side; yellowish throat.

HABITAT AND DISTRIBUTION: Wet meadows, pond and stream borders, marshes, damp litter; found only in western PA in our region.

HABITS: Hibernates/aestivates; forages mostly under leaf litter, rotten logs, or other surface debris in damp, open areas; flattens itself dramatically when threatened.

DIET: Feeds mostly on earthworms, slugs, and leeches.

REPRODUCTION: Mating occurs Feb–Jul; 4–15 live young are born Jul–Sep.

CONSERVATION STATUS: SH—PA; little is known about this very rare water snake.

RANGE: Northeastern MO, IL, IN, OH, southern MI, and western PA.

KEY REFERENCES: Van Dam and Hammerson (2005a).

Eastern Racer *Coluber constrictor*

DESCRIPTION: L = 0.9–1.6 m (35–63 in); a long, slim, cylindrical snake (not flattened on the belly like the rat snakes [*Elaphe*]); entirely black above,

somewhat paler below with a white chin; juveniles are brightly patterned with reddish, brown, or gray blotches.

HABITAT AND DISTRIBUTION: Fields, meadows, agricultural areas, open woodlands; found throughout the region.

HABITS: Diurnal; bimodal summer activity period with peaks in May and Sep; hibernates from Oct–Mar.

DIET: Feeds by capturing and swallowing whole a variety of invertebrates and small vertebrates, especially grasshoppers, mice, bird eggs, frogs, lizards, and other snakes; racers are not constrictors, but may use a body loop to pin larger prey.

REPRODUCTION: Mating takes place in spring after emergence from hibernation; female lays a clutch of 5–28 eggs under rocks, litter, logs, or similar sites, in Jun–Jul; the same sites may be used year after year and by more than one female; hatching occurs in 6–9 weeks; sexual maturity is reached in 2–3 years.

RANGE: North America from southern Canada throughout most of the United States.

KEY REFERENCES: Hulse et al. (2001).

Ring-necked Snake *Diadophis punctatus*

DESCRIPTION: L = 25–51 cm (10–20 in); a small snake; black above with a yellowish collar; reddish or orange below.

HABITAT AND DISTRIBUTION: Woodlands, fields, rocky hillsides, and stream borders with abun-

dant rocks or logs for cover; found throughout the region.

HABITS: Active day or night; fossorial; bimodal summer activity period with peaks in Jun and Sep; hibernates from Nov–Mar.

DIET: Feeds on a variety of small vertebrates and invertebrates, but especially salamanders and earthworms.

REPRODUCTION: Mating occurs in spring; female lays a clutch of 2–10 eggs in Jun–Jul under a rock or rotting log or in a burrow, which hatch in 4–8 weeks; more than one female may use the same nest.

RANGE: North America from Canada to northern Mexico.

KEY REFERENCES: DeGraaf and Yamasaki (2001).

Cornsnake *Elaphe guttata*

DESCRIPTION: L = 0.7–1.2 m (28–47 in); a medium-sized snake, red, orange, gray, or brown above with prominent red bands on the back outlined

in black; whitish below checkered or banded with black; belly is noticeably flattened, as in all rat snakes (*Elaphe*).

HABITAT AND DISTRIBUTION: Mesic forest with loose soil and litter; also old fields, second growth, and agricultural areas; southern NJ, MD, parts of DE, most of VA, and a few localities in the Eastern Panhandle of WV.

HABITS: Mostly crepuscular or nocturnal; hibernates Oct–Mar; can climb relatively well; vibrates tail when alarmed.

DIET: Constrictor-dispatches prey by coiling around it and suffocating it; main prey items are small mammals, but birds, bird eggs, lizards, insects, frogs, and other snakes are also taken.

REPRODUCTION: Female lays a clutch of 6–31 eggs in Jun–Jul under a rotted log, forest litter, or in a burrow; hatching occurs in 8–10 weeks; sexual maturity is reached in 2–3 years.

CONSERVATION STATUS: S1—WV, DE; restricted range makes the species vulnerable.

RANGE: Eastern and central United States from NJ west to UT and south to FL and NM; also northern Mexico and introduced into several Caribbean islands.

KEY REFERENCES: Mitchell (1994), Hammerson (2005a).

Eastern Ratsnake *Elaphe obsoleta*

DESCRIPTION: L = 1–2 m (39–78 in); a long snake, black above and white below; belly is noticeably flattened as in all rat snakes (*Elaphe*); juveniles are gray with dark blotches on the back, a "Y"-shaped blotch on the back of the head, and a dark bar across the snout between the eyes.

HABITAT AND DISTRIBUTION: Woodland, fields, agricultural areas; found throughout the Mid-Atlantic region.

HABITS: Diurnal; good climber; hibernates Oct–Mar in holes, caves, cellars, or crevices, sometimes

with other snakes including other species; rat snakes assume a coiled, aggressive posture with raised head and vibrating tail when alarmed.

DIET: Constrictor-dispatches prey by coiling around it and suffocating it; main prey items are small mammals, birds, and bird eggs; also takes lizards, insects, frogs, and other snakes.

REPRODUCTION: Mates in May–Jun; female lays a clutch of 5–14 eggs in Jun–Jul under a rotten log, stump, or similar piles of decaying litter, sometimes communally; eggs hatch in about 2 months; sexually mature in 4 years.

RANGE: Eastern North America from southern Canada south to FL and west to TX.

KEY REFERENCES: Hulse et al. (2001).

Red-bellied Mudsnake *Farancia abacura*

DESCRIPTION: L = 1–1.9m (39–74 in); a large, stout, shiny snake with blunt snout and spine-tipped tail; black back with red or pinkish blotches along the sides; reddish belly with black checks.

HABITAT AND DISTRIBUTION: Blackwater swamps, cypress bayous, quiet rivers and streams; Coastal Plain of southeastern VA.

HABITS: Mostly nocturnal; good swimmer; burrows in mud during inactive periods; hibernates Oct–Apr.

DIET: Principal prey is aquatic salamanders (e.g., *Amphiuma* and *Siren*), but also eats other amphibians and fish.

REPRODUCTION: Female digs a nest burrow in Jul–Aug under a log, stump, or other detritus, and deposits 30–50 eggs with which she remains until hatching in Aug–Oct; reproduction data are few.

RANGE: Coastal Plain of the southeastern United States from VA to east TX; also up the Mississippi drainage to IL.

KEY REFERENCES: Mitchell (1994).

Rainbow Snake *Farancia erytrogramma*

DESCRIPTION: L = 0.9–1.7 m (35–66 in); a large, brightly colored snake; black back with three red stripes running from head to tail; belly is red or yellow with two longitudinal black stripes; tail tipped with a spine.

HABITAT AND DISTRIBUTION: Rivers, streams, bayous, freshwater and brackish marshes, and adjacent sandy soiled uplands; Charles County, MD, and Coastal Plain of eastern VA.

HABITS: Mostly nocturnal and apparently active throughout the year; good swimmer.

DIET: Freshwater eels (*Anguilla*) are a major prey item, as well as other aquatic vertebrates and invertebrates.

REPRODUCTION: Female lays 20–40 eggs in Jun–Jul in a burrow dug in sandy soil, and remains with the eggs until the young hatch in Sep–Oct; young may remain in a burrow in or near the nest until spring; sexual maturity is reached at 2–3 years.

CONSERVATION STATUS: S1—MD; S3-VA; small popu-

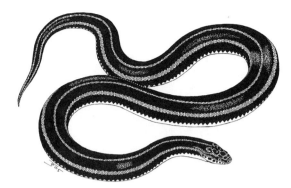

lation size and limited distribution in wetlands threatened by development.

RANGE: Coastal Plain of the southeastern United States from MD to LA.

KEY REFERENCES: Mitchell (1994), Hammerson (2005a).

Eastern Hog-nosed Snake *Heterodon platirhinos*

DESCRIPTION: L = 0.4–1.2 m (16–47 in); a medium-sized to large snake; variable in color pattern from solid black or gray to alternating broad dorsal bands of black, gray, reddish, brown, or orange; upturned snout ("hognose").

HABITAT AND DISTRIBUTION: Woodlands, old fields, and grasslands with loose, sandy soils, usually near water; found nearly throughout the region except northern and western PA.

HABITS: Diurnal; may be mildly venomous (salivary extract kills amphibians); hibernates Oct–Mar; bizarre defense behavior includes initial cobra-like expansion of the head and neck in an erect, threatening posture accompanied by hissing, but if the attacker persists, the snake will roll over on its back with its mouth open and "play dead."

DIET: Feeds mostly on toads, but will take other small vertebrates and invertebrates.

REPRODUCTION: Mating occurs in Apr–May; female lays 4–61 eggs in Jun–Jul in a burrow or under decaying logs or other vegetation; hatching occurs in 1.5–2 months; sexual maturity reached in second year.

CONSERVATION STATUS: S3—PA, WV; small local populations are vulnerable to extirpation.

RANGE: Eastern North America from southern Canada to FL and TX.

KEY REFERENCES: DeGraaf and Yamasaki (2001), Hulse et al. (2001), Hammerson (2005a).

Yellow-bellied Kingsnake *Lampropeltis calligaster*

DESCRIPTION: L = 0.7–1.2 m (28–47 in); a handsome, medium-sized to large snake, tan or brown with dark irregular diamond blotches on the back outlined with black; older individuals may be uniformly brown; small head.

HABITAT AND DISTRIBUTION: Woodlands and fields with dry, sandy soil; Piedmont and parts of the Coastal Plain of MD and VA.

HABITS: Nocturnal; fossorial; hibernates Oct–Mar.

DIET: Feeds on lizards, snakes, small mammals, birds, insects, grasshoppers, and frogs killed by constriction.

REPRODUCTION: Female lays a clutch of 5–17 eggs in

Jun–Jul in a burrow; hatching occurs in Aug–Sep; sexual maturity is reached in 2–3 years.

RANGE: Eastern United States from MD west to NE and south to FL and east TX; absent from Appalachians and most of southern GA, AL, and peninsular FL.

KEY REFERENCES: Mitchell (1994).

Common Kingsnake *Lampropeltis getula*

DESCRIPTION: L = 0.9–1.8 m (35–71 in); a large shiny snake, black with white or yellow bands.

HABITAT AND DISTRIBUTION: Woodlands, fields, agricultural areas, swamps, marshes; southern NJ, DE, VA, and MD mostly east of highlands; also southwestern WV and VA (Black Kingsnake, *L. g. nigra*).

HABITS: Diurnal or crepuscular; hibernates Oct–Apr; vibrates tail and discharges musk from anal glands when alarmed.

DIET: Feeds on snakes, including venomous species to whose venom it appears to be immune; also takes lizards, amphibians, small mammals, birds, and insects.

REPRODUCTION: Female lays a clutch of 2–24 eggs in Jun–Jul under rotting logs, litter, or stumps; hatching occurs in 2–3 months.

CONSERVATION STATUS: SX—PA; S2—DE, small population with limited distribution.

RANGE: North America from NJ west to OR and south to northern Mexico.

KEY REFERENCES: Hulse et al. (2001), Hammerson (2005a).

Milksnake *Lampropeltis triangulum*

DESCRIPTION: L = 0.6–1.1 m (24–43 in); a medium-sized shiny snake; this species occurs as two very different forms in the Mid-Atlantic region: 1) the "Scarlet Kingsnake (*L. t. elapsoides*)," which has a beautiful red body with black and white bands; and 2) the "Eastern Milksnake (*L. t. triangulum*)," which has a brown body with silver and black bands.

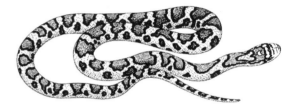

HABITAT AND DISTRIBUTION: Woodlands, fields, agricultural areas; the "Eastern Milksnake" form occurs throughout most of the Mid-Atlantic region except the Piedmont and Coastal Plain of VA, while the "Scarlet Kingsnake" form occurs only in the Coastal Plain of southeastern VA.

HABITS: Mostly crepuscular or nocturnal; hibernates Oct–Apr; vibrates tail and discharges musk from anal glands when alarmed.

DIET: Feeds on small mammals, birds, other snakes, lizards, amphibians, and insects.

REPRODUCTION: Female lays a clutch of 6–24 eggs in Jun–Jul under rotting logs, litter, or stumps; hatching occurs in 1.5–3 months; sexual maturity is reached in 3–4 years.

CONSERVATION STATUS: S2—DE; small population with limited distribution.

RANGE: Western Hemisphere from ME west to MT and south to northern South America.

KEY REFERENCES: Mitchell (1994), Hulse et al. (2001), DeGraaf and Yamasaki (2001), Hammerson (2005a).

Plain-bellied Watersnake *Nerodia erythrogaster*

DESCRIPTION: L = 0.8–1.5 m (32–59 in); a medium-sized to large snake; the form found in the Mid-Atlantic region is known as the "Redbelly Watersnake," which is reddish-brown or brown above and orange or red below; juveniles are tan or reddish with black diamonds on the back.

HABITAT AND DISTRIBUTION: Sloughs, blackwater swamps, cypress bays, quiet streams and rivers, and nearby uplands; southeastern VA and parts of the southern Delmarva Peninsula.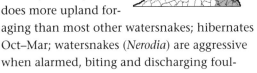

HABITS: Mostly aquatic, although this species does more upland foraging than most other watersnakes; hibernates Oct–Mar; watersnakes (*Nerodia*) are aggressive when alarmed, biting and discharging foul-smelling musk from anal glands.

DIET: Feeds on fish, frogs, salamanders, crayfish, and aquatic insects.

REPRODUCTION: Mating takes place in spring; Plain-bellied Watersnakes are viviparous, giving birth to 5–37 live young in Aug–Oct.

CONSERVATION STATUS: S1—DE; small population with limited distribution.

RANGE: Eastern and central North America from DE west to KS and south to northern FL and central Mexico.

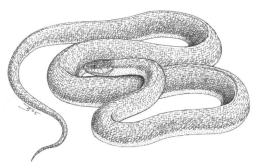

KEY REFERENCES: Mitchell (1994), Hammerson (2005a).

Northern Watersnake *Nerodia sipedon*

DESCRIPTION: L = 0.6–1.3 m (24–51 in); a medium-sized to large snake, variable in pattern; brown, reddish, or gray above with dark blotches down the center of the back alternating with dark blotches on the sides; yellowish, brown, or reddish below with dark spots.

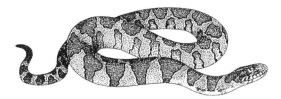

HABITAT AND DISTRIBUTION: Rivers, streams, lakes, ponds, swamps, marshes, bogs, often with exposed snags and logs (for basking) and rocky shores (for hibernating); found throughout the Mid-Atlantic region.

HABITS: Mostly aquatic; hibernates Oct–Mar in rocky crevices or burrows in stream banks; watersnakes (*Nerodia*) are aggressive when alarmed, biting and discharging foul-smelling musk from anal glands.

DIET: Feeds mostly on fish, frogs, and salamanders, but takes other aquatic vertebrates and invertebrates as well.

REPRODUCTION: Mating occurs in Apr–May or Oct; viviparous, with 11–36 live young born Aug–Oct; sexual maturity reached at 2–3 years.

RANGE: Eastern North America from ME west to SD and south to GA and east TX.

KEY REFERENCES: DeGraaf and Yamasaki (2001).

Brown Watersnake *Nerodia taxispilota*

DESCRIPTION: L = 0.8–1.6 m (32–64 in); a large, stout snake; brown above with black blotches down the center of the back and alternating black blotches down the side; yellowish below with dark spots; wide triangular head.

HABITAT AND DISTRIBUTION: Lakes, rivers, fresh and brackish water marshes; Coastal Plain of southeastern VA.

HABITS: Aquatic; mostly diurnal; arboreal, climbing snags and branches overhanging water bodies for basking and mating; hibernates Nov–Mar; watersnakes (*Nerodia*) are aggressive when alarmed, biting and discharging foul-smelling musk from anal glands.

DIET: Feeds mostly on fish, especially catfish (*Ictalurus punctatus*).

REPRODUCTION: Mating occurs in spring; viviparous, giving birth to 20–40 live young in Aug–Sep.

RANGE: Coastal Plain of the southeastern United States from VA to AL.

KEY REFERENCES: Mitchell (1994).

Rough Greensnake *Opheodrys aestivus*

DESCRIPTION: L = 60–90 cm (24–35 in); a long, thin snake; green above, yellow below; dorsal scales are keeled (the shorter, thicker-bodied, Smooth Greensnake has unkeeled dorsal scales).

HABITAT AND DISTRIBUTION: Riparian woodlands, forest, thickets, old fields, gardens, parks; preferred habitat is shrubs overhanging water bodies, but also forages on forest floor or through grassy understory; southern NJ, a few localities in southeastern and southwestern PA, DE, eastern MD, VA except highlands, western WV and Jefferson County, WV, in the Eastern Panhandle.

HABITS: Largely arboreal; hibernates Nov–Mar.

DIET: Forages at the lower and mid- level of shrubs, often bordering or overhanging streams and ponds, for caterpillars, spiders, grasshoppers, crickets, and other arthropods.

REPRODUCTION: Female lays a clutch of 3–12 eggs in Jun–Jul under rotten logs, decaying vegetation, rocks, or other forest litter; eggs hatch in 3–6 weeks; sexual maturity is reached in 2–3 years.

CONSERVATION STATUS: S1—PA; S2—DE; S3—WV; small populations with limited distribution.

RANGE: Eastern United States from NJ west to KS south to FL and TX.

KEY REFERENCES: Mitchell (1994), Hammerson (2005a).

Smooth Greensnake *Opheodrys vernalis*

DESCRIPTION: L = 40–60 cm (16–24 in); a small to medium length snake; green above, yellowish below; dorsal scales are unkeeled (the longer, slimmer Rough Greensnake has keeled dorsal scales); placed in the genus *Liochlorophis* by some authors.

HABITAT AND DISTRIBUTION: Old fields, grasslands, pasture, meadows, open woodlands; highlands of VA, parts of WV, most of PA, northwestern NJ, and western MD.

HABITS: Terrestrial; diurnal; hibernates Nov–Mar, often in communal hibernacula.

DIET: Feeds mainly on spiders, caterpillars, crickets, grasshoppers, and other terrestrial arthropods.

REPRODUCTION: Female lays a clutch of 3–12 eggs in Jul–Sep in a nest under a rotted log or rock or in a burrow; hatching occurs in 4–12 days (females retain eggs throughout most of the early developmental period of the young); nests communally at times; sexual maturity reached in 2 years.

CONSERVATION STATUS: S3—PA, VA, NJ; populations declining in many parts of the range, perhaps due to pesticide use (caterpillars are a major food item) and reversion of farm land to forest.

RANGE: Disjunct distribution across eastern and central North America from eastern Canada west to MT and south to VA and NM.

KEY REFERENCES: DeGraaf and Yamasaki (2001), Hammerson (2005a).

Pinesnake *Pituophis melanoleucus*

DESCRIPTION: L = 1–1.8 m (39–71 in); a large, heavy bodied snake, strikingly patterned with black and white blotches above; white below, sometimes spotted with black; head is similar in width to the body; face is white spotted with black; formerly included the Gophersnake (*P. catenifer*), Bullsnake (*P. c. sayi*) and Louisiana Pinesnake (*P. ruthveni*), which are now recognized as separate taxa.

HABITAT AND DISTRIBUTION: Rocky slopes with loose soil in the highlands of western VA; pine-oak woodlands with sandy soil in the Pine Barrens of southern NJ.

HABITS: Crepuscular; fossorial; hibernates Oct–Apr in burrows built by the snake; impressive defense displays include rising on coils with erect head, hissing, lunging and striking, tail vibrating, and, if captured, biting vigorously.

DIET: Feeds mainly on small mammals killed by constriction; also takes birds and bird eggs.

REPRODUCTION: Mating occurs in spring; female digs a burrow in sandy soil or under a log or stump or in a mammal burrow in Jun–Aug and lays a clutch of 2–24 eggs, which hatch in 2–4 months; sometimes nests communally.

CONSERVATION STATUS: SH—WV, MD; S2-VA; probably extirpated from WV and MD; small populations in limited range about which there is little information are a problem; also destruction of preferred sandy woodland habitat.

RANGE: Eastern United States from NJ west to KY and south to FL and LA; disjunct populations in IL, IN, KY, and TN.

KEY REFERENCES: Mitchell (1994), Hammerson (2005a).

Glossy Crayfish Snake *Regina rigida*

DESCRIPTION: L = 36–78 cm (14–31 in); a small to medium-sized snake, dark and shiny above and yellow below with black half moons in two rows along the belly; dark lips contrast with yellow cheeks and throat.

HABITAT AND DISTRIBUTION: Cypress bays, swamps, ponds, marshes; found only in New Kent County, VA, in the Mid-Atlantic region.

HABITS: Nocturnal; aquatic; hibernates Oct–Apr.

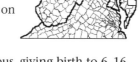

DIET: Adults feed mostly on crayfish and other aquatic arthropods.

REPRODUCTION: Viviparous, giving birth to 6–16 young Jul–Sep.

CONSERVATION STATUS: S1—VA; limited range in VA makes the species vulnerable to extirpation.

RANGE: Coastal Plain of the southeastern United States from VA to east TX; also north to OK and AR.

KEY REFERENCES: Mitchell (1994), Hammerson (2005a).

Queen Snake *Regina septemvittata*

DESCRIPTION: L = 38–86 cm (15–34 in); a slender, medium-sized snake with narrow head; brown above with 3 dark lines on the back and 2 yellow lines on the side; yellow below with 4 dark stripes or rows of mottling; yellow cheeks, lips, and throat.

HABITAT AND DISTRIBUTION: Rocky cataracts, streams, and rivers of the Piedmont and highlands; parts of VA, PA, WV, DE, and MD.

HABITS: Aquatic; mostly diurnal; often basks on limbs above water; spends inactive periods under rocks or stones bordering waterway; hibernates Oct–Apr; wriggles vigorously and emits foul anal secretions when attacked.

DIET: Feeds mostly on soft-shelled (recently molted) crayfish; also takes small fish, amphibians, and aquatic insects.

REPRODUCTION: Viviparous, giving birth to 5–15 young Jul–Sep; sexual maturity reached in 2–3 years.

CONSERVATION STATUS: SH—NJ; S1—DE; S3-PA; possibly extirpated from NJ; threatened by small population size, limited distribution, and stream degradation.

RANGE: Patchily distributed in eastern North America from NJ and western NY west to WI and south to the Florida Panhandle and MS.

KEY REFERENCES: Mitchell (1994), Hammerson (2005a).

DeKay's Brownsnake *Storeria dekayi*

DESCRIPTION: L = 23–46 cm (9–18 in); a small snake; brown, gray, or reddish above with parallel rows of dark checks along the back; whitish or pinkish below.

HABITAT AND DISTRIBUTION: Woodlands and open areas with abundant ground cover; found throughout most of VA, MD, NJ, and DE, and parts of WV and PA.

HABITS: Nocturnal; hibernates Oct–Apr, often sharing communal hibernacula in mammal burrows, ant mounds, and rock crevices.

DIET: Feeds mostly on slugs and snails.

REPRODUCTION: Mates in spring; viviparous, giving birth to 3–31 live young in summer; sexual maturity reached in 2 years.

CONSERVATION STATUS: S3—DE; small population size and limited range make the species vulnerable to extirpation in DE.

RANGE: Eastern and central North America from ME west to SD and south through the eastern United States, Mexico, and Central America to Honduras.

KEY REFERENCES: DeGraaf and Yamasaki (2001), Hulse et al. (2001), Hammerson (2005a).

Red-bellied Snake *Storeria occipitomaculata*

DESCRIPTION: L = 17–31 cm (7–12 in); a small snake; brown or gray above with two rows of dark checks down the back, often bordering a lighter central stripe; belly can be yellowish, pink, or red; 3 yellowish blotches on the neck, sometimes more or less connected; small white spot at the corner of the mouth.

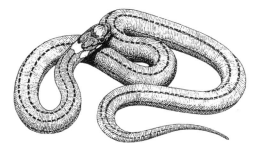

HABITAT AND DISTRIBUTION: Woodlands and old fields with abundant ground cover; most often encountered under rocks, logs, and litter; found throughout most of the region except western VA and northwestern WV.

HABITS: Terrestrial; fossorial; mostly nocturnal, but can be active day or night; hibernates Oct–Apr in burrows, rock crevices, ant mounds, and similar sites, often communally; when attacked they wriggle vigorously, discharge an offensive, anal secretion, and sometimes gape or curl their lips; sometimes feign death.

DIET: Feeds mostly on slugs, although other litter invertebrates are taken as well.

REPRODUCTION: Viviparous, giving birth to 1–21 live young Jul–Sep; sexual maturity reached in 2 years.

CONSERVATION STATUS: S1— DE; small population size and limited range make the species vulnerable to extirpation in DE.

RANGE: Eastern North America from New Brunswick to Saskatchewan in Canada south to FL and east TX; disjunct populations in SD, WY, and NE.

KEY REFERENCES: DeGraaf and Yamasaki (2001), Hulse et al. (2001), Hammerson (2005a).

Southeastern Crowned Snake
Tantilla coronata

DESCRIPTION: L = 20–30 cm (8–12 in); a small snake; head and neck are black dorsally separated by a whitish collar; otherwise brown or tan above; whitish or yellowish below.

HABITAT AND DISTRIBUTION: Pine, pine-oak, and oak woodlands, dunes, sand hills, and slopes with loose soil and abundant ground cover of rotting logs, stumps, litter, and rocks; found only in 6 Piedmont counties and one Coastal Plain county in VA.

HABITS: Nocturnal; terrestrial; fossorial; hibernates Nov–Mar; has an enlarged, grooved, rear tooth and weak venom.

DIET: Feeds on litter and soil invertebrates, e.g., centipedes, slug, earthworms, and termites.

REPRODUCTION: Mating occurs in fall and spring; female lays a clutch of 1–4 eggs in a nest in a rotten log or stump in Jun–Jul, which hatch in Aug; sexual maturity is reached in 2–3 years.

CONSERVATION STATUS: S2—VA; small population size and limited range make the species vulnerable to extirpation in VA.

RANGE: Southeastern United States from VA west to KY and south to northern FL and eastern LA.

KEY REFERENCES: Martof et al. (1980), Mitchell (1994), Hammerson (2005a).

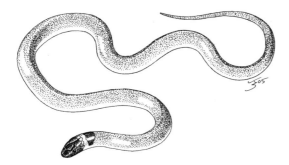

Short-headed Gartersnake
Thamnophis brachystoma

DESCRIPTION: L = 25–35 cm (10–14 in); a small to medium-sized snake; brown with dark checks above with a white or yellowish central stripe and 2 buff lateral stripes; belly is tan or grayish green; chin and throat are whitish or yellowish; head appears too small for body; unlike the Common Gartersnake (*T. sirtalis*), the Short-headed Gartersnake usually lacks a row of dark spots between the central and lateral stripes.

HABITAT AND DISTRIBUTION: Old fields, pastures, meadows, and woodlands, often under logs, rocks, and litter near creeks; northwestern PA.

HABITS: Terrestrial; fossorial; gregarious, often congregating under suitable cover objects; hibernates Oct–Apr; wriggles vigorously when attacked and exudes foul anal secretions.

DIET: Feeds mainly or exclusively on earthworms.

REPRODUCTION: Mating occurs in spring; viviparous, giving birth to 5–14 live young, Jun–Sep.

CONSERVATION STATUS: S3—PA; small population size and limited range make the species vulnerable to extirpation in PA; reversion of farm land to forest may impact Short-headed Gartersnakes.

RANGE: Unglaciated portions of northwestern PA and southwestern NY; successfully introduced into areas bordering the original range.

KEY REFERENCES: Hulse et al. (2001), Hammerson (2005a).

Eastern Ribbonsnake *Thamnophis sauritus*

DESCRIPTION: L = 46–97 cm (18–38 in); a small to medium-sized, slim snake with a long tail; brown above with a white or yellowish central stripe and 2 lateral stripes; belly is yellowish or grayish green; usually has a white spot in front of the eye.

HABITAT AND DISTRIBUTION: Wetland borders; found throughout much of the Mid-Atlantic region but absent from parts of central and western PA and WV.

HABITS: Mostly diurnal; terrestrial; arboreal, using shrubs over water for basking; aquatic in pursuit of prey; hibernates Oct–Mar.

DIET: Feeds mainly on frogs and salamanders.

REPRODUCTION: Mating occurs in spring; viviparous, giving birth to 3–26 live young Jul–Aug; sexual maturity reached in 2–3 years.

CONSERVATION STATUS: S2—DE, WV; S3—PA; small population size and limited range make the species vulnerable to extirpation in parts of the Mid-Atlantic where wetland drainage or pollution are problems.

RANGE: Eastern North America from ME west to MI and south to FL and LA.

KEY REFERENCES: DeGraaf and Yamasaki (2001), Hulse et al. (2001), Hammerson (2005a).

Common Gartersnake *Thamnophis sirtalis*

DESCRIPTION: L = 0.5–1.1 m (20–43 in); a relatively slim, medium-sized snake; pattern is highly variable, but often is brown with dark checks and a central light stripe above; yellowish below with dark checks.

HABITAT AND DISTRIBUTION: Forest or open areas with dense understory, usually near water; found throughout the Mid-Atlantic region.

HABITS: Diurnal; terrestrial; hibernates Oct–Mar

DIET: Feeds on litter and soil invertebrates and vertebrates, including earthworms, salamanders, toads, spiders, insects, and millipedes.

REPRODUCTION: Mating occurs in spring; viviparous, giving birth to 3–85 live young Jul–Sep; sexual maturity reached at 2 years.

RANGE: North America from central Canada south to FL, TX, and southern CA; mostly absent from western mountain and desert regions.

KEY REFERENCES: DeGraaf and Yamasaki (2001), Hulse et al. (2001).

Rough Earthsnake *Virginia striatula*

DESCRIPTION: L = 18–32 cm (7–13 in); a small, plain snake with pointed snout and strongly keeled dorsal scales; brown above and shiny white, pinkish or yellowish below; sometimes with a light band across the head.

HABITAT AND DISTRIBUTION: Open woodlands and fields with abundant logs, litter, or other debris;

found only along the Coastal Plain of southeastern VA in our region.

HABITS: Terrestrial; fossorial; hibernates Oct–Mar; emits foul secretions or feigns death when attacked.

DIET: Feeds mostly on earthworms.

REPRODUCTION: Mating occurs in spring; viviparous, giving birth to 2–13 live young Jun–Sep; sexual maturity reached at 1–2 years.

RANGE: Southeastern United States from VA west to KS and south to northern FL and east TX.

KEY REFERENCES: Martof et al. (1980), Mitchell (1994).

Smooth Earthsnake *Virginia valeriae*

DESCRIPTION: L = 18–33 cm (7–13 in); a small, plain snake with pointed snout and little or no keel on dorsal scales (i.e., "smooth"); brown or gray

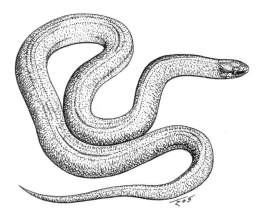

above with black flecks; whitish below. This taxon was split into 2 species in our region by Collins (1990) although the herpetological community has yet to recognize these taxa: the Smooth Earth Snake (*V. valeriae*) in the eastern portion and western WV, and the Mountain Earthsnake (*V. pulchra*) in the mountains of western PA, eastern WV, and Highland County in western VA.

HABITAT AND DISTRIBUTION: Woodlands, old fields, and pastures with abundant logs, litter, or other debris; patchy distribution throughout the Mid-Atlantic.

HABITS: Terrestrial; fossorial; hibernates Nov–Apr; emits foul secretions when attacked.

DIET: Feeds mostly on earthworms.

REPRODUCTION: Viviparous, giving birth to 4–14 live young Aug–Sep; sexual maturity reached at 1–2 years.

CONSERVATION STATUS: S1—DE; S2—PA; S3—WV; small population size, poor knowledge of the life history, and limited range make the species vulnerable to extirpation in parts of the Mid-Atlantic, especially the "Mountain Earthsnake" form, which is endemic.

RANGE: Patchy, local distribution in eastern North America from NJ west to KS and south to FL and east TX.

KEY REFERENCES: Mitchell (1994), Hulse et al. (2001), Hammerson (2005a).

FAMILY VIPERIDAE

Copperhead *Agkistrodon contortrix*

DESCRIPTION: L = 0.6–1.1 m (24–43 in); a medium-sized to large, thick-bodied, venomous snake; note heavy, triangular head and pit between eye and nostril; brown, gray, tan, or pinkish dorsal color with reddish hour-glass shaped blotches on the back.

HABITAT AND DISTRIBUTION: Woodlands and open areas, especially with abundant rocks, litter, logs, and detritus; wood piles are favorite resting sites; found nearly throughout the region except northern PA and southern NJ.

HABITS: Terrestrial; diurnal in spring but crepuscular or nocturnal in summer; hibernates Nov–Mar; venomous; able to inject venom into predators or prey with retractable front fangs.

DIET: Adults eat mostly small mammals, some of which, e.g., the wood rats (*Neotoma*), have developed resistance to venom; juvenile copperheads include invertebrates in the diet.

REPRODUCTION: Mating occurs mainly in spring; viviparous, giving birth to 1–21 live young Aug–Sep; sexual maturity reached in 2–3 years.

CONSERVATION STATUS: S1—DE; small population size and limited range make the species vulnerable to extirpation in DE.

RANGE: Eastern North America from MA west to NE and south to peninsular FL and northern Mexico.

KEY REFERENCES: DeGraaf and Yamasaki (2001), Hulse et al. (2001), Hammerson (2005a).

Cottonmouth *Agkistrodon piscivorus*

DESCRIPTION: L = 0.8–1.8 m (32–71 in); a large, heavy bodied, venomous snake; note triangular head and pit between eye and nostril; dark brown or olive dorsal color with alternating thick dark cross bands and thin lighter cross bands; some individuals are almost entirely dark on the back; light line above eye; dark line through eye; whitish or yellowish below with dark streaks or blotches.

HABITAT AND DISTRIBUTION: Swamps, bayous, freshwater and brackish marshes, quiet streams, ponds, and rivers; Coastal Plain of southeastern VA.

HABITS: Semi-aquatic; more active at night; basks on logs, banks, and rocks along watercourses; hibernates Oct–Apr in rock crevices or mammal burrows; when threatened, it will attempt escape, or coil, vibrate its tail, and open its mouth exposing the white, cottony lining and long front fangs; doesn't hesitate to bite if pressed.

DIET: Feeds mostly on fish, amphibians, and reptiles, for which its largely neurotoxic venom is best adapted.

REPRODUCTION: Females give birth to 1–16 live young in Aug–Sep; sexual maturity is reached in 3–4 years.

CONSERVATION STATUS: S3—VA; threatened by drainage and development of limited wetland habitat.

RANGE: Southeastern United States from VA west to KS and south to FL and TX

KEY REFERENCES: Mitchell (1994), Hammerson (2005a).

Timber (Canebrake) Rattlesnake
Crotalus horridus

DESCRIPTION: L = 0.9–1.9 m (35–74 in); a large, heavy bodied, venomous snake; note triangular head and pit between eye and nostril; end of tail with a few to several hard, dry skin segments forming a rattle. The Timber Rattlesnake has at least three color phases, two of which were formerly recognized as subspecies: the Timber Rattlesnake (formerly *C. h. horridus*), found mostly in the highlands, which has a black phase in which the dorsal color is dark with yellowish blotches down the center of the back, and a yellow phase in which the head and dorsal color are yellowish and there are dark central cross bands edged in yellow; the Canebrake Rattlesnake (formerly *C. h. atricaudatus*) from the Coastal Plain and adjacent Piedmont, which has tan or pinkish dorsal color anteriorly, darkening posteriorly, with a rusty central stripe and dark cross bands.

HABITAT AND DISTRIBUTION: The *C. h. horridus* forms frequent highland forests, especially rocky areas; they have a patchy distribution in highlands across most of the Mid-Atlantic region, as well as in the Pine Barrens of Southern NJ. The Canebrake Rattlesnake  form is found in woodlands and wetlands of the Coastal Plain of southeastern VA.

HABITS: Terrestrial; hibernates Oct–Apr, often in communal hibernacula in caves or crevices.

DIET: Feeds mostly on small mammals, for which its largely hemotoxic venom is best adapted.

REPRODUCTION: Female gives birth to 4–19 young in late summer or fall; sexual maturity is reached at 5–11 years; a long-lived snake with low reproductive rate.

CONSERVATION STATUS: S3—PA, WV, MD; human efforts at extirpation have been successful in many areas due to low reproductive rate.

RANGE: Eastern North America from ME west to MN and south to northern FL and east TX.

KEY REFERENCES: Mitchell (1994), Hulse et al. (2001), Hammerson (2005a).

Massasauga *Sistrurus catentatus*

DESCRIPTION: L = 0.5–0.8 m (20–32 in); a thick, medium-sized, venomous rattlesnake; note triangular head and pit between eye and nostril; end of tail with a few to several hard, dry skin segments forming a rattle; light brown above with dark blotches outlined in yellow or white; black belly; tail appears ringed.

HABITAT AND DISTRIBUTION: Old fields, meadows, grasslands, marshes; prefers drier habitats in summer, wetter areas in fall and spring; found only in western PA in the Mid-Atlantic region.

HABITS: Mostly diurnal during spring and fall, but more nocturnal during warm periods; hibernates in burrows or crevices in marshy sites from Oct–Apr.

DIET: Main prey is small mammals.

REPRODUCTION: Females give birth to 8–10 live young in the summer; sexual maturity is reached in 3 years; females produce offspring every 2 years.

CONSERVATION STATUS: S1—PA; wetland habitat for this species is threatened by drainage and pollution.

RANGE: Central North America from southern Ontario, NY, and PA west to MN and KS, south to AZ and TX.

KEY REFERENCES: Hulse et al. (2001).

Class Aves—Birds

Birds first appear in the fossil record about 150 million years ago during the late Jurassic Period of the Mesozoic Era, and have their evolutionary origins from Reptilia, although there is considerable debate concerning which group of reptiles is ancestral. They are characterized by a 4-chambered heart, feathers, ability to thermoregulate (i.e., they are "warm blooded"), and flight. There are about 9,000 species of birds in 23 orders; 331 species are found regularly in the Mid-Atlantic during one or more seasons of the year.

ORDER ANSERIFORMES; FAMILY ANATIDAE

Fulvous Whistling-Duck
Dendrocygna bicolor

DESCRIPTION: L = 51 cm (20 in); W = 92 cm (36 in); a long-necked, long-legged duck; body a rich, tawny buff; mottled black and tan on back; dark streak running from crown down nape to back; whitish streaking on throat; white on flanks; gray bill and legs; dark wings; dark tail with white base.

HABITAT AND DISTRIBUTION: Rare summer visitor (Mar–Oct) in extreme southeastern VA: Lakes, marshes, impoundments, flooded grasslands.

DIET: Forages by wading or swimming in shallow water (< 0.5 m), surface dabbling, tipping up, or shallow diving to feed on aquatic plant seeds (>96%) and some aquatic invertebrates.

RANGE: Breeds locally from southern CA, AZ, TX, LA, and FL south in coastal regions to Peru and central Argentina, West Indies; also in the Old World in East Africa, Madagascar, India, Sri Lanka, and Myanmar.

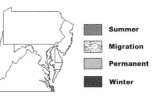

Summer
Migration
Permanent
Winter

KEY REFERENCES: Hohman and Lee (2001).

Greater White-fronted Goose *Anser albifrons*

DESCRIPTION: L = 71 cm (28 in); W = 144 cm (57 in); grayish-brown body; barred with buff on back; speckled with black on breast and belly; white lower belly and undertail coverts; pinkish bill with white feathering at base, edged in black; orange legs; in flight white rump and gray wings are key. **Immature** Lacks white feathering at base of bill and speckling on breast and belly; pale legs and bill.

HABITAT AND DISTRIBUTION: Rare winter resident (Oct–Mar), mainly along the coast of DE, MD, and VA: Lakes, marshes, grasslands, croplands.

DIET: Roosts often in large mixed-species flocks of geese in open water, and flies to foraging areas in mixed-species goose flocks; foraging habitats are agricultural fields and marshes where birds feed, mostly in daylight hours, by grazing, picking, and probing for seeds, grains, grasses, bulbs, and tubers.

RANGE: Breeds in Arctic regions of AK and western Canada; win-

Summer
Migration
Permanent
Winter

ters in temperate regions of the Northern Hemisphere, mainly in CA, TX, LA, and Mexico in North America.

KEY REFERENCES: Ely and Dzubin (1994).

Snow Goose *Chen caerulescens*

DESCRIPTION: L = 71 cm (28 in); W =144 cm (57 in); a medium-sized white goose with black primaries; orange-red bill with dark line bordering mandibles; orangish or reddish legs. **Immature** Patterned like adult but with gray bill and legs and grayish wash on back. **Blue Phase** Dark gray body with white head, neck, and belly; orange-red legs and bill. **Immature** Dark brownish-gray throughout with white chin and belly; dark legs and bill.

HABITAT AND DISTRIBUTION: Common winter resident (Nov–Mar) along the coast; rare to casual transient and winter resident inland; most WV and PA records are from the west (Ohio River Valley and Lake Erie respectively), presumably stragglers from the Mississippi Valley flyway: Bays, estuarine marshes, lakes, wet grasslands, croplands.

DIET: Forages, often in large flocks, by grazing plant shoots and leaves or probing in mud for tubers, rhizomes and roots.

RANGE: Breeds in Arctic Canada, AK, and north-eastern Siberia; win-

Summer
Migration
Permanent
Winter

ters in the west along the Pacific Coast from southwestern Canada to central Mexico, and in the east from Chesapeake Bay south through the southeastern United States, especially the lower Mississippi drainage, to Veracruz, Mexico.

KEY REFERENCES: Mowbray et al. (2000).

Brant *Branta bernicla*

DESCRIPTION: L = 61 cm (24 in); W = 100 cm (39 in); a small goose; dark above, barred with brown; white upper tail coverts nearly obscure dark tail; white below with gray barred flanks; black breast, neck, and head with white bars on side of throat; dark bill and legs: **Immature** Lacks barring on throat.

HABITAT AND DISTRIBUTION: Common winter resident (Oct–Apr) along the coast; rare in summer; uncommon to rare transient (Oct–Nov, Apr–May) in inland NJ and eastern PA; rare to casual transient and winter visitor elsewhere in the region: Bay shores, mudflats, tidal pools, lagoons, estuaries, golf courses, agricultural fields.

DIET: Forages by walking on exposed tidal beds and picking eelgrass, green algae, widgeon grass, or sea lettuce or swimming in shallow water by dipping or tipping up to harvest these or similar vegetation.

RANGE: Breeds in high Arctic; winters along northern coasts of Northern Hemisphere.

KEY REFERENCES: Reed et al. (1998).

Canada Goose *Branta canadensis*

DESCRIPTION: L = 66–122 cm (26–48 in); W = 137–213 cm (54–84 in); black neck and head with white chin strap; grayish-brown back and breast; white rump, belly, and undertail coverts; black tail; subspecies of this goose vary considerably in size.

HABITAT AND DISTRIBUTION: Common winter resident (Oct–Mar) and increasingly common summer resident* (Apr–Sep) nearly throughout; mainly a transient (Feb–Mar, Oct–Nov) in WV, but widely introduced (22 counties) and increasing as a resident: Lakes, ponds, rivers, marshes, estuaries, grasslands, croplands.

DIET: Canada Geese are almost entirely herbivorous. Foods vary seasonally from green native and domestic grasses and sedges in winter and spring to berries and seeds in late summer and fall. In coastal regions they feed on eelgrass and salt-marsh grasses. They will also take submerged rhizomes and bulbs in wetlands as available, and readily forage on crops, e.g., wheat, soybeans, sorghum, corn, and clover.

REPRODUCTION: Monogamous, apparently pairing for life; in the Mid-Atlantic region, nesting begins in March and April; open platform nest (0.7 m dia.) (28 in) usually is placed on the ground on islets, peninsulas or shorelines in dense vegetation and built (female) of materials gathered from the nest site vicinity, e.g., grasses, sedges, twigs, lichens, mosses, and cattail blades, and lined with contour feathers and down; clutch—4–7; incubation (female)—25–28 days; fledging—young are precocial, able to leave nest and swim within 24 h post-hatching, but cannot fly until 6–7 weeks later; post-fledging parental care—young birds may remain with parents throughout the first year; single brood per season.

CONSERVATION STATUS: Resident goose populations have increased dramatically in recent years, becoming a problem in terms of water sanitation and crop depredation in some areas. The source of resident goose populations is assumed to be game department introductions, but they likely derive from range expansion of wild populations as well.

RANGE: Breeds across northern half of North America; winters from northern United States south to northern Mexico, farther north along coasts; also introduced in various Old World localities.

KEY REFERENCES: Mowbray et al. (2002).

Mute Swan *Cygnus olor*

DESCRIPTION: L = 152 cm (60 in); W = 224 cm (88 in); white body; orange bill with large, black knob at base; black face from base of bill to eye; neck usually held in a graceful curve while swimming. **Immature** Brownish body with grayish bill. Young birds begin molt into white plumage by midwinter; bill becomes pinkish.

HABITAT AND DISTRIBUTION: Uncommon to rare and local resident* along the coast and locally inland (domesticated): Coastal ponds, inlets, rivers, bogs.

DIET: Forages in shallow water primarily on benthic vegetation that it can reach by dipping or tipping up; some aquatic invertebrates and vertebrates are also taken, especially during molt.

REPRODUCTION: Monogamous, apparently pairing for life; open mound nest, 1.5 m dia. (60 in), 0.5 m high (20 in) is placed on islets, peninsulas, or shorelines in dense vegetation and built by both pair members of grass, sedge, and cattail blades; clutch—5–7; incubation (female, fed by male)—36 days; fledging—young are precocial, able to leave nest and swim within 24 h post-hatching, but cannot fly until 120–150 days later; post-fledging parental care—young birds may remain with parents through the first winter; single brood per season.

CONSERVATION STATUS: Mute Swans are an Old World species introduced into the Western Hemisphere multiple times over the past two or three centuries. Since the 1970s, feral breeding populations have undergone rapid increase, rais-ing concerns regarding potential damage to some aquatic environments.

RANGE: Breeds in northern and temperate Eurasia; winters in southern portions of breeding range south to the Mediterranean and Caspian seas, northern India, and eastern China. Introduced widely throughout North America, but scarce in most parts except along the coast of the northeastern United States.

KEY REFERENCES: Ciaranca et al. (1997).

Tundra Swan *Cygnus columbianus*

DESCRIPTION: L = 132 cm (52 in); W = 206 cm (81 in); very large, entirely white bird; rounded head; black bill, often with yellow preorbital spot. **Immature** Brownish-gray with orangish bill.

HABITAT AND DISTRIBUTION: Uncommon to rare transient (Feb–Mar, Nov–Dec) nearly throughout, scarcer in western VA and WV; common winter resident in Chesapeake Bay: Lakes, bays, ponds, marshes, agricultural fields.

DIET: Forages in shallow water by using long neck to reach benthic aquatic plants, seeds, and tubers; foods include pondweed, sedge, bulrush, widgeongrass, wild celery, and shoalgrass; also feeds on crops, e.g., wheat, corn, rice, and sorghum.

CONSERVATION STATUS: Populations were decimated by hunting during the nineteenth century, but

with protection have recovered significantly since.

Summer
Migration
Permanent
Winter

RANGE: Breeds in Arctic regions of Northern Hemisphere; winters in coastal boreal and north temperate areas.

KEY REFERENCES: Rosenberg and Roth (1994).

Wood Duck *Aix sponsa*

DESCRIPTION: L = 48 cm (19 in); W = 74 cm (29 in); green head and crest streaked with white; red eye, face plate, and bill; white throat; purplish breast; iridescent dark bluish back; beige belly with white flank stripes. **Female** Deep iridescent blue on back; brownish flanks; grayish belly; brownish-gray head and crest; white eyering and postorbital stripe; white chin.

HABITAT AND DISTRIBUTION: Common to uncommon summer resident* (Mar–Oct) throughout; uncommon (Coastal Plain) to rare or casual (inland) in winter (Nov–Feb): Rivers, swamps, ponds, marshes.

DIET: Forages in shallow wetlands with emergent vegetation, pecking, dabbling, or tipping up to capture prey; feeds on both plants (e.g., seeds, roots, tubers, shoots, fruits) and animals (e.g., aquatic insects, crustacea, isopods), the proportion varying by seasonal availability.

REPRODUCTION: Monogamous; pairs form during fall and winter, and arrive at breeding sites in late March or April; nest is placed in a natural cavity, abandoned Pileated Woodpecker hole, or nest box, and built by the female with down plucked from herself; clutch—10–12; incubation (female)—30 days; fledging—precocial young jump from nest about 24 h post-hatching, but cannot fly until 8–10 weeks of age; parental care—female cares for

Summer
Migration
Permanent
Winter

young 4–8 weeks post-hatching; two broods per season are common.

RANGE: Breeds across southeastern Canada and the eastern half of the United States, and in the west from southwestern Canada to central CA, also Cuba and Bahamas; winters in southeastern United States south through northeastern Mexico, and in the west from Oregon, CA, and NM south through northwestern Mexico.

KEY REFERENCES: Hepp and Bellrose (1995).

Gadwall *Anas strepera*

DESCRIPTION: L = 51 cm (20 in); W = 86 cm (34 in); a medium-sized duck; gray above; scalloped gray, black, and white on breast and flanks; brownish head; white belly; black hindquarters; chestnut, black, and white patches on wing. **Female** Brown body mottled with buff and dark brown; orange bill marked with black; white wing patch, which, when visible, is distinctive.

HABITAT AND DISTRIBUTION: Common to uncommon transient (Oct–Nov, Mar–Apr) throughout; common to rare winter resident (Nov–Feb) and locally uncommon summer resident*, mainly

along the coast: Lakes, marshes, estuaries, ponds, bays.

DIET: Forages both diurnally and nocturnally in shallow water (30 cm) (12 in) by dabbling, dipping, or tipping up, mostly for aquatic vegetation (e.g., pond weed, widgeon grass, bulrush, and spike rush), but aquatic invertebrates can form a significant portion of the breeding and molting season diets.

CONSERVATION STATUS: S2—MD, VA; S3—DE; limited breeding distribution threatened by wetland drainage and pollution.

REPRODUCTION: Monogamous; pairs form in winter flocks in November and arrive on breeding territory in April (often the same territory used by the female in a previous year); nest , 20–30 cm dia. (8–12 in), is placed on the ground in a grassy area, often on a marsh islet; female digs out a nest bowl and lines it with grasses and down feathers plucked from her breast; clutch—8–12; incubation (female—male leaves territory for molting ground)—26 days; fledging—young are precocial, able to leave the nest and swim 24–36 h post-hatching, but cannot fly until 50 days post-hatching; post-fledging care by female—4 weeks; single brood per season.

RANGE: Breeds in boreal and north temperate grassland, prairie and steppe

Summer
Migration
Permanent
Winter

regions, parts of the Great Lakes, the Mid-Atlantic coast, and western U.S.; winters in temperate and northern tropical areas of the Northern Hemisphere.

KEY REFERENCES: Leschack et al. (1997).

Eurasian Wigeon *Anas penelope*

DESCRIPTION: L = 48 cm (19 in); W = 81 cm (32 in); medium-sized duck; gray above and on flanks; pinkish breast; chestnut head with creamy forehead and crown; white belly; black hindquarters; white and black patches on wing; green specu-

lum. **Female** Brown body mottled with buff and dark brown; grayish bill; black wing patch; green speculum.

HABITAT AND DISTRIBUTION: Rare winter resident (Oct–Mar) along the coast: Estuaries, lakes, marshes, crops, fields.

DIET: Forages in shallow water by tipping up and plucking vegetative parts of aquatic vegetation; also grazes in agricultural fields on vegetative parts and seeds of crops.

RANGE: Breeds in northern Eurasia; winters in temperate and subtropical Eurasia, but

Summer
Migration
Permanent
Winter

regular along coasts of North America.

KEY REFERENCES: Cannings (2005).

American Wigeon *Anas americana*

DESCRIPTION: L = 48 cm (19 in); W = 84 cm (33 in); medium-sized duck; gray above; purplish-brown on breast and flanks; white crown and forehead; broad, iridescent green stripe through and past eye; densely mottled black and white on cheek, chin, and throat; pale blue bill with black tip; white belly; black hindquarters; green speculum. **Female** Brown body mottled with buff and dark brown; grayish bill; black wing patch; green speculum.

HABITAT AND DISTRIBUTION: Common to uncommon transient (Oct–Nov, Mar–Apr) throughout; common (coast) to uncommon or rare (inland) winter resident (Oct–Apr); rare to casual summer resident* along the coast: Lakes, estuaries, bays, ponds, crops, fields.

DIET: Forages in shallow water by tipping up and plucking vegetative parts of aquatic vegetation; also grazes in agricultural fields on vegetative parts and seeds of crops; aquatic invertebrates form a significant portion of the breeding season diet.

RANGE: Breeds from northern North America south to the northern

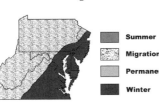

Summer
Migration
Permanent
Winter

United States; winters along Atlantic and Pacific coasts, and inland from southern United States to northwestern Colombia; West Indies.

KEY REFERENCES: Mowbray (1999).

American Black Duck *Anas rubripes*

DESCRIPTION: L = 59 cm (23 in); W = 92 cm (36 in); a large duck; dark brown body, mottled with light brown; violet speculum; reddish legs; yellowish or greenish bill; light brown head and neck finely streaked with dark brown.

HABITAT AND DISTRIBUTION: Common to uncommon winter resident (Oct–Mar) throughout; uncommon to rare summer resident* (Apr–Sep): Rivers, lakes, ponds, marshes, bogs, coastal marshes, estuaries.

DIET: Forages in shallow water, often in emergent vegetation, by tipping up, surface dabbling, or occasionally diving (2–4 m) (7–13 ft), and by picking on land for both plant (e.g., tubers, seeds, shoots) and animal (e.g., Ephemeroptera, Odonata, Diptera, Isopoda, molluscs) prey, with more animal prey taken during the early breeding season.

REPRODUCTION: Monogamous, with pairs forming during the wintering period; pairs arrive on breeding territory as early as February in the Mid-Atlantic states, but the peak of nest initiation is in April; open ground nest is placed in low dense vegetation, sometimes on a stump, trunk, or similar platform, often on a wooded island, and is built (female) by excavation of a shallow bowl, which is then lined with debris from the nest vicinity (e.g., leaves, sticks, grass, pine needles); eggs are covered with breast down by the time incubation begins; clutch—8–10; incubation (female—male leaves for molting site in mid-incubation)—25 days; fledging—young are precocial, and leave the nest within 24 h after hatching, walking 1–2 km (0.6–1.2 km) to aquatic rearing site, and are able to fly about 6 weeks post-hatching; post-fledging care by female—female leaves young at 6–7 weeks post-hatching; usually one brood per season.

CONSERVATION STATUS: S2—WV; American Black Duck populations plummeted from 1950 (800,000) to 1990 (300,000) due to overhunting with breeding and wintering habitat loss. Restrictions on hunting instituted in 1983 have

resulted in modest increases; limited breeding distribution in WV makes the bird vulnerable.

RANGE: Breeds across northeastern North America south to NC; winters in the eastern United States south to the Gulf states.

Summer
Migration
Permanent
Winter

KEY REFERENCES: Longcore et al. (2000).

Mallard *Anas platyrhynchos*

DESCRIPTION: L = 59 cm (23 in); W = 92 cm (36 in); a large duck; iridescent green head; yellow bill; white collar; rusty breast; gray scapulars and belly; brownish back; purple speculum; black rump and undertail coverts; curling black feathers at tail; white tail. **Female** Brown body mottled with buff; orange bill marked with black; white outer tail feathers; blue speculum.

HABITAT AND DISTRIBUTION: Common resident* throughout; more common during migration and in winter (Oct–Mar): Ponds, lakes, streams, marshes, flooded fields.

DIET: Forages in shallow water by surface dabbling, dipping, and tipping up to take in plant and animal matter, which is filtered out from water or mud medium by blade-like lamellae of the bill; foods include aquatic insects, molluscs, annelids, crustacea, duckweed, tubers, and seeds.

REPRODUCTION: Monogamous; pairs form during wintering period, usually by November, and return to female's home breeding area by March or April; ground nest (0.3 m dia.) is placed in dense vegetation, usually within 1 km (0.6 mi) of feeding pond or lake, and built by the female who creates a small depression by digging with her feet and pressing down with her body, and then lines it with vegetation pulled in from the immediate vicinity, including grass, leaves, and twigs; down plucked from her breast is used to line the nest and cover the eggs once incubation has begun; clutch—5–11; incubation (female)—28 days; male normally leaves for molting site late in incubation or once the young hatch; young are precocial and able to follow the female to water a day or so after hatching; female cares for young at least until they are able to fly, at about 7 weeks, or sometimes 1–2 weeks later; single brood per season.

RANGE: Breeds across boreal and temperate regions of the Northern Hemisphere; winters in temperate and subtropical regions.

Summer
Migration
Permanent
Winter

KEY REFERENCES: Drilling et al. (2002).

Blue-winged Teal *Anas discors*

DESCRIPTION: L = 41 cm (16 in); W = 64 cm (25 in); a small duck; brown mottled with dark brown above; tan marked with spots below; light blue patch on wing; green speculum; dark gray head with white crescent at base of bill. **Female** Mottled brown and dark brown above and below; tan undertail coverts spotted with brown; yellowish legs.

HABITAT AND DISTRIBUTION: Common transient (Apr–May, Aug–Sep) throughout; uncommon to rare and local in summer* (May–Aug) and winter (Oct–Mar), mainly along the coast: Ponds, lakes, estuaries, marshes.

DIET: Forages in shallow water (< 30 cm) (12 in) by surface dabbling and tipping up to capture aquatic invertebrates or take aquatic plants.

HABITAT AND DISTRIBUTION: Common (Coastal Plain) to uncommon or rare (inland) transient (Sep–Oct, Mar–Apr); common to uncommon winter resident (Nov–Feb) and rare summer resident* (May–Aug) along the coast: Lakes, marshes, estuaries, bays, ponds.

REPRODUCTION: Monogamous; birds pair in winter and migrate to female's home area to breed, arriving in April or May; open platform nest is placed on the ground, and built (female) of grasses and other plant parts gathered in the immediate vicinity of the nest site, and lined with feathers and down plucked from the female's breast; clutch—10; incubation (female)—19–29 days; young are precocial and able to follow the female to water within a day or so post-hatching; young are able to fly 40 days post-hatching, but female leaves them before then; single brood per season.

CONSERVATION STATUS: S1—VA; S2—MD, WV; S3—PA, DE; limited breeding distribution threatened by wetland drainage and pollution.

RANGE: Breeds across boreal and temperate North America south to central United States; winters southern United States to northern South America, West Indies.

KEY REFERENCES: Rohwer et al. (2002).

Northern Shoveler *Anas clypeata*

DESCRIPTION: L = 48 cm (19 in); W = 79 cm (31 in); medium-sized duck with large, spatulate bill; green head; golden eye; black back; rusty sides; white breast and belly; black rump and undertail coverts; blue patch on wing; green speculum. **Female** Mottled brown and dark brown above and below; brown eye; orange "lips" on dark bill.

DIET: Forages in open, shallow water mostly by surface or subsurface dabbling, followed by straining of prey from mud and water through the specially adapted bill; main prey are crustacea, aquatic insects, plankton, and plant seeds.

RANGE: Breeds across boreal and north temperate regions (mainly in west in North America); winters along temperate coasts south to subtropics and tropics of Northern Hemisphere.

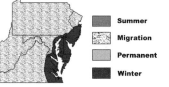

KEY REFERENCES: Dubowy (1996).

Northern Pintail *Anas acuta*

DESCRIPTION: L = 66 cm (26 in); W = 92 cm (36 in); a long-necked, long-tailed duck; gray back and sides; brown head; white neck, breast and belly; black rump, undertail coverts and tail with extremely long central feathers; speculum iridescent brown. **Female** Brownish mottled with dark brown throughout; grayish bill; pointed tail.

HABITAT AND DISTRIBUTION: Common to uncommon transient (Nov, Mar) throughout; common to uncommon winter resident (Dec–Feb) along

coastal regions and the Piedmont. Rare in summer* (May–Aug), mainly along the coast: Flooded fields, swales, shallow ponds, bays.

DIET: Forages both diurnally and nocturnally, usually in flocks, by walking and picking (fields) or wading or swimming and tipping up (shallow wetlands) for plant matter (seeds, tubers, shoots); also feeds on invertebrates, mostly aquatic insects; roosts in large flocks in open water, often some distance from foraging sites.

RANGE: Breeds in Arctic, boreal, and temperate grasslands and tundra; winters in temperate, subtropical, and tropical areas of Northern Hemisphere.

KEY REFERENCES: Austin and Miller (1995).

Green-winged Teal *Anas crecca*

DESCRIPTION: L = 38 cm (15 in); W = 61 cm (24 in); a small, fast-flying duck; chestnut head with broad iridescent green stripe above and behind eye; gray body; beige breast with black spots; white bar on side of breast; white tail; black rump and undertail coverts. **Female** Mottled brown and white above and below; whitish undertail coverts; green speculum.

HABITAT AND DISTRIBUTION: Common to uncommon transient (Mar–Apr, Aug–Oct) throughout; common (along coast) to rare (inland) winter resident (Nov–Feb); rare summer resident*

(May–Aug), mainly on Virginia's eastern shore: Lakes, estuaries, marshes, ponds.

DIET: Forages mostly by dabbling on surface or shallow dipping below in shallow water (< 12 cm) for seeds of aquatic plants (e.g., sedge, pondweed) and aquatic invertebrates (e.g., insects, molluscs, crustacea).

RANGE: Breeds in boreal and Arctic areas; winters in temperate and subtropical regions of Northern Hemisphere.

KEY REFERENCES: Johnson (1995).

Canvasback *Aythya valisineria*

DESCRIPTION: L = 53 cm (21 in); W = 84 cm (33 in); a medium-sized, heavy bodied duck with steeply, sloping forehead; rusty head; red eye; black breast; gray back and belly; black hindquarters. **Female** Grayish body, brownish neck and head.

HABITAT AND DISTRIBUTION: Common transient (Nov, Feb–Mar) throughout; common to uncom-

mon winter resident (Nov–Mar) in the Piedmont, Coastal Plain, and Lake Erie: Estuaries, lagoons, marshes, bays, lakes.

DIET: Forages, often in flocks, in shallow water (usually < 2 m) to harvest benthic and epibenthic plant parts (leaves, rhizomes, tubers), aquatic insects, and invertebrates; a major winter food in the Chesapeake Bay region is the bay bottom-growing plant, wild celery (*Vallisneria americana*).

RANGE: Breeds across north-western North America south to CA

and IA; winters locally from southern Canada south through the United States to southern Mexico.

KEY REFERENCES: Mowbray (2002).

Redhead *Aythya americana*

DESCRIPTION: L = 51 cm (20 in); W = 84 cm (33 in); rusty head; red eye; bluish bill with white ring and black tip; black breast; gray back and belly; dark brown hindquarters. **Female** Brownish throughout; bluish bill with black tip.

HABITAT AND DISTRIBUTION: Common to uncommon transient (Oct–Nov, Mar–Apr) throughout; common to uncommon winter resident (Nov–Feb) along the coast and on Lake Erie; rare to casual summer resident* (May–Sep) along the coast: Bays, lakes, rivers, ponds, lagoons, estuaries.

DIET: Forages in shallow water, usually < 1 m (3.3 ft) by diving, tipping up, dipping, or surface gleaning for floating or submerged vegetation (e.g., shoalgrass, manateegrass, and widgeongrass); also takes some aquatic invertebrates.

RANGE: Breeds in the prairie pothole region of the north-central United States

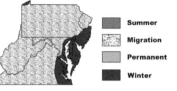

and Canada, AK, western Canada, the northwestern United States, and locally in the Great Lakes region; winters in the central and southern United States south to Guatemala, with a large portion wintering in the Laguna Madre of south TX and northern Mexico; also Greater Antilles.

KEY REFERENCES: Woodin and Michot (2002).

Ring-necked Duck *Aythya collaris*

DESCRIPTION: L = 43 cm (17 in); W = 71 cm (28 in); a smallish duck with characteristically pointed (not rounded) head; black back, breast, and hindquarters; dark head with iridescent purple sheen; gray flanks with white bar edging breast; golden eye; white feather edging at base of bill; white band across dark bill. **Female** Brown body and head; white eyering; bill with whitish band.

HABITAT AND DISTRIBUTION: Common transient and common to uncommon winter resident (Oct–Apr) throughout; rare in summer along the coast: Marshes, swamps, lakes, ponds—mostly in freshwater environments.

DIET: Forages by diving in shallow water (< 1.5 m) (4.9 ft) for aquatic seeds, tubers, and benthic invertebrates.

RANGE: Breeds across central and southern Canada, northern United States;  winters along both United States coasts, southern United States south to Panama, West Indies.

KEY REFERENCES: Hohman and Eberhardt (1998).

Greater Scaup *Aythya marila*

DESCRIPTION: L = 48 cm (19 in); W = 79 cm (31 in); dark, rounded head with iridescent green sheen; gray mottled with black back; black breast and hindquarters; gray flanks; golden eye; bluish bill with dark tip. **Female** Brown body and head; white patch at base of bill; bill bluish-gray.

HABITAT AND DISTRIBUTION: Common transient and winter resident (Oct–Apr) along the Coastal Plain and Lake Erie; rare transient (Nov, Mar) inland: Large lakes, bays, inlets, in-shore marine.

DIET: Forages in flocks during nonbreeding periods, diving in shallow water (usually < 2m deep) (6.6 ft) to feed in soft benthic substrates, mostly on invertebrates (e.g., aquatic insects [lakes], molluscs, crustacea), but vegetation and algae (e.g., sea lettuce) are also taken, depending on site, season, and availability.

RANGE: Breeds in Old and New World Arctic and subarctic regions; win-  ters along temperate and northern coasts and large lakes in the Northern Hemisphere.

KEY REFERENCES: Kessel et al. (2002).

Lesser Scaup *Aythya affinis*

DESCRIPTION: L = 43 cm (17 in); W = 71 cm (28 in); a smallish duck; gray back; black breast and hindquarters; dark head with iridescent purple sheen (sometimes greenish); gray flanks; golden eye; bluish bill with dark tip. **Female** Brown body and head; white patch at base of bill; bill grayish.

HABITAT AND DISTRIBUTION: Common to uncommon transient (Nov, Mar–Apr) throughout; common to uncommon (Coastal Plain) to rare (inland) winter resident (Oct–Apr); rare in summer along coast: Bays, estuaries, lakes.

DIET: Forages by diving in shallow water to harvest benthic invertebrates (e.g., molluscs, arthropods, annelids) or vegetation (e.g., waterweed, pondweed, and bulrush seeds).

RANGE: Breeds in AK, western and central Canada, and northern United States;

winters in coastal and central inland United States south to northern South America, West Indies.

KEY REFERENCES: Austin et al. (1998).

King Eider *Somateria spectabilis*

DESCRIPTION: L = 56 cm (22 in); W = 92 cm (36 in); a large duck with distinct orange protuberance (frontal shield) from forehead (male); pale blue crown and nape; blood orange bill; beige cheek and chin; white neck and breast; black back and sides; white crissum; black hindquarters. **Female** Entirely brown mottled with dark brown; dark brown markings are wedge-shaped on sides; feathering on side of bill extends nearly to nares.

HABITAT AND DISTRIBUTION: Rare winter resident (Nov–Mar) along the immediate coast: Coastal waters, often at rocky shores and jetties.

DIET: Forages mostly on the bottom in coastal marine waters 15–25 m (50–82 ft) in depth; main prey items include molluscs, crustacea, echinoderms, and algae.

RANGE: Breeds in Old and New World high Arctic; winters along northern coasts, south to OR and NY in the United States.

	Summer
	Migration
	Permanent
	Winter

KEY REFERENCES: Suydam (2000).

Common Eider *Somateria mollissima*

DESCRIPTION: L = 61 cm (24 in); W = 97 cm (38 in); a large duck with white back and creamy breast; black sides and tail; peculiar, wedge-shaped head; greenish stripe along bill and forehead; black patch through and behind eye; greenish back of neck and white throat. **Male—First Winter** Brownish head and body with white breast. **Female** Brownish-gray head and neck; striking dark brown and white scalloping on back and sides.

HABITAT AND DISTRIBUTION: Rare winter resident (Oct–Mar) along the immediate coast: Coastal marine waters, often near shoals, jetties, and rocky shorelines or islands.

DIET: Forages shallow water (< 20 m) (66 ft) over kelp beds and rocky offshore or intertidal areas by diving for benthic invertebrates (e.g., molluscs, crustacea, sea urchins); often in large flocks.

RANGE: Breeds in the high Arctic of the Old and New World. Winters in Arctic, North Atlantic, and North Pacific oceans and coasts.

	Summer
	Migration
	Permanent
	Winter

KEY REFERENCES: Goudie et al. (2000).

Harlequin Duck *Histrionicus histrionicus*

DESCRIPTION: L = 43 cm (17 in); W = 53 cm (21 in); dark slate-gray head with striking white, black, and chestnut markings; slate-gray back, neck, breast, and belly; white markings on back and wings; white collar and shoulder bar; chestnut flanks. **Female** Brown with whitish belly; white patches on ear, forehead, and base of bill.

HABITAT AND DISTRIBUTION: Rare winter resident (Nov–Mar) along the immediate coast: Winter habitat is rocky coastlines where birds forage in-shore or in the surf, and rest on rocks.

DIET: Forages by diving in shallow water (< 20 m) (66 ft) to capture fish or pick invertebrates from rocks, stones, and pebbles, e.g., crustacea, amphipods, and molluscs.

CONSERVATION STATUS: Eastern populations of the Harlequin Duck have been designated as "Endangered" by the Committee on Status of Endangered Wildlife in Canada; current eastern populations are estimated at < 1000 birds, and may still be declining for reasons that are not well understood.

RANGE: Breeds in northwest-ern North America from AK to ID; in the north-east south to

Quebec; also Iceland, Greenland, and Siberia; winters mainly along northern coasts, south to CA and NY in the United States.

KEY REFERENCES: Robertson and Goudie (1999).

Surf Scoter *Melanitta perspicillata*

DESCRIPTION: L = 48 cm (19 in); W = 86 cm (34 in); black with white patches on nape and forehead; orange bill with bull's-eye on side (white circle, black center); white eye. **Female** Entirely brown with white patches in front of and behind eye; dark bill.

HABITAT AND DISTRIBUTION: Common winter resi-dent (Oct–Apr) along coast: Estuaries, bays, inshore ocean areas.

DIET: Forages by diving in shallow water (1–3m) (3–10 ft), mostly for molluscs (60–70%), but other aquatic invertebrate prey are taken as well includ-ing crustacea and aquatic insects (summer, fresh water).

RANGE: Breeds in northern North America south to cen-tral Canada;

winters mainly along coasts from AK to north-western Mexico, and Nova Scotia to FL and the northern Gulf Coast, also Great Lakes.

KEY REFERENCES: Savard et al. (1998).

White-winged Scoter *Melanitta fusca*

DESCRIPTION: L = 56 cm (22 in); W = 100 cm (39 in); black with white eye and wing patches; dark bill with black knob at base and orange tip. **Female** Entirely dark brown (sometimes with whitish pre- and postorbital patches); bill dark orange with black markings; feathering on bill extends nearly

to nostrils; white secondaries sometimes visible on swimming bird.

HABITAT AND DISTRIBUTION: Common winter resident (Oct–Apr) along coast: Bays, estuaries, inlets, lakes.

DIET: Forages by diving in shallow water (usually < 5 m) (16 ft) for benthic invertebrates (e.g., molluscs, crustacea).

RANGE: Breeds in boreal and Arctic regions of the Old and New World; winters mainly along northern coasts, south to northwestern Mexico and FL in North America.

KEY REFERENCES: Brown and Fredrickson (1997).

Black Scoter *Melanitta nigra*

DESCRIPTION: L = 48 cm (19 in); W = 84 cm (33 in); entirely black with orange knob at base of bill.

Female Dark brown with whitish cheek and throat contrasting with dark crown and nape.

Immature Patterned similarly to female but whitish belly.

HABITAT AND DISTRIBUTION: Common winter resident (Oct–Apr) along coast: Marine, bays, lakes.

DIET: Forages by surface diving in open water, < 10 m (33 ft) in depth normally, for invertebrate prey (e.g., molluscs, crustacea), often in conspecific flocks.

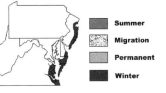

RANGE: Breeds locally in tundra regions of Eurasia and North America; winters in northern and temperate coastal waters of Northern Hemisphere, south to CA and FL in the United States.

CONSERVATION STATUS: Black Scoter populations have declined in recent years, perhaps because the take allowed hunters exceeds replacement capability of the population.

KEY REFERENCES: Bordage and Savard (1995).

Long-tailed Duck *Clangula hyemalis*

DESCRIPTION: L = 48 cm (19 in); W = 74 cm (29 in); dark head with white face; dark neck and breast; gray and brown on back; dark wings; white belly and undertail coverts; black, extremely long pointed central tail feathers; pink patch on bill.

Female and Winter Male White head with dark brown cheek patch and crown; white neck; grayish-brown breast; white belly and undertail coverts; short, sharply pointed tail.

HABITAT AND DISTRIBUTION: Common (along coast and Lake Erie) to rare (inland) transient and winter resident (Nov–Mar): Coastal marine waters, bays, large lakes, often near rocky headlands, cliffs, and ledges.

DIET: Forages by diving to or near the bottom at depths of 5–15 m (16–49 ft), but occasionally much deeper (up to 66 m) (217 ft) to harvest fish and epibenthic and benthic invertebrates (e.g., crustacea, aquatic insects [lakes and ponds], and molluscs) from cobble, rocky, or sandy bottoms; also takes pondweed, eel grass, and other vegetable foods on occasion.

RANGE: Breeds in Arctic and sub-Arctic regions of the Old and New World; winters mainly along northern coasts of Northern Hemisphere, in the United States south to CA and GA.

KEY REFERENCES: Robertson and Savard (2002).

Bufflehead *Bucephala albeola*

DESCRIPTION: L = 36 cm (14 in); W = 59 cm (23 in); a small, plump, short-billed duck; head white from top of crown to nape, the rest iridescent purple; black back; white breast, belly and sides; gray bill; pink legs. **Female and Immature Male** Dark back; grayish-white below; dark head with large white patch extending below and behind eye.

HABITAT AND DISTRIBUTION: Common to uncommon transient (Nov–Dec, Feb–Apr) throughout; common to uncommon winter resident

(Nov–Apr) along the Coastal Plain and Piedmont, scarce inland except along the Ohio River and Lake Erie where it can be locally common; rare or casual in summer along the coast: Bays, lakes, estuaries.

DIET: Feeds by diving for aquatic invertebrates, mainly insects in fresh water, e.g., dragonfly and midge larvae; crustacea and molluscs in salt water; occasionally other invertebrates, fish, and seeds.

RANGE: Breeds across Canada and extreme northern United States;

winters from sub-Arctic along both coasts of North America, and inland from central United States south to central Mexico.

KEY REFERENCES: Gauthier (1993).

Common Goldeneye *Bucephala clangula*

DESCRIPTION: L = 46 cm (18 in); W = 76 cm (30 in); iridescent green head (sometimes purplish); golden eye; white patch at base of bill; black back and hindquarters; white breast, sides, and belly; black and white scapulars; white wing patch visible in flight. **Female** Gray body; brown head; white neck; golden eye; gray bill yellowish at tip (mostly yellow in some birds).

HABITAT AND DISTRIBUTION: Common to uncommon transient (Nov–Dec, Feb–Mar) throughout; common to uncommon winter resident (Nov–Mar) along the Coastal Plain and Piedmont, scarce inland except along the Ohio River and Lake Erie where it can be locally common; rare or casual in summer along the coast: Bays, lakes.

DIET: Mostly diurnal forager, diving for prey in shallow water, < 4 m (13 ft) deep; feeds mainly (75%) on aquatic invertebrates (e.g., crustacea, insects, molluscs) and plant material (e.g., seeds).

RANGE: Breeds across boreal regions of Old and New World; winters along northern coasts south to temperate and subtropical regions of the Northern Hemisphere.

KEY REFERENCES: Eadie et al. (1995).

Hooded Merganser *Lophodytes cucullatus*

DESCRIPTION: L = 46 cm (18 in); W = 66 cm (26 in); head with white crest from top of crown to nape broadly edged in black, the rest black; golden eye; black back and tail; white breast with prominent black bar; rusty sides; sharp, black bill. **Female and Immature Male** Body brownish; head a pale orange with dusky crown; pale orange crest off back of crown and nape; upper mandible dark; lower mandible orangish.

HABITAT AND DISTRIBUTION: Common transient (Oct–Dec, Feb–Mar) throughout; common winter

resident (Oct–Apr) along the Coastal Plain and Piedmont; uncommon to rare and local in summer*, mainly in northern PA and the Ohio River Valley in western WV: Ponds, lakes, estuaries, bays; for breeding sites, the bird prefers cavities located in heavily wooded bottomlands with swift, clear-running streams nearby.

DIET: Uses visual foraging to capture crayfish, fish, and aquatic insects by diving from surface in clear, shallow (usually < 1.5 m) (5 ft) water; additional prey items include other arthropods, amphibians, molluscs, and vegetation.

REPRODUCTION: Monogamous; pair formation may occur on the wintering ground; pair arrives at the breeding territory in March or April; nest is placed in natural tree cavity, abandoned woodpecker hole, or nest box, usually near water; female lines bottom of nest cavity with her own down; clutch—8–10 ("dumping," i.e., placement of eggs in the nest by other females, makes calculation of normal clutch size difficult); incubation—26–41 days; male leaves territory after female begins incubation; fledging—young are precocial and leave the nest within 24 h of hatching, but are not able to fly until 70 days posthatching; female cares for young for several weeks after departure from the nest, but exact duration of parental care is not known; single brood per season.

CONSERVATION STATUS: S1—MD, WV, DE; S3—PA; limited breeding distribution threatened by wetland drainage and pollution.

RANGE: Breeds across central and southern Canada and northern United States,

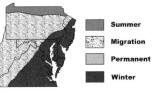

farther south in Rockies and Appalachians; winters mainly along coasts from southern Canada to northern Mexico, West Indies.

KEY REFERENCES: Dugger et al. (1994).

Common Merganser *Mergus merganser*

DESCRIPTION: L = 64 cm (25 in); W = 92 cm (36 in); iridescent green head; sharp, red-orange bill; black back; gray rump and tail; white breast, sides and belly. **Female and Immature Male** Rufous, crested head; white chin; rufous throat and neck ending abruptly at white breast; gray back and sides; orange bill.

HABITAT AND DISTRIBUTION: Common winter resident (Nov–Mar) along the coast and on Lake Erie; uncommon to rare and local transient (Nov–Dec, Feb–Mar) and rare or casual winter resident inland; uncommon and local summer resident* (Apr–Oct) in northern PA: Lakes, rivers, bays, estuaries.

DIET: Forages by surface diving or head-down probing of benthos for prey, mostly fish (10–30 cm long) (4–12 in).

REPRODUCTION: Monogamous; males arrive on breeding territories in late March or April followed by females; cavity nest is placed in a natural hole, abandoned Pileated Woodpecker nest, or artificial nest box located 1–30 m (3–100 ft) up in a tree or stub, and lined by the female with breast down; clutch—9–12; incubation (female)—28–35 days; fledging—young are precocial and leave the nest 1–2 days post-hatching, but are unable to fly until about 9 weeks; post-fledging parental care—females abandon young 4–6 weeks post-hatching, before they can fly; single brood per season.

CONSERVATION STATUS: S1— VA; S3—PA; Common Mergansers have been subject to population control efforts in the past due to perceived negative effects on native fisheries; limited wetland breeding habitat threatened by drainage and pollution.

RANGE: Breeds in Old and New World sub-Arctic and boreal regions south in mountains into temperate areas; winters from northern coasts south inland through temperate and sub-tropical zones of the Northern Hemisphere.

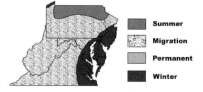

Summer
Migration
Permanent
Winter

KEY REFERENCES: Mallory and Metz (1999).

Red-breasted Merganser *Mergus serrator*

DESCRIPTION: L = 56 cm (22 in); W = 81 cm (32 in); iridescent green, crested head; sharp, red-orange bill; red eye; white collar; buffy breast scalloped with dark brown; black back; gray rump; black tail; black shoulder with white chevrons; white scapulars; grayish sides. **Female and Immature Male** Rufous, crested head; white chin and throat; gray back and sides; white wing patch.

HABITAT AND DISTRIBUTION: Common winter resident (Nov–Mar) and rare summer resident (has bred) along the coast and on Lake Erie; uncommon to rare and local transient (Nov, Feb–Mar) and rare or casual winter resident inland: Bays, estuaries, in-shore marine, lakes.

DIET: Forages by surface-diving for fish, marine

invertebrates, or, at inland sites, aquatic insects; sometimes works in groups to herd prey.

CONSERVATION STATUS: Like the Common Merganser, the Red-breasted Merganser has been subjected to population control efforts due to perceived negative effects on native fisheries.

RANGE: Breeds in Arctic and boreal regions of Old and New World; winters 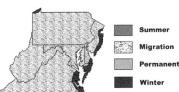 mainly along coasts in southern boreal and temperate areas.

KEY REFERENCES: Titman (1999).

Ruddy Duck *Oxyura jamaicensis*

DESCRIPTION: L = 38 cm (15 in); W = 58 cm (23 in); a small duck; chestnut body; stiff black tail held at a 45-degree angle; black cap; white cheek; sky-blue bill with dark tip. **Winter** Grayish-brown body; dark cap; white cheek. **Female** Mottled grayish and white body; stiff black tail; dark cap and dark line below eye.

HABITAT AND DISTRIBUTION: Common to uncommon winter resident (Oct–Apr) and rare to casual summer resident* along the coast and on Lake Erie; uncommon to rare and local transient (Oct–Nov, Mar–Apr) and rare or casual winter resident inland: Lakes, marshes, ponds, bays, estuaries.

DIET: Forages mostly by diving in shallow water (< 1 m) (3.3 ft) for aquatic invertebrates (60–70% of

migration and winter diet, including mollusca, aquatic insects, and crustacea) and benthic plants (e.g., wild celery, muskgrass, and widgeongrass).

RANGE: Breeds mainly in the prairie pothole region of northcentral United States  and Canada; also locally from northern Canada through the United States to central Mexico, West Indies, and South America; winters in coastal and southern United States south to Nicaragua and elsewhere in tropical breeding range.

KEY REFERENCES: Brua (2002).

ORDER GALLIFORMES; FAMILY PHASIANIDAE

Ring-necked Pheasant *Phasianus colchicus*

DESCRIPTION: L = 84 cm (33 in); W = 79 cm (31 in); a chicken-sized bird with long legs and extremely long, pointed tail; green head with naked red skin on face; white collar; body rich chestnuts, grays, golds and bronzes, spotted with white and brown; grayish-brown, long, pointed tail barred with brown; has spurs on tarsi. **Female** Pale grayish-brown spotted with dark brown; long, pointed tail.

HABITAT AND DISTRIBUTION: Introduced repeatedly throughout the region. Uncommon resident* in most areas, though rare in the mountains and absent from WV (except northern and eastern panhandles where still present): Pastures, hedgerows, old fields, marshes, agricultural areas.

DIET: Forages by walking on the ground and picking seeds, nuts, fruits, leaves, shoots, and invertebrates from litter and surrounding vegetation; domestic grains, e.g., corn, wheat, soybeans, rice, and barley, are preferred.

REPRODUCTION: Harem polygyny—several females on territory of a single male; harem formation

begins in March or April; all nesting and brood-rearing activities are performed by the female; ground nest (18 cm dia.) (7 in) is placed in a depression in dense vegetation, and built of grasses, twigs, plant stems, and other materials gathered from the immediate vicinity; clutch—7–15; incubation—23–28 days; young are precocial, able to leave the nest within a few hours of hatching, and remain with the female until 10–11 weeks post-hatching; some double-brooding occurs.

RANGE: Native of central Asia. Introduced throughout North America. Self-sustaining populations exist in cool temperate regions with extensive grain crops and moderate hunting.

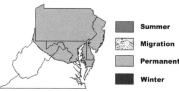

KEY REFERENCES: Giudice and Ratti (2001).

Ruffed Grouse *Bonasa umbellus*

DESCRIPTION: L = 43 cm (17 in); W = 71 cm (28 in); body mottled brown and white; head with a crest; white line through eye; black shoulder patches; tail chestnut barred with black and white with a broad black subterminal band and white terminal band. **Female** Similar to male but black and white terminal bands are incomplete (central tail feathers grayish at tip).

HABITAT AND DISTRIBUTION: Uncommon resident* inland; scarce or absent from the Coastal Plain, though found in the southern NJ Pine Barrens: Coniferous, deciduous, and mixed woodlands, especially near aspen groves (feeding) and dense conifer stands (roosting).

DIET: Forages on tree twigs, catkins, and buds, especially poplar (*Populus*), in winter; feeds by eating seeds, fruits, nuts, acorns, leaves, and shoots from the forest floor at other times of the year; adults are almost entirely herbivorous, while newly hatched young feed primarily on forest-litter invertebrates.

REPRODUCTION: Promiscuous; males begin defense of small territories in April, where they give both visual and audio displays; audio displays consist of increasingly rapid wing flapping ("booming"); females visit male territories for copulation; male has no role in care of eggs or young; nest is a scrape in forest-floor litter; clutch—9–14; incubation—23–24 days; young are precocial and able to follow the hen from the nest at 2–3 h post-hatching; young remain with female 12–15 weeks post-hatching; single brood per season.

RANGE: Boreal and north temperate regions of North America, south in mountains to northern CA and Utah in the west and north GA in the east.

KEY REFERENCES: Rusch et al. (2000).

Wild Turkey *Meleagris gallopavo*

DESCRIPTION: L = 119 cm (47 in); W = 160 cm (63 in); dark brown body with iridescent bronze highlights; naked red head and neck; blue skin on face with dangling red wattles; hairy beard hanging from breast; tail dark barred with buff; tarsi with spurs. **Female** Naked facial skin is grayish; lacks beard and wattles. Adult males associate in groups of 2–3, females and immature males in larger flocks.

HABITAT AND DISTRIBUTION: Uncommon resident*, mainly west of the Coastal Plain; apparently absent from some southwestern WV counties: Deciduous forest, oak woodlands.

DIET: Mast (e.g., acorns, beech nuts), plant seeds, buds, fruits, occasionally insects and small vertebrates; young birds during the first few weeks post-hatching eat mostly insects.

REPRODUCTION: Polygynous, with no apparent participation by the male in care for the brood at any stage; males begin gobbling in Feb with copulation in Mar–Apr; eggs laid Apr–Jun; nest, 20 cm dia. (8 in) is a shallow depression scraped in the ground by the female, lined with plant material, often placed at the base of a tree or in dense brush; clutch—12–14; incubation—25–30 days; young are precocial, capable of following hen 12–24 h after hatching; brood remains together for up to eight weeks post-hatching; one brood per season.

RANGE: Formerly from southern Canada to central Mexico, extirpated from many areas but being reintroduced; populations are expanding rapidly to occupy former range.

KEY REFERENCES: Eaton (1992).

Legend:
- Summer
- Migration
- Permanent
- Winter

FAMILY ODONTOPHORIDAE

Northern Bobwhite *Colinus virginianus*

DESCRIPTION: L = 26 cm (10 in); W = 38 cm (15 in); brown and gray above mottled with dark brown and white; chestnut sides spotted with white; white breast and belly scalloped with black; chestnut crown with short, ragged crest; white eyebrow and throat. **Female** Tawny eyebrow and throat.

HABITAT AND DISTRIBUTION: Uncommon and declining resident* of grasslands throughout the region; scarce in central WV, extreme northern PA, and highlands throughout: Brushy fields, pastures, grasslands, savanna, agricultural areas.

DIET: Forages in coveys by scratching and pecking through litter and vegetation on the ground for seeds, shoots, and arthropods; more arthropods are eaten during the breeding season, while seeds predominate in the diet during the nonbreeding period.

REPRODUCTION: Monogamy, polygyny, or polyandry, apparently depending on the circumstances of each population; pair formation begins in March or April and first clutches are laid in May; ground nest, 15 cm dia. (6 in), is a scrape located in low, dense vegetation and lined by both sexes with grasses and plant stems, sometimes with a dome; clutch—12–14; incubation (mostly by female)—23 days; fledging—precocial young are able to leave the nest and follow the female within 24 h of hatching and are able to fly at 14 days post-hatching; post-fledging parental care—young remain in covey with adults until late winter; two or more broods per season are common.

CONSERVATION STATUS: SZS3—PA; S3—WV; Mid-Atlantic populations of the Northern Bobwhite have shown long-term declines, presumably resulting from reversion of open habitats to forest and conversion of grasslands to intensive agriculture.

RANGE: Eastern and central United States south to Guatemala; isolated populations in AZ and Sonora.

KEY REFERENCES: Lehmann (1984), Guthery (1986), Brennan (1999).

ORDER GAVIIFORMES; FAMILY GAVIIDAE

Red-throated Loon *Gavia stellata*

DESCRIPTION: L = 64 cm (25 in); W = 107 cm (42 in); black mottled with gray on back; gray, rounded head; gray neck striped with white on nape; rufous throat; red eye; white below with barring on flanks; relatively thin, slightly upturned bill. **Winter** Black back with indistinct white spots; head grayish with some white below eye; throat and underparts white.

HABITAT AND DISTRIBUTION: Fairly common winter resident (Oct–Apr) in coastal areas where they frequent in-shore marine sites and bays.

DIET: Forages by surface diving, mostly for fish, but also for aquatic invertebrates.

RANGE: Breeds in the high Arctic; winters in coastal temperate and boreal regions of the Northern Hemisphere.

KEY REFERENCES: Barr et al. 2000.

Common Loon *Gavia immer*

DESCRIPTION: L = 79 cm (31 in); W = 137 cm (54 in); checked black and white on back; black head with a white streaked collar; white below with black streaking on breast and flanks; thick, heavy bill; red eye. **Winter** Dark gray above; white

below; dark gray crown and nape; white face and throat with a partial collar of white around neck.

HABITAT AND DISTRIBUTION: Common transient and winter resident (Oct–May) along the coast; uncommon to rare transient and rare winter resident inland; some individuals remain through the summer; has bred*: Bays, lakes, rivers.

DIET: Dives for fish, usually within 5 m (16 ft) of surface, but as deep as 60 m (200 ft); also eats other aquatic vertebrates and invertebrates.

RANGE: Breeds in northern North America south to northern

United States; winters coastal North America and on large lakes inland south to Baja California, Sonora, and south TX.

KEY REFERENCES: McIntyre and Barr (1997).

ORDER PODICIPEDIFORMES; FAMILY PODICIPEDIDAE

Pied-billed Grebe *Podilymbus podiceps*

DISTRIBUTION: L = 33 cm (13 in); W = 56 cm (22 in); grayish-brown body; short pale bill with dark black ring; black throat; dark brown eye with white eyering. **Winter** Whitish throat; no black ring on bill. **Juvenile** Prominently striped in dark brown and white (Mar to as late as Oct).

HABITAT AND DISTRIBUTION: Common to uncommon transient and winter resident (Aug–Apr) throughout; uncommon to rare and local summer resident*. Ponds, lakes, rivers, swales, marshes, estuaries, bays. This bird has the curious facility of simply sinking when disturbed.

DIET: Forages by diving in shallow water for aquatic invertebrates (e.g., crayfish, insects) and vertebrates (e.g., fish, amphibians); casts pellets of indigestible material.

CONSERVATION STATUS: S1—NJ, DE; S2—MD, VA, WV; S3—PA; limited breeding distribution threatened by wetland drainage and pollution.

REPRODUCTION: Monogamous; pairs may form during migration, in winter, or at the breeding site, and some birds remain paired throughout the year; breeding begins in late March or early April; floating platform nest, 11–13 cm dia. (4–5 in), is placed in emergent vegetation or open water, and built by both sexes of buoyant plant stems (e.g., bulrush, cattail, water-lily); clutch—4–8; incubation (both sexes) 23–27 days; fledging—young are semi-precocial, able to leave the nest on the adult's back or by swimming within 1 h after hatching; newly-hatched young are often carried on the parents' back; parental care—continues for 25–62 days after hatching; second broods are common.

RANGE: Breeds across most of Western Hemisphere from central Canada south

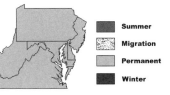

to southern Argentina; West Indies; winters in temperate and tropical portions of breeding range.

KEY REFERENCES: Muller and Storer (1999).

Horned Grebe *Podiceps auritus*

DESCRIPTION: L = 36 cm (14 in); W = 59 cm (23 in); dark back; rusty neck and sides; whitish breast; black head with buffy orange ear patch and eyebrow; red eye. **Winter** Upper half of head dark with white spot in front of eye; cheek, throat and breast white; nape and back dark.

HABITAT AND DISTRIBUTION: Common transient and winter resident (Nov–Apr) along the coast; uncommon transient (Nov, Mar–Apr) and rare to casual winter resident (Dec–Feb) inland: Bays, estuaries; larger lakes.

DIET: Forages by diving in shallow water (< 6 m) (20 ft) for fish, crustacea, and arthropods.

RANGE: Breeds in boreal and north temperate freshwater wetlands;

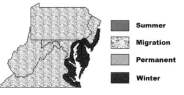

Summer
Migration
Permanent
Winter

winters mainly along coast in boreal and temperate regions of the Northern Hemisphere.

KEY REFERENCES: Stedman (2000).

Red-necked Grebe *Podiceps grisegena*

DESCRIPTION: L = 51 cm (20 in); W = 81 cm (32 in); a large, long-necked grebe; dark back and upper half of head; rusty neck; lower half of head and chin grayish-white; dark brown eye; breast grayish-white tinged with rust; yellowish lower mandible. **Winter** Dark back, posterior portion of neck, and crown; white chin and partial collar; grayish white anterior portion of neck, breast and belly.

HABITAT AND DISTRIBUTION: Rare to casual transient (Nov, Mar) throughout, scarcer inland; rare winter resident along coast; more numerous in some years: Lakes, rivers, marshes, ponds, inlets, estuaries, bays, in-shore marine.

DIET: Forages by surface diving and pursuing prey visually under water at depths from 3–55 m (10–180 ft); prey includes fish, aquatic insects, and other aquatic invertebrates (e.g., crustacea, annelids, molluscs).

RANGE: Breeds locally in boreal and north temperate areas; winters along

Summer
Migration
Permanent
Winter

temperate, and boreal coasts of the Northern Hemisphere.

KEY REFERENCES: Stout and Nuechterlein (1999).

Eared Grebe *Podiceps nigricollis*

DESCRIPTION: L = 31 cm (12 in); W = 57 cm (22 in); looks mostly black with a red eye when swimming at a distance; dark back; black neck; dark

rusty sides; white belly; black head with buffy orange ear tufts; red eye. **Winter** Dark gray head and back; light grayish chin and ear patch; whitish on breast, throat, and sides.

HABITAT AND DISTRIBUTION: Rare to casual winter resident (Dec–Mar) along the coast: prefers hypersaline environments in winter, e.g., in-shore marine areas, lagoons, and bays.

DIET: Forages by swimming and skimming or pecking prey from the surface or diving to capture prey (e.g., brine shrimp and other aquatic invertebrates).

RANGE: Breeds locally in southern boreal, temperate and tropical

regions of the world; in North America mainly in the west; winters mostly along the Pacific Coast, southwestern United States, Mexico, and Guatemala.

KEY REFERENCES: Cullen et al. (1999).

FAMILY PELECANIDAE

American White Pelican *Pelecanus erythrorhynchos*

DESCRIPTION: L = 157 cm (62 in); W = 267 cm (105 in); white body; black primaries and secondaries; enormous orange bill and gular pouch; often

with a horn-like growth on upper mandible (breeding); orange-yellow feet.

HABITAT AND DISTRIBUTION: Rare (VA) to casual (elsewhere) visitor along the coast in fall, winter, and spring (Sep–May): Large lakes, impoundments, coastal waters. Migrates in large flocks.

DIET: Does not dive from the air for fish like the Brown Pelican; forages by dipping for prey from the surface of the water, often in groups; principal prey taken is fish, usually in water depths of 0.3–2.5 m (3.3–6.6 ft); crayfish, salamanders, and other aquatic organisms also taken.

RANGE: Breeds locally at large lakes and marshes in central and western

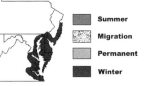

Canada south through western half of the United States; winters southwestern United States to Nicaragua, and from FL around the Gulf of Mexico to Yucatan.

KEY REFERENCES: Evans and Knopf (1993).

Brown Pelican *Pelecanus occidentalis*

DESCRIPTION: L = 122 cm (48 in); W = 198 cm (78 in); grayish above; brownish below with dark chestnut or black nape and neck (breeding—neck and nape are white in nonbreeding period); whitish head tinged with yellow; enormous bill

and gular pouch (reddish in breeding period). Immature Entirely brownish-gray.

HABITAT AND DISTRIBUTION: Uncommon to rare summer resident* (Apr–Oct) along the immediate coast north to Cape May. A few individuals remain through the winter: Coastal marine environments.

DIET: Forages by flying over marine in-shore waters, 10–20 m above the surface, and plunging into the water to a depth of 1–2 m (3.3–6.6 ft) to capture small, surface-schooling fish in its gular pouch.

REPRODUCTION: Monogamous; males arrive at breeding colonies in April followed within a few days by females; colony sites are mainly on coastal islands in the Mid-Atlantic region; nest (1–2 m dia.) (3.3–6.6 ft) is a mound of sticks on the ground, built by the female using material brought by the male; nest cup is made by trampling; clutch—3; incubation (both sexes—using feet rather than belly)—29–32 days; fledging—young are able to fly and are independent at 11—12 weeks post-hatching; single brood per season.

CONSERVATION STATUS: S1—MD, VA; the Brown Pelican was nearly extirpated from its United States breeding range during the mid-twentieth century due to poor reproductive rates resulting from pesticide poisoning. Populations have recovered in many areas. Breeding colony disturbance and coastal devel-

	Summer
	Migration
	Permanent
	Winter

opment are the most imminent threats to regional populations.

RANGE: Coastal Western Hemisphere from southern NJ to eastern Brazil, including West Indies, in east and CA to southern Chile in west.

KEY REFERENCES: Shields (2001).

FAMILY PHALACROCORACIDAE

Double-crested Cormorant
Phalacrocorax auritus

DESCRIPTION: L = 81 cm (32 in); W = 130 cm (51 in); entirely iridescent black; whitish or dark ear tufts during breeding season; gular pouch of bare orange skin; green eye. **Immature** Brown with buffy head and neck. Often sits on snags or posts with wings spread to dry.

HABITAT AND DISTRIBUTION: Common transient (Apr, Sep–Oct) along the coast, rare inland; uncommon to rare and local summer* and winter resident along the coast: Estuaries, lagoons, bays, in-shore marine, lakes, ponds.

DIET: Forages in shallow water (< 8 m) by swimming low in water, sometimes with just the neck protruding, and surface-diving to capture prey, mostly fish; casts pellets of indigestible materials.

REPRODUCTION: Monogamous; males arrive at colony sites, often located in a grove of shrubs or

trees, or in an open, unvegetated area, on coastal islands, in April, followed shortly thereafter by females; open platform nest (0.5 m dia.) (20 in) is built by both sexes, mostly of sticks, though other beach debris is often included; clutch—4; incubation (both sexes)—25–28 days; fledging—young may leave ground nests at 3–4 weeks, but remain in tree nests until able to fly at 6–8 weeks post-hatching; post-fledging parental care—young are essentially independent once they are able to fly long distances; single brood per season is usual.

CONSERVATION STATUS: S1—NJ, DE, MD; S3—VA; cormorants are subject to intensive population control efforts in some parts of their range due to actual or perceived impacts on aquaculture sites and native fisheries; limited breeding populations in the region are vulnerable to coastal development and disturbance.

RANGE: Breeds locally across central and southern Canada and northern and central United States, along both coasts from AK to southern Mexico on the west, and Newfoundland to east TX on the east, also Cuba; winters in the southeastern United States and along both coasts to southern Mexico, Greater Antilles, Yucatan Peninsula, and Belize.

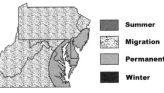

KEY REFERENCES: Hatch and Weseloh (1999).

Great Cormorant *Phalacrocorax carbo*

DESCRIPTION: L = 92 cm (36 in); W = 160 cm (63 in); **Breeding** Black body with short black tail; yellowish gular pouch; white throat; white flank patches. **Winter** Lacks white flank patches. Immature Dark brown above; brown neck; lighter belly.

HABITAT AND DISTRIBUTION: Rare winter resident (Oct–May) along the immediate coast: Rocky inlets, islands, cliffs, ledges, bays, and coastal waters.

DIET: Forages by diving in shallow marine waters (< 20 m) (66 ft) for fish.

RANGE: Resident in temperate, tropical, and sub-Arctic coastal regions throughout much of the Old World; breeds in southwestern Greenland and northeastern North America; winters along northeastern coast of North America (Newfoundland to SC).

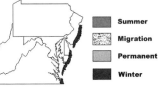

KEY REFERENCES: Hatch et al. (2000).

FAMILY ANHINGIDAE

Anhinga *Anhinga anhinga*

DESCRIPTION: L = 89 cm (35 in); W = 114 cm (45 in); snake-like neck and long triangular tail with terminal buffy band; long, sharp, yellow-orange bill; black body with iridescent green sheen; mottled white on shoulders and upper wing coverts. **Female and Immature** Light buff head, neck, and breast; lacks white wing spotting.

HABITAT AND DISTRIBUTION: Rare summer visitor in southeastern VA: Rivers, lakes, ponds.

DIET: Forages by swimming slowly underwater with

wings partially opened, and spearing prey, mostly small fish, with long, dagger-like bill; often forages with only the sinuous neck and head protruding above water; sits on snags along waterways with wings spread to dry.

RANGE: Breeds from the southeastern United States south through the lowlands of the subtropics and tropics to southern Brazil, Cuba; winters throughout breeding range except inland in Gulf states.

KEY REFERENCES: Frederick and Siegel-Causey (2000).

ORDER CICONIIFORMES; FAMILY ARDEIDAE

American Bittern *Botaurus lentiginosus*

DESCRIPTION: L = 66 cm (26 in); W = 100 cm (39 in); a chunky, relatively short-legged heron; buffy brown above and below streaked with white and brown on throat, neck, and breast; dark brown streak on side of neck; white chin; greenish-yellow bill and legs; yellow eyes.

HABITAT AND DISTRIBUTION: Common to uncommon transient (Apr, Oct–Nov) throughout; rare and local summer* (May–Sep) and winter (Nov–Mar) resident, mainly along the coast: Marshes. A secretive bird; often, rather than fly when approached, it will "freeze" with its neck extended in an attempt to blend in to the reeds and rushes of the marsh.

DIET: Aquatic insects, fish, crustacea, frogs, salamanders, snakes, small mammals.

REPRODUCTION: Apparently monogamous, but perhaps with some polygyny; pair formation occurs in late April to early May; open platform nest, 25–40 cm (10–16 in) in diameter, is built in dense emergent vegetation, 8–20 cm (3–8 in) above water, and constructed of plant stems (e.g., reeds, cattails, sedges), and lined with grasses; clutch— 3–5; incubation 24–28 days; chicks remain on or near the nest for about 14 days post-hatching; one brood per season.

CONSERVATION STATUS: S1—PA, WV, VA, DE, MD; S2—NJ; disappearance of freshwater wetland habitats appears to be the main threat for this species.

RANGE: Breeds central and southern Canada south to southern United States

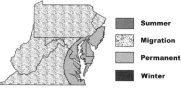

and central Mexico; winters southern United States to southern Mexico and Cuba.

KEY REFERENCES: Gibbs et al. (1992a).

Least Bittern *Ixobrychus exilis*

DESCRIPTION: L = 36 cm (14 in); W = 43 cm (17 in); the smallest of our herons; dark brown with white streaking on back; tan wings, head, and neck; dark brown crown; white chin and throat streaked with tan; white belly; yellowish legs and bill; yellow eyes; extended wings are half tan (basally) and half dark brown.

HABITAT AND DISTRIBUTION: Uncommon transient and summer resident* (May–Sep), mainly along the Coastal Plain; rare and local inland. A few can be found in winter in coastal marshes: Marshes. Like the American Bittern, this bird will often "freeze" with neck extended when approached.

DIET: Small fish mostly, but also aquatic insects, crustacea, snakes, frogs, salamanders, small mammals, and occasional vegetable matter.

REPRODUCTION: Presumed monogamous; pair formation in late April or early May; canopy nest is built mainly by the male, 15–80 cm (6–31 ft) above water in marsh vegetation (e.g., *Typha, Carex, Scirpus*); nest is 15–20 cm (6–8 in) diameter; base of nest is constructed of sticks and stems with top built by inter-weaving living support stems with dead plant material; clutch—4–5; incubation—17–20 days; fledging—13–15 days post-hatching; post-fledging parental care is not well known, but may persist for up to 30 days. One brood is produced per season so far as is known.

CONSERVATION STATUS: S1—PA, WV, DE; S2—MD; S3—NJ; wetland breeding habitats becoming scarce as a result of drainage and pollution.

RANGE: Breeds in the eastern half of the United States and southeastern

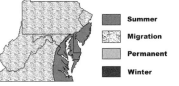

Summer
Migration
Permanent
Winter

Canada; locally in the western United States, south through lowlands to southern Brazil, West Indies; winters through breeding range from the southern United States southward.

KEY REFERENCES: Gibbs et al. (1992b).

Great Blue Heron *Ardea herodias*

DESCRIPTION: L = 122 cm (48 in); W = 183 cm (72 in); a very large heron; slate-gray above and on neck; white crown bordered with black stripes that extend as plumes (breeding); white chin and throat streaked with black; gray breast and back plumes; white below streaked with chestnut; chestnut thighs; orange-yellow bill; dark legs. **Immature** Dark cap; brownish-gray back; buffy neck. **White Phase** ("Great White Heron" of some authors) Entirely white or mixed white and blue with yellow bill and legs; this phase is extremely rare except in southern FL and the Caribbean.

HABITAT AND DISTRIBUTION: Common to uncommon summer resident* (Mar–Sep), scarcer inland especially in the mountains; uncommon (Coastal Plain) to rare or casual (inland) winter resident (Oct–Feb): Lakes, rivers, marshes, bays, estuaries.

DIET: Predominantly fish, but also insects, crustaceans, amphibians, reptiles, small mammals, and birds (eggs or nestlings mostly), taken by slowly wading through or standing in shallow water.

REPRODUCTION: Mostly monogamous; adults return to colony sites and form pairs in Jan–Mar; colonies are in low shrubs or trees along the coast or on coastal islands or in groves of trees in swamps or upland forest inland; nests are placed as high as 30 m (98 ft) above ground; the bulky, open platform nests, 0.5–1.2 m (20–47 in) in diameter, are built of sticks, often brought by the male, and placed by the female, and lined with grasses, twigs, moss, or other finer plant materials; clutch—3–4; incubation—27 days; fledging—50–80 days post-hatching; young are essentially independent once they leave the nest; single brood per season.

CONSERVATION STATUS: S2—WV, NJ, DE; S3—VA, PA; colonial breeding sites vulnerable to disturbance.

RANGE: Breeds from central and southern Canada south to coastal Colombia

■	**Summer**
▨	**Migration**
▣	**Permanent**
■	**Winter**

and Venezuela, West Indies; winters southern United States southward through breeding range.

KEY REFERENCES: Butler (1992).

Great Egret *Ardea alba*

DESCRIPTION: L = 100 cm (39 in); W = 145 cm (57 in); entirely white, with shaggy plumes on breast and back in breeding season; long yellow bill; long dark legs that extend well beyond tail in flight; yellow eye; greenish lores (breeding).

HABITAT AND DISTRIBUTION: Common summer resident* (Apr–Sep) along the Coastal Plain; rare to casual and local inland, mainly as a transient or

post-breeding wanderer (Apr–May, Aug–Oct) inland. A few winter on the coast: Marshes, swamps, lagoons, impoundments, lakes, ponds, rivers, estuaries, bays.

DIET: Forages by walking slowly or standing and peering in shallow water, 0.2–0.4 m (8–16 in), and lunging to capture aquatic prey (e.g., fish, crustacea, amphibians, insects).

REPRODUCTION: Monogamous; males arrive at breeding colony sites in April; colonies typically are located on coastal islands with low, shrubby vegetation or in groves of tall trees in or near wetlands in the Mid-Atlantic region; open platform nest, 0.5 m dia. (20 in), is placed in a shrub, tree, or on the ground, and is built (both sexes) of sticks and lined with green twigs, leaves, and shoots; clutch—3; incubation (both sexes)—23–27 days; young can clamber off and on the nest by about 3 weeks post-hatching, are able to fly at 7–8 weeks, and are independent at 9–10 weeks; single brood per season.

CONSERVATION STATUS: S1—PA, DE; S2—VA; coastal wetland breeding habitat threatened by development.

RANGE: Breeds in temperate and tropical regions of the world; winters mainly in sub-

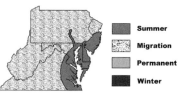

■	**Summer**
▨	**Migration**
▣	**Permanent**
■	**Winter**

tropical and tropical portions of breeding range.

KEY REFERENCES: McCrimmon et al. (2001).

Snowy Egret *Egretta thula*

DESCRIPTION: L = 59 cm (23 in); W = 114 cm (45 in); small, entirely white heron; black legs with yellow feet; black bill; yellow lores; white plumes off neck and breast (breeding). **Immature** has yellow line up back of leg.

HABITAT AND DISTRIBUTION: Common summer resident* (Apr–Sep) along the Coastal Plain; rare transient (Apr–May, Aug–Oct) and late summer wanderer in Piedmont, casual farther inland. Rare in winter on the coast: Marshes, ponds, lakes, estuaries, bays.

DIET: Forages by walking slowly in shallow pools; occasionally uses canopy feeding; also puts foot forward and shakes it on bottom substrate; main foods are fish and crustacea.

REPRODUCTION: Monogamous; pairs arrive at colonies, often located on islands, in March or April; open platform nest is placed about 2 m (6.6 ft) up in as shrub or low tree and built mostly by the female of twigs and sticks, 41 cm dia. (16 in), and lined with grasses, cattails, and Spanish moss; clutch—3–5; incubation (both sexes)—23 days; fledging—30–40 days post-hatching, but young are able to climb off the nest within 10 days post-hatching; post-fledging parental care—young leave the colony 7–8 weeks post-hatching; single brood per season.

CONSERVATION STATUS: S1—DE; S2—VA; S3—MD, NJ; coastal wetlands habitats threatened by development.

RANGE: Breeds locally across United States and extreme southern

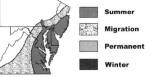

Canada south in lowlands to southern South America, West Indies; winters from coastal southern United States southward through breeding range.

KEY REFERENCES: Parsons and Master (2000).

Little Blue Heron *Egretta caerulea*

DESCRIPTION: L = 59 cm (23 in); W = 100 cm (39 in); a smallish heron; dark blue body; maroon neck and head; two-tone bill, black at tip, pale gray or greenish at base; plumes on neck, breast and head; dark legs. **Winter** Mainly navy blue on neck with maroon tinge; no plumes. **Immature** Almost entirely white in first year with some gray smudging and blue-gray wing tips; greenish legs; two-tone bill; more gray smudging in second year.

HABITAT AND DISTRIBUTION: Common to uncommon summer resident* (Apr–Sep) along the Coastal Plain; rare to casual transient (Apr–May,

Aug–Oct) and late summer wanderer inland: Freshwater marshes, lakes, ponds, swamps, estuaries.

DIET: Usually hunts by walking slowly in shallow water, pausing to scan, and then grasping prey, mainly small fish, aquatic insects, crustacea, and amphibians, from at or below the surface with a sudden jab of the bill.

REPRODUCTION: Monogamous; male arrives at breeding colony in mid to late April with female arriving some days later; colonies normally include pairs of other ardeid species, and are often located in stands of shrubs or trees over water or on coastal islands; the open platform nest, 0.3–0.5 m dia. (122–20 in) is placed in a shrub or tree, and built mostly by the female with sticks brought by the male; clutch—3–4; incubation (both sexes)—22–23 days; fledging—35 days post-hatching, but young are able to climb off the nest several days before they are able to fly; post-fledging parental care—young leave the colony at 45 days post-hatching; single brood per season.

CONSERVATION STATUS: S1—DE; S2—VA, NJ; S3—MD; Little Blue Herons have undergone a significant long-term (1966–1994) decline in their eastern U.S. breeding populations, probably as a result of disturbance or loss of breeding colony habitat.

RANGE: Breeds along Coastal Plain of eastern United States from ME to TX

Summer
Migration
Permanent
Winter

south through lowlands to Peru and southern Brazil, West Indies; winters from southern United States south through breeding range.

KEY REFERENCES: Rodgers and Smith (1995).

Tricolored Heron *Egretta tricolor*

DESCRIPTION: L = 66 cm (26 in); W = 92 cm (36 in); slate-blue body; red eye; bluish lores; rusty throat; white central stripe down neck streaked with dark blue; maroon on breast; white belly and thighs;

buffy plumes on back (breeding). **Immature** Rusty and slate above, whitish below.

HABITAT AND DISTRIBUTION: Common to uncommon summer resident* (Apr–Sep) along the Coastal Plain, scarcer northward (NJ); rare to casual transient (Apr–May, Aug–Oct) and late summer wanderer inland. Rare in winter on the coast: Marshes, bays, estuaries, lakes, ponds.

DIET: Forages mostly by standing or walking slowly along shoreline or in shallow water and jabbing small fish (90%) or other small animals (insects, vertebrates, crustacea); sometimes uses canopy feeding or chases food items.

REPRODUCTION: Monogamous; pairs arrive at colony sites, often on islands, in late April or early May; open platform nests, 30 cm dia. (12 in) are placed 1–2 m (40–80 in) up in shrubs or small trees and built by both sexes (male brings material to female) of sticks and twigs and lined with twigs and grass; clutch—3–4; incubation (both sexes)—21–24 days; fledging—young are able to clamber off the nest at 17 days onto neighboring branches, and are able to fly at 25–30 days post-hatching; post-fledging parental care—up to 60 days post-hatching; single brood per season.

CONSERVATION STATUS: S1—DE; S2—VA; S3—NJ, MD; loss of suitable estuarine breeding habitat may have caused declines.

RANGE: Breeds in the Coastal Plain of the eastern United States from ME to TX, south through coastal lowlands to Peru and northern Brazil, West Indies; win-

Summer
Migration
Permanent
Winter

ters from Gulf states south through breeding range.

KEY REFERENCES: Frederick (1997).

Cattle Egret *Bubulcus ibis*

DESCRIPTION: L = 51 cm (20 in); W = 92 cm (36 in); a small, entirely white heron; yellow-orange bill and legs; buff coloration and plumes on crest, breast, and back (breeding). **Immature** Like adult but lacks buff coloration, and legs are dark.

HABITAT AND DISTRIBUTION: Common to uncommon summer resident* (Apr–Sep) along the Coastal Plain, scarcer northward (NJ); rare transient (Apr–May, Aug–Oct) and rare late summer wanderer inland in the Piedmont, casual farther inland. Rare to casual in winter along coast: Pastures, fields, grasslands, parks, lawns (foraging); woodlands, swamps, wooded islands, coastal islands (breeding, roosting).

DIET: Forages often by following grazing animals, farm machinery, or other Cattle Egrets to capture prey disturbed by their passage; feeds on a wide variety of invertebrates and small vertebrates including Orthoptera, Lepidoptera, spiders, fish, annelids, birds, crustacea, and reptiles.

REPRODUCTION: Monogamous; Cattle Egrets nest later on average than other herons, and tend to choose nest sites in already established breeding colonies with other ardeids; colony locations include woodland groves, swamps, or wooded islands (inland) and coastal islands with low

shrubs or trees; breeding begins in May; the nest, 0.3 m dia. (12 in) is placed in the crotch of a shrub or tree, and is built mostly by the female with sticks provided by the male; clutch—3–4; incubation—24 days; fledging—young clamber off nest at about 20 days post-hatching, but are not able to fly for another 10 days or so; post-fledging parental care—15 days; single brood per season.

CONSERVATION STATUS: S1—DE; S3—MD; disturbance of coastal breeding colonies and destruction of coastal habitat threaten populations.

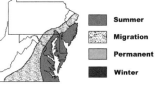

Summer
Migration
Permanent
Winter

RANGE: Formerly strictly an Old World species, the Cattle Egret appeared in South America in the late 1800s and has expanded steadily northward, reaching VA in 1953. Current distribution includes most temperate and tropical regions of the world. Northern populations are migratory.

KEY REFERENCES: Telfair (1994).

Green Heron *Butorides virescens*

DESCRIPTION: L = 48 cm (19 in); W = 71 cm (28 in); a small, dark heron; olive back; black cap; chestnut neck; white throat with chestnut striping; yellow eye and yellow lores; white malar stripe; grayish belly; greenish legs; bill dark above, yellowish below. **Immature** Heavily streaked below.

HABITAT AND DISTRIBUTION: Common to uncommon summer resident * (May–Sep) throughout: Swamps, streams, rivers, lakes, marshes.

DIET: Forages mainly by standing in or near water, scanning for prey, and then capturing prey with a quick jab; mostly diurnal, but also crepuscular or nocturnal; sometimes uses a lure or bait to attract prey; main prey items are small fish, but also feeds on a variety of invertebrates (e.g., aquatic insects, annelids, crustacea) and other small vertebrates (e.g., frogs, snakes).

REPRODUCTION: Monogamous so far as known; pair formation may occur during migration; pairs arrive at breeding territory in April or early May; open platform nest, 20–30 cm dia. (8–12 in) built of sticks and twigs is placed in shrubs or trees, often over or near water; both sexes work on the nest with the male bringing sticks to the female who works them into the nest structure; clutch—3–5; incubation (both sexes)—19–21 days; fledging—young clamber off the nest at 16–17 days post-hatching, and are able to fly 5–6 days later; post-fledging parental care—not well known, but may persist for 15–20 days.

RANGE: Breeds in the eastern half of North America from southern Canada

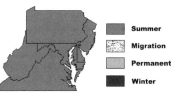

▓	Summer
░	Migration
░	Permanent
■	Winter

southward, and along the Pacific coast from southern British Columbia southward, to northern Argentina (including Green-backed Heron form, *B. striatus*); West Indies; winters from southern United States southward through breeding range.

KEY REFERENCES: Davis and Kushlan (1994).

Black-crowned Night-Heron
Nycticorax nycticorax

DESCRIPTION: L = 66 cm (26 in); W = 114 cm (45 in); a rather squat heron; black on back and crown with long, trailing white plumes (breeding); gray wings; pale gray breast and belly; white forehead,

cheek and chin; red eye; dark beak; pale legs.

Immature Dark brown above, heavily streaked with white; whitish below streaked with brown; red eye; bluish lores; pale legs.

HABITAT AND DISTRIBUTION: Common to uncommon summer resident* (Apr–Sep) along the immediate coast, rare to casual inland; uncommon to rare transient (Apr–May, Aug–Oct) and late summer wanderer inland; uncommon to rare in winter along the coast: Bays, lakes, marshes, swamps, streams, rivers, ponds, canals, ditches, wet fields.

DIET: Mostly nocturnal, but also diurnal when feeding nestlings; feeds mainly on fish captured by making sudden thrusts to grasp prey in the bill while standing or walking quietly in shallow water with emergent vegetation, but includes a wide variety of other vertebrates (e.g., amphibians, reptiles, mammals, birds) and invertebrates (e.g., insects, crustacea, molluscs), depending on availability.

REPRODUCTION: Presumed monogamous; breeds in colonies at sites with dense shrub or tree growth, often over water; male arrives at colony, selects nest site, and begins nest construction in April or May; pair formation occurs over the next 2–3 weeks; female completes open platform nest, 30–45 cm dia. (12–18 in), with sticks brought by the male; clutch—3–5; incubation—24–26 days; fledging—young clamber away from nest at 29–34 days and are able to fly at 6–7 weeks; post-fledging parental care—not known; single brood per season.

CONSERVATION STATUS: SH-WV; S1—DE; S2—PA; S3—MD, NJ; principal threat to populations in the Mid-Atlantic appears to be wetland loss to agriculture and development.

RANGE: Breeds locally in temperate and tropical regions of the world; with-

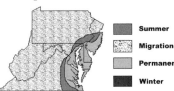

draws from seasonally cold portions of breeding range in winter.

KEY REFERENCES: Davis (1993).

Yellow-crowned Night-Heron
Nyctanassa violacea

DESCRIPTION: L = 61 cm (24 in); W = 107 cm (42 in); gray body, streaked with black above and on wings; black head with white cheek patch and creamy crown and plumes (breeding); red eye; dark bill; pale legs. **Immature** Dark brown above, finely spotted with white; whitish below streaked with brown; dark bill; yellow legs; red eye.

HABITAT AND DISTRIBUTION: Uncommon summer resident* (Apr–Sep) along the immediate coast, scarcer northward (NJ); rare to casual transient (Apr–May, Aug–Oct) and late summer wanderer inland. Rare in winter on the coast: Wetlands.

DIET: Forages both diurnally and nocturnally along water margins in estuaries, tidal pools, beaches,

creeks, and swamps, walking slowly or standing and then lunging to grasp prey, mainly crustacea.

REPRODUCTION: Monogamous (?); usually, but not always, in colonies; inland-breeding pairs are often solitary, nesting in swamps or along rivers or streams; typical colony sites are in trees near water (mainland) or shrubs (coastal islands); birds either arrive paired at colony or nesting sites in late March or early April, or pair shortly after arrival; open platform nest, 0.5–1.3 m dia. (20–55 in), is placed in a shrub or tree (loblolly pine is a common nest tree in VA), and built (both sexes) of sticks, often lined with twigs, Spanish moss, or leaves; clutch—3–4; incubation (both sexes)—24–25 days; fledging—37 days post-hatching; post-fledging parental care—at least 20 days and perhaps longer; single brood per season.

CONSERVATION STATUS: S1—PA, DE; S2—VA, MD, NJ; destruction and degradation of coastal wetlands constitute the main threats to this species.

RANGE: From the south-eastern United States south in coastal

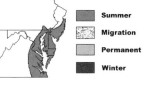

regions to Peru and southern Brazil, West Indies; winters from coastal Gulf states south through breeding range.

KEY REFERENCES: Watts (1995).

FAMILY THRESKIORNITHIDAE

White Ibis *Eudocimus albus*

DESCRIPTION: L = 64 cm (25 in); W = 100 cm (39 in); entirely white with long, pink, decurved bill; pink facial skin; yellow eye; pink legs. **Immature** Brown above; white below; neck and head mottled brown and whitish; pinkish bill and legs.

HABITAT AND DISTRIBUTION: Uncommon to rare summer visitor* (Apr–Sep) to coastal VA, especially the Eastern Shore; rare to casual elsewhere: Bays, rivers, and estuaries, especially in *Spartina* marshes and mangroves.

DIET: Feeds by benthic tactile or visual probing in or on soft mud while wading in shallow water; main prey are crustaceans, aquatic insects, annelids, and molluscs (snails), but small fish, reptiles, and amphibians are occasionally taken.

REPRODUCTION: White Ibis are monogamous with pair formation occurring shortly after their April arrival in the region; pairs nest in mixed colonies with other ciconiiforms where nests are placed in dense growths of low to mid-sized shrubs located on the coast or on coastal islands; the female selects the nest site and builds an open platform nest, 30 cm dia. (12 in), of sticks and twigs, many contributed by the male, and lined with grasses, leaves, and moss; clutch size 2–3; incubation 21 days; young altricial, beginning to fly at 4 weeks; they leave the colony and become independent at 40–60 days post-hatching; usually only one brood is produced per season.

CONSERVATION STATUS: S1—VA; principal threat to populations is loss of coastal wetlands for foraging and breeding.

RANGE: Coastal southeastern United States south to French Guiana and Peru, West Indies; withdraws from northern portions in winter.

Summer
Migration
Permanent
Winter

KEY REFERENCES: Kushlan and Bildstein (1992).

Glossy Ibis *Plegadis falcinellus*

DESCRIPTION: L = 56 cm (22 in); W = 92 cm (36 in); body entirely dark purplish-brown with green sheen on wings and back; bare dark skin on face, edged with bluish skin (breeding); long, dark, decurved bill; dark legs; brown eyes. **Immature** Brownish throughout with white streaking on head and neck; dark bill and legs.

HABITAT AND DISTRIBUTION: Common to uncommon summer resident* (Mar–Sep) along the Coastal Plain, scarcer northward (NJ); rare to casual transient (Mar–May, Aug–Oct) and late summer wanderer inland—not recorded from WV. Rare in winter on the VA coast: Fresh, brackish, and saltwater marshes, tidal flats, and lagoons.

DIET: Forages by walking in shallow water and probing substrate for invertebrates (e.g., aquatic insects, annelids, crustacea, molluscs); the bill tip is sensitive and useful for detecting prey swimming in murky water or buried in mud.

REPRODUCTION: Monogamous; pairs arrive at breeding colonies in April; platform nest, 30–40 cm dia. (12–16 in), is placed on the ground, in low shrubs, or in trees, and is built (both sexes) of sticks, twigs, rushes, plant fibers, grasses, or reeds, depending on the habitat in which the colony is located (e.g., freshwater marsh, saltmarsh, swamp, woodland grove); clutch—3–4; incubation (both sexes)—21 days; fledging—young clamber off nest at 8 days post-hatching, but can't fly until about 28 days; young are independent about 7 weeks post-hatching; single brood per season.

CONSERVATION STATUS: S1—DE; S2—VA; S3—NJ; coastal colonies threatened by development.

RANGE: Breeds along coast of eastern United States from ME to LA south through West Indies to northern Venezuela and in Old World temperate and tropical regions; withdraws from colder portions of breeding range in winter.

Summer
Migration
Permanent
Winter

KEY REFERENCES: Davis and Kricher (2000).

FAMILY CICONIIDAE

Wood Stork *Mycteria americana*

DESCRIPTION: L = 104 cm (41 in); W = 168 cm (66 in); a large white-bodied bird with long, heavy, bill, down-turned towards the tip; black primaries and secondaries; naked, black-skinned head and neck; pale legs with pinkish feet. **Immature** Patterned like adult but with grayish feathering on neck and head; yellowish bill.

HABITAT AND DISTRIBUTION: Rare summer and fall visitor (Jun–Sep) along the southern VA coast; casual elsewhere: Estuaries, coastal marshes, swamps, bays.

DIET: Forages by walking through shallow water, head down, and bill open 7–8 cm and submerged to the nares; bill is used to grope from side to side, capturing prey by touch; most common foods are fish, but other aquatic organisms are also taken.

CONSERVATION STATUS: "Endangered"—USFWS; destruction of marshland and swamp habitats has decimated populations.

RANGE: SC, GA, and the Gulf states south through Mexico, Central, and South America to Argentina; West Indies.

Summer
Migration
Permanent
Winter

KEY REFERENCES: Coulter et al. (1999).

FAMILY CATHARTIDAE

Black Vulture *Coragyps atratus*

DESCRIPTION: L = 66 cm (26 in); W = 145 cm (57 in); entirely black with naked black-skinned head; primaries silvery from below. Black Vultures do not normally soar; they alternate flapping and gliding, usually at low levels, and with their wings held horizontally, not in a "V", like Turkey Vultures.

HABITAT AND DISTRIBUTION: Uncommon to rare resident*, scarcer in winter throughout except northern and western WV, northern and western PA, northern NJ, and eastern shore of VA, where rare to casual or absent: Mainly open

areas for foraging, but nests at times in woodlands.

DIET: Forages by flying over open areas, searching visually for carrion, and watching behavior of other vultures, both Black and Turkey, for signs of prey detection. Turkey Vultures are able to locate prey both visually and by olfaction, while Black Vultures must depend on their vision alone, and therefore often follow Turkey Vultures, or other Black Vultures, to carcasses. Occasionally, they forage by walking along shorelines or by taking live prey (e.g., nestlings, newborn livestock—attacked by groups of birds, fish, small mammals).

REPRODUCTION: Monogamous, perhaps pairing for life, at least in resident populations; breeding begins in March; no nest is built; eggs are placed in a natural cavity (e.g., cave, crevice, stump, or hollow log), brush pile, thicket, or even an abandoned building; clutch—2; incubation (both sexes)—38–39 days; fledging—75–80 days post-hatching; post-fledging parental care—several months; vulture families remain together for long periods, often occupying communal roosts with other vulture families, which can also serve as "information centers" for prey location.

CONSERVATION STATUS: S2—DE; S3—WV; limited breeding population at the edge of the range; state and federal fish and wildlife agencies have issued permits allowing killing of entire vulture populations for some Mid-Atlantic states (e.g., VA) because of presumptive damage to domestic stock. These programs have been challenged, and no massive removal has occurred to date.

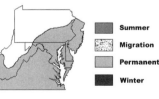

Summer
Migration
Permanent
Winter

RANGE: Breeds from eastern (NJ) and southwestern (AZ) United States south to central Argentina; winters from central and southern United States south through breeding range.

KEY REFERENCES: Buckley (1999).

Turkey Vulture *Cathartes aura*

DESCRIPTION: L = 66 cm (26 in); W = 175 cm (69 in); black with naked, red-skinned head; relatively long tail; silvery flight feathers (outlining black wing lining). **Immature** Black head.

HABITAT AND DISTRIBUTION: Common summer resident* (Mar–Oct) nearly throughout, though scarcer in northwestern WV; uncommon in winter except northern and western WV, northern and western PA, northern NJ, and mountains throughout, where rare or absent: Nearly ubiquitous except in extensive agricultural or residential areas.

DIET: Forages by soaring and searching visually, and through use of olfactory clues, for carrion; regurgitates pellets of indigestible material (e.g., bones, feathers); will take live prey on occasion.

REPRODUCTION: Monogamous, possibly mating for life; pairs arrive on breeding territory in March; nests in a variety of sites, e.g., caves, rock outcrops, abandoned hawk or heron nests, thickets, and hollow stumps, trees, or logs; no nest is actually constructed, and eggs are laid directly on the substrate; clutch—2; incubation (both sexes)—28 days; fledging—60–80 days post-hatching; little evidence of post-fledging parental care; single brood per season.

RANGE: Breeds from southern Canada south to southern South America, Bahamas and Cuba; winters from

Summer
Migration
Permanent
Winter

central and southern United States south through breeding range.

KEY REFERENCES: Stager (1964), Kirk and Mossman (1998).

ORDER FALCONIFORMES; FAMILY ACCIPITRIDAE

Osprey *Pandion haliaetus*

DESCRIPTION: L = 59 cm (23 in); W = 160 cm (63 in); dark brown above; white below, often with dark streaking (females); white crown with ragged crest from occiput; broad dark line extending behind yellow eye; white chin and cheek; extended wings are white finely barred with brown with prominent dark patches at wrist.

HABITAT AND DISTRIBUTION: Common to uncommon transient (Mar–May, Sep–Oct) throughout, more common in spring; common to uncommon and local summer resident* (May–Sep) along the coast, in eastern PA, and in the southern Ohio River in WV; rare but increasing as a summer resident at scattered inland localities along lakes and rivers nearly throughout the region; rare to casual winter resident along the coast: Estuaries, lakes, rivers, ponds, bays, marshes, lagoons; occasionally offshore for surface-feeding fish.

DIET: Forages mostly over shallow water by flapping, gliding, and hovering 10–40 m (30–130 ft) above the surface, and diving, feet first, to grasp fish,

10–30 cm (4–12 in) in length, up to 1 m (40 in) below the water surface.

REPRODUCTION: Usually monogamous; some polygyny; adults arrive on breeding territories in March or April; open nest, 0.7 m dia. (28 in), is built of sticks by both sexes on a wide variety of platforms, often over or near water, e.g., trees, cliffs, channel markers, buoys, telephone poles, and rocks; clutch—3; incubation (both sexes)—37 days; fledging—50–55 days post-hatching; post-fledging parental care—up to 8–10 weeks; single brood per season.

CONSERVATION STATUS: S2—PA, NJ, WV; S3—DE; decimated during the early twentieth century by hunting as pests, especially inland; in the mid-twentieth century most remaining North American populations were sharply reduced by pesticide poisoning; combinations of hacking, prohibition of hunting, and reduction of pesticide use have promoted a rapid recovery in many areas, especially of coastal populations.

RANGE: Breeds in boreal, temperate, and some tropical localities of Old and New World, particularly along coasts; winters mainly in tropical and subtropical zones.

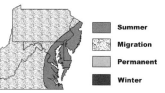

■	**Summer**
▨	**Migration**
▨	**Permanent**
■	**Winter**

KEY REFERENCES: Poole et al. (2002).

Mississippi Kite *Ictinia mississippiensis*

DESCRIPTION: L = 38 cm (15 in); W = 89 cm (35 in); dark gray above; pale gray below with black primaries and tail; red eye; orange-yellow legs; gray cere; in flight note pointed wings, uniform gray underparts and dark, slightly forked tail.

Immature Streaked rusty below; barred tail; red or yellow eye.

HABITAT AND DISTRIBUTION: Rare summer resident* (Apr.-Sep.) in southern VA, especially along the Meherrin River in Greensville County; Riparian

and oak woodlands, deciduous forest and swamps, savanna.

DIET: Forages mainly by sallying or hawking from perches, but also by soaring and hovering; feeds mostly on flying insects (e.g., Coleoptera, Homoptera, Orthoptera, Odonata), but also takes birds, reptiles, amphibians, small mammals; casts pellets of bone, feathers, and other indigestible parts of prey; very little known concerning nonbreeding period habits or habitat.

REPRODUCTION: Monogamous with some assistance from "helpers" (nonbreeding birds); birds arrive at breeding sites paired in late April or early May; there is no breeding territory, and pairs often nest in proximity with other pairs; open platform nest, 27 cm dia. (in), is placed in the crotch or fork of a tree, and built by both pair members of sticks and twigs lined with leaves and Spanish moss; clutch—2; incubation (both sexes)—30 days; fledging—30–35 days post-hatching; post-fledging group care—at least 50 days and perhaps until migration; single brood per season.

CONSERVATION STATUS: S1—VA; limited breeding habitat at the periphery of the range.

RANGE: Breeds in the southern United States; winters south to central South America, but little is known of the species during this period.

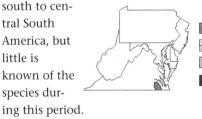

Summer
Migration
Permanent
Winter

KEY REFERENCES: Parker (1999).

Bald Eagle *Haliaeetus leucocephalus*

DESCRIPTION: L = 89 cm (35 in); W = 213 cm (84 in); huge size; dark brown body; white head and tail; yellow beak and feet. **Immature** Entirely brown with whitish wing linings and base of tail.

HABITAT AND DISTRIBUTION: Uncommon and local resident* along the coast and in northwestern PA; rare to casual elsewhere: Forested lakes, rivers, estuaries.

DIET: Forages by sallying from perches or soaring over water and grasping prey (fish, mammals, birds) from water surface or shoreline; also feeds on carrion.

REPRODUCTION: Monogamous, perhaps mating for life; nest-building begins 1–3 months before egglaying, usually in February in Mid-Atlantic region; open platform nest, 1–2 m dia. (3.3–6.6 ft), is placed in the top of a large tree and built (both sexes) of sticks and lined with grasses, moss, green leaves, and down; clutch—2; incubation (mostly female)—24 days; fledging—10–11 weeks post-hatching; post-fledging parental care—4–11 weeks; single brood per season.

CONSERVATION STATUS: S1—NJ; S2—PA, WV, VA, MD, DE; "Endangered"-USFWS; populations were decimated by effects of DDT and other pesticides on reproduction; however, pesticide controls and reintroduction programs have fostered recovery in many areas.

RANGE: Breeds across Canada and northern United States, south along coasts to FL, CA, and TX; winters throughout breeding range from southern Canada southward to Mexico, particularly along

the coasts and at larger inland lakes.

KEY REFERENCES: Buehler (2000).

Summer
Migration
Permanent
Winter

Northern Harrier *Circus cyaneus*

DESCRIPTION: L = 48 cm (19 in); W = 107 cm (42 in); a slim, long-tailed, long-winged hawk; gray above; pale below with dark spots; white rump; yellow eye and cere; long, yellow legs. **Female** 50% heavier and 13% longer than male (54 cm, 21 in); streaked dark brown and tan above; whitish below heavily streaked with brown; yellow eyes; pale yellowish cere; yellow legs; barred tail; white rump.

HABITAT AND DISTRIBUTION: Uncommon to rare and local resident* in PA, the Delmarva Peninsula, and scattered localities elsewhere throughout the region; uncommon transient (Oct, Apr) throughout; common to uncommon winter resident (Nov–Mar), mainly along the coast and Piedmont; birds congregate in terrestrial, communal roosts of up to 20 individuals in open areas in winter: Marshes, grasslands, estuaries, agricultural areas.

DIET: Forages by flying low over open areas, usually within a few meters of the ground, alternately flapping and gliding; wings at an angle during glides; often hovers just above the ground; depends on both visual and auditory cues for prey detection; prey items are captured with talons, and include voles, mice, shrews, rabbits, and small birds.

REPRODUCTION: Mostly monogamous, but with some polygyny, apparently depending on prey abundance; males arrive on breeding territories in March or April, 5–10 days before females; open platform nest, 0.4–0.6 m dia. (16–24 in), is placed on the ground, generally in low, dense vegetation, in open grassland; both sexes build the nest of plant stems, stalks, and grasses lined with grasses and sedges; clutch—4–5; incubation (female)—30–32 days; fledging—young are able to walk off from nest at 14 days, but cannot fly until 4–5 weeks post-hatching; post-fledging parental care—7–8 weeks post-hatching; single brood per season.

CONSERVATION STATUS: S1—WV, VA, NJ, DE; S2—MD; S3—PA; breeding populations of Northern Harriers, like several other open-country birds, have declined in the past few decades as abandoned farms have reverted to forest or been developed for residential areas or intensive agriculture; in addition to this form of habitat loss, harriers have suffered from loss of wetland (marsh) habitats as well.

RANGE: Breeds across boreal and temperate regions of Northern Hemisphere; winters in temperate and tropical zones.

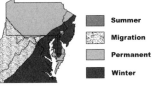

Summer
Migration
Permanent
Winter

KEY REFERENCES: MacWhirter and Bildstein (1996).

Sharp-shinned Hawk *Accipiter striatus*

DESCRIPTION: L = 30 cm (12 in); W = 60 cm (24 in); slate-gray above; barred rusty and white below; gray crown with rusty face; red eye; yellow cere; barred tail; long yellow legs; as in other accipiters,

northern Argentina (except prairie regions and most of southern United States), also in Greater Antilles; winters from northern coastal regions and southern Canada south through breeding range.

KEY REFERENCES: Bildstein and Meyer (2000).

the female is much larger than the male.
Immature Brown above; streaked brown and white on head, breast, and belly. All 3 accipiters are distinguished from other hawks by their relatively short, broad wings, and long, square-tipped tails, and by their behavior in flight, which is characterized by a series of rapid wing beats followed by a short, flat-winged glide. They seldom soar except during migration. All 3 species are forest bird hunters.

HABITAT AND DISTRIBUTION: Uncommon resident* in highlands except central WV where scarce; uncommon winter resident (Sep–Apr) along coast and Piedmont, rare in summer: Forests.

DIET: Forages by perching quietly in forest or edge mid-story, and launching to pursue and capture prey, usually small birds, although small mammals, reptiles, amphibians, and insects are also taken.

REPRODUCTION: Monogamous; pairs arrive on breeding territory, and begin courtship flights near nest site in late April or early May; open platform nest, 35–60 cm dia. (14–24 in), is placed against the trunk on a horizontal branch about 8 m up (26 ft), often in a conifer, and is built (female) of twigs and sticks lined with bark; clutch—4–5; incubation (female)—32 days; fledging—28–32 days post-hatching; post-fledging parental care—3–4 weeks; single brood per season.

CONSERVATION STATUS: S1—MD; S2—NJ; S3—VA, WV; highland forest breeding habitat for this species is scarce in parts of the region.

RANGE: Breeds from sub-Arctic AK and Canada to

Cooper's Hawk *Accipiter cooperii*

DESCRIPTION: L = 46 cm (18 in); W = 81 cm (32 in); slate-gray above; barred rusty and white below; gray crown with rusty face; red eye; yellow cere; barred tail; long, yellow legs. **Immature** Brown above; streaked brown and white on head, breast, and belly.

HABITAT AND DISTRIBUTION: Uncommon to rare resident* nearly throughout; less numerous along Coastal Plain in summer: Forests.

DIET: Hunts by making short flights to inconspicuous perches, scanning for prey, giving chase, and then grasping prey with both feet; small to medium-sized birds are the main prey, although mammals, reptiles, amphibians, insects, and fish are also eaten; prey parts in order of consumption are head, viscera, and muscle.

REPRODUCTION: Monogamous; pairs begin nesting in April; open platform nest, 0.6–0.8 m dia. (24–32 in), is placed in a tree crotch or large hori-

zontal limb 8–15 m (24–49 ft) above ground, and built of sticks with a lining of bark shards, mistletoe, and green sprigs; clutch—3–4; incubation—34–36 days; fledging—30–34 days post-hatching; post-fledging parental care—up to seven weeks; single brood per season.

CONSERVATION STATUS: S1—DE; S3—WV, VA, NJ; original declines, 1950–1970, are attributed to reduced reproductive success resulting from pesticide-induced eggshell thinning; however, many populations have recovered since the banning of DDT and related compounds, and the cause of current low population numbers in some Mid-Atlantic states is unknown.

RANGE: Breeds from southern Canada to northern Mexico; winters from northern United States to Honduras.

KEY REFERENCES: Rosenfield and Bielefeldt (1993).

Northern Goshawk *Accipiter gentilis*

DESCRIPTION: L = 59 cm (23 in); W = 100 cm (39 in); slate-gray above; barred gray and white below; gray crown with prominent white eyebrow; dark patch behind eye; red eye; yellow cere; unevenly barred tail; long yellow legs. **Immature** Brown above; streaked brown and white on head, breast, and belly; white eyebrow; uneven tail barring.

HABITAT AND DISTRIBUTION: Uncommon to rare resident* in northern PA and scattered localities in northwestern NJ, western MD, and eastern WV; rare transient and winter visitor (Nov–Apr) elsewhere: Northern mixed hardwood and coniferous forests.

DIET: Forages by making short flights through or over forest to perches in forest mid-story, and then launching out to chase and capture prey; main food items are birds, e.g., grouse, woodpeckers, corvids, and mammals, including squirrels and rabbits.

REPRODUCTION: Monogamous; pair arrive on breeding territory in March; open platform nest, 1 m dia. (40 in), is placed in a large coniferous or deciduous tree, often against the trunk on a horizontal branch, and built (female) of sticks lined with bark and conifer sprigs; clutch—2–4; incubation (female)—28–30 days; fledging—35–36 days (males), 40–42 days (females) post hatching; post-fledging parental care—30 days; single brood per season.

CONSERVATION STATUS: S1—WV, MD, NJ; mature highland forest habitat is scarce for much of the Mid-Atlantic region.

RANGE: Breeds in boreal regions of Northern Hemisphere, south in mountains to temperate and tropical zones; winters in breeding range and irregularly southward and in lowlands.

KEY REFERENCES: Squires and Reynolds (1997).

Red-shouldered Hawk *Buteo lineatus*

DESCRIPTION: L = 48 cm (19 in); W = 100 cm (39 in); mottled dark brown and white above with rusty shoulders; barred rusty and white below; tail dark with 3 or 4 narrow, whitish bars; brown eye; pale yellowish cere; yellow legs; in flight note brown and white barring on flight feathers, pale white patch at base of primaries ("window"), and rusty wing linings. **Immature** Brown mottled with white above; buff streaked with brown on breast, barred on belly; rusty shoulders.

HABITAT AND DISTRIBUTION: Uncommon to rare resident* throughout, more common along the Coastal Plain: Riparian forest, woodlands, swamps.

DIET: Forages usually by scanning from a perch in a forest tree, and dropping or swooping onto prey; main prey items are small mammals (e.g., voles, chipmunks, deer mice), but birds, reptiles, amphibians, invertebrates, and fish can also be part of diet.

REPRODUCTION: Monogamous; pair formation begins in February or March; open platform nest, 0.4–0.6 m dia. (16–24 in), is placed in the crotch of a large tree (12–19 m up) (40–60 ft); both pair members build the nest of sticks, leaves, bark, and often live conifer sprigs, lined with bark shreds, moss, lichen, and conifer sprigs; clutch—3–4; incubation—33 days; fledging—six weeks post-hatching; post-fledging parental care—8–10 weeks?: single brood per season.

CONSERVATION STATUS: S1—NJ; S2—DE; significant long-term declines have occurred in some populations despite increasing amounts of forest.

RANGE: Breeds from southeastern Canada and the eastern United States

	Summer
	Migration
	Permanent
	Winter

south to central Mexico, also CA; winters from central United States south through breeding range.

KEY REFERENCES: Crocoll (1994).

Broad-winged Hawk *Buteo platypterus*

DESCRIPTION: L = 41 cm (16 in); W = 89 cm (35 in); dark brown above; barred rusty and white below; dark tail with 2 white bands equal in width to dark bands; brown eye; yellowish cere; yellow legs; in flight note whitish flight feathers with black tips; buffy wing linings. **Immature** Brown above; streaked brown and white below; tail narrowly barred with brown and white.

HABITAT AND DISTRIBUTION: Common to uncommon transient and summer resident* (Apr–Sep) nearly throughout; uncommon to rare as a summer resident along the coast. This species is the most abundant transient at hawk watch sites in the region: Deciduous or mixed forests, but seen in migration kettles over any habitat.

DIET: Forages by swooping from perches 10–20 m (30–70 ft) up to capture prey with talons; prey items include small mammals (e.g., voles, chipmunks, shrews), amphibians (e.g., toads), reptiles, birds, especially nestlings and fledglings, and insects.

REPRODUCTION: Monogamous; pairs arrive on breeding territory in April; open platform nest, 0.3–0.5 m dia. (12–20 in), is placed in the first main crotch of a deciduous tree or on horizontal branches next to the trunk of conifers; both sexes build nest of sticks and twigs lined with bark shreds, lichen, moss, pine needles, feathers, and green sprigs; clutch—2–3; incubation (female, male brings food for her)—28–31 days; fledging—5–6 weeks post-hatching; post-fledging parental care—8 weeks; single brood per season.

CONSERVATION STATUS: S2—DE; S3—NJ; perhaps loss of wintering forest habitat in the Neotropics.

RANGE: Breeds across southern Canada from Alberta eastward and the eastern half of the  United States south to TX and FL; winters from southern Mexico south to Brazil; resident in West Indies.

KEY REFERENCES: Goodrich et al. (1996).

Red-tailed Hawk *Buteo jamaicensis*

DESCRIPTION: L = 56 cm (22 in); W = 135 cm (53 in); a large hawk, extremely variable in plumage; most common adult plumage is mottled brown and white above; white below with dark streaks and speckling across belly; rusty tail (appears whitish from below); in flight, dark forewing lining contrasts with generally light underwing.
Immature Mottled brown and white above; streaked brown and white below; brown tail finely barred with grayish-white. **Light Phase (Krider's)** Much paler; pale orange tail. **Dark Phase (Harlan's)**  Dark throughout with some white speckling; dark tail, whitish at base, darker at tip with rusty wash.

HABITAT AND DISTRIBUTION: Common resident* nearly throughout; uncommon to rare in summer along the coast: Open areas, woodlands.

DIET: Normally a sit-and-wait predator, scanning for prey from an elevated perch, although cruising and soaring are also used; small to medium-sized mammals (e.g., rabbits, squirrels, mice, and gophers) are the principal prey, but birds, reptiles, insects, and crabs are also taken in varying amounts based on availability.

REPRODUCTION: Monogamous, often mating for life in nonmigratory populations; both members of the pair select the nest site, usually the top of a tall tree, and build the open platform nest, diameter 0.7–0.8 m (28–32 in), of sticks and twigs lined with strips of bark, stems, and leaves; egglaying begins in mid to late March; clutch—2–3; incubation—28–35 days; fledging—42–46 days post-hatching; parental care post-fledging—4–6 weeks; one brood per season.

RANGE: Breeds from the sub-Arctic of AK and Canada to Panama; winters 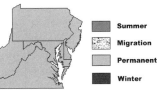 northern United States south through the breeding range; resident in the West Indies.

KEY REFERENCES: Preston and Beane (1993).

Rough-legged Hawk *Buteo lagopus*

DESCRIPTION: L = 56 cm (22 in); W = 130 cm (51 in); a large hawk with legs feathered all the way to the toes; mottled brown and white above; buffy with brown streaks on breast; dark brown belly; white tail with dark subterminal band.
Dark Phase Entirely dark except whitish flight feathers and tail with broad, dark, subterminal band. **Immature** Similar to adult—whitish or finely barred tail with broad, dark,

subterminal band; base of primaries show white from above in flight.

HABITAT AND DISTRIBUTION: Uncommon to rare winter resident (Nov–Mar) in PA and NJ, scarcer in WV and the more southern portions of the region: Open areas, e.g., grassland, pasture, old fields, marshes, and farmland.

DIET: Forages from the air, mostly for small and medium-sized mammals and birds (e.g., voles, rabbits, mice, muskrats, weasels, longspurs, and sparrows), by soaring, alternate flapping and gliding, or hovering, and then dropping to seize prey in talons; also hunts from perches, sitting and waiting until prey is observed, and then launching out to pounce.

RANGE: Breeds in Arctic and sub-Arctic regions of Old and New World; winters mainly in temperate zone.

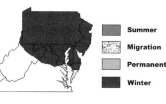

Summer

Migration

Permanent

Winter

KEY REFERENCES: Bechard and Swem (2002).

Golden Eagle *Aquila chrysaetos*

DESCRIPTION: L = 86 cm (34 in); W = 180 cm (71 in); huge; dark brown with golden wash on head and shaggy neck; legs feathered to feet; appears entirely dark in flight. **Immature** Dark brown with white patch at base of primaries and basal half of tail white.

HABITAT AND DISTRIBUTION: Rare transient (Nov, Mar), more common in fall, and rare winter visitor (Dec–Feb) throughout: open areas including grasslands, parklands, shrubby second growth, savanna, and open woodlands.

DIET: Mammals comprise most of prey taken (80–90%) including rabbits, woodchucks, squirrels, and domestic livestock (e.g., lambs and kids).

CONSERVATION STATUS: Various human activities account for 70% of recorded eagle deaths.

RANGE: Breeds primarily in open and mountainous regions of boreal and temperate zones in Northern Hemisphere; winters in central and southern portions of breeding range.

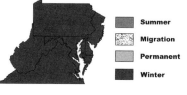

Summer

Migration

Permanent

Winter

KEY REFERENCES: Kochert et al. (2002).

FAMILY FALCONIDAE

American Kestrel *Falco sparverius*

DESCRIPTION: L = 26 cm (10 in); W = 56 cm (22 in); a small falcon; tercel (male) has rusty crown with gray borders; black and white facial pattern; orange buff nape with black spot; orange buff back and underparts spotted with brown; blue-gray wings with dark spots; orange tail with black subterminal bar edged in white. **Female** Rusty back and wings barred with brown.

HABITAT AND DISTRIBUTION: Common resident* throughout, somewhat more common in winter: Open areas.

DIET: Forages by dropping on prey from perches in open grassland, savanna, fields, or hedgerows, or flying and hovering to drop on prey; main prey items are large insects (e.g., grasshoppers, cicadas, beetles, and dragonflies), voles, mice, bats, and small birds.

REPRODUCTION: Mostly monogamous; pairs arrive on breeding territories in April; nests in abandoned woodpecker hole or natural cavity in a pole or snag; clutch—4–5; incubation (both sexes)—27–29 days; fledging— 28–31 days post-hatching; post-fledging parental care— 2 weeks; single brood per season.

CONSERVATION

STATUS: S3—NJ, DE; perhaps declining as a result of conversion of open habitats to forest or intensive developments.

RANGE: Breeds nearly throughout Western Hemisphere from sub-

Arctic AK and Canada to southern Argentina, West Indies; winters from north temperate regions south through breeding range.

KEY REFERENCES: Smallwood and Bird (2001).

Merlin *Falco columbarius*

DESCRIPTION: L = 28 cm (11 in); W = 64 cm (25 in); a small falcon; slate above, buffy below with dark brown spots and streaks; white throat; brown eyes, yellow cere and legs; black tail, white at base, with two white bars and white terminal edging. **Female** Dark brown above.

HABITAT AND DISTRIBUTION: Uncommon to rare transient (Sep–Oct, Mar–Apr) throughout; rare winter visitor (Nov–Feb) along the coast and Piedmont: Open woodlands, hedgerows, second growth; often hunts small birds in trees or shrubs bordering water in winter.

DIET: Principally a bird hunter; sallies from low perches, using surprise and speed to capture prey, most of which is taken on the wing; sometimes cooperative hunting is involved where pairs, usually a male and a female, but sometimes involving even a member of a different species (e.g., Sharp-shinned Hawk), work together to surprise and capture prey; insects, small mammals, and reptiles taken occasionally.

RANGE: Breeds across boreal and north temperate regions of Old and New

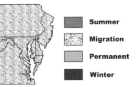

World; winters in south temperate and tropical zones.

KEY REFERENCES: Sodhi et al. (1993).

Peregrine Falcon *Falco peregrinus*

DESCRIPTION: L = 46 cm (18 in); W = 102 cm (40 in); a large falcon; dark gray above; white below with spotting and barring on belly and thighs; black crown and cheek with white neck patch; brown eye; yellow eye-ring, cere, and legs; in flight note large size, long pointed wings, long tail, whitish under-parts finely barred and spotted; black and white facial pattern. **Immature** Dark brown above; buffy

below spotted and streaked with brown; has facial pattern of adult.

HABITAT AND DISTRIBUTION: Uncommon to rare transient (Sep–Oct, Mar–Apr) and rare winter visitor (Nov–Feb) along the coast, scarcer inland. Rare and local summer resident* at reintroduction sites in NJ, VA, MD, and PA: Open areas, usually near water, during the nonbreeding period; breeding habitat seems limited mainly by nest site availability (ledges on vertical walls).

DIET: Forages by locating prey from a perch or while flying, then chasing and/or stooping on prey, usually birds (>80%), and either grasping them with the talons, or striking them and then following injured prey to the ground either to pick them up and carry them to a perch (smaller birds), or to consume them in place; caches surplus prey or prey parts.

REPRODUCTION: Mostly monogamous; breeding pairs arrive at eyries in late February or March; nest, if any, is a crude scrape on a horizontal ledge against a vertical wall, usually a cliff face, but also uses abandoned hawk, osprey, raven, or eagle nests and artificial nest boxes or ledges on buildings in metropolitan areas; clutch—3–4; incubation (both sexes)—33–35 days; fledging—6 weeks post-hatching; post-fledging parental care—from 5–6 weeks for migratory populations up to 9–10 weeks for residents; single brood per season.

CONSERVATION STATUS: S1—PA, WV, VA, NJ; S2—MD; "Threatened"-USFWS; Peregrine Falcon breeding populations were extirpated in the Mid-Atlantic region in the 1950s and 1960s due to reproductive failure caused by pesticide poisoning; hacking programs have achieved some success, but there is not, as yet, a self-sustaining population in the region.

RANGE: Breeds (at least formerly) in boreal and temperate regions of Northern Hemisphere; winters mainly in the tropics.

Summer
Migration
Permanent
Winter

KEY REFERENCES: White et al. (2002).

Yellow Rail *Coturnicops noveboracensis*

DESCRIPTION: L = 18 cm (7 in); W = 33 cm (13 in); a small (cowbird-sized) rail; dark brown with tawny stripes above; dark brown crown; broad tawny eyebrow; dark brown mask; whitish chin; tawny underparts barred with black on the flanks; short, yellowish bill and pale legs; in flight (a rarely observed event) shows white secondaries.

HABITAT AND DISTRIBUTION: Rare to casual transient (Oct, Apr–May), mainly in fall, along the coast. Some summer and winter records as well: Wet prairies, meadows, marshes.

DIET: Forages by walking in shallow water or on wet ground and picking invertebrates (e.g., snails, crustacea, aquatic insects) from benthos, soil surface, or vegetation; also eats plant seeds on occasion.

RANGE: Breeds in east and central Canada and northeastern and north central United States; winters along the southern Atlantic and Gulf coast of United States; resident in central Mexico.

Summer
Migration
Permanent
Winter

KEY REFERENCES: Stalheim (1974), Bookhout (1994).

Black Rail *Laterallus jamaicensis*

DESCRIPTION: L = 15 cm (6 in); W = 26 cm (10 in); a tiny (sparrow-sized) rail; dark grayish-black with chestnut nape and black and white barring on flanks; short black bill.

HABITAT AND DISTRIBUTION: Uncommon to rare transient and local summer resident* (Apr–Oct) along the immediate coast; some winter records as well: Wet prairies and meadows, marshes; preferred coastal habitat is moist marshland with scattered pools.

DIET: Forages mostly on invertebrates picked from the ground, vegetation surfaces, or shallow water while walking or wading; occasionally feeds on plant seeds.

REPRODUCTION: Monogamous?; pair formation—May?; domed nest, 11–15 cm dia. (4–6 in), is placed in low, dense marsh vegetation, 0–0.5 m (0–20 in) above the ground, and built with a thick base of woven grasses with a canopy of dead and living grasses woven over the top and sides; ramp of dead vegetation leads to side entrance; clutch 5–7; incubation—19–20 days, evidently by both male and female; fledging—young are semiprecocial and able to walk from the nest within a few days post-hatching, but age at first flight is unknown; post-fledging parental care—? number of broods per season—?

CONSERVATION STATUS: S1—DE; S2—VA, MD, NJ; the status of breeding populations of this species in the Mid-Atlantic region is not well known, but its coastal marshland habitat is threatened by development.

RANGE: Breeds locally in CA, KS, and along east coast from NY to TX, also West

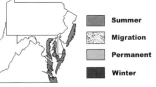

Indies, Central and South America; winters along Gulf coast and in tropical breeding range.

KEY REFERENCES: Eddleman et al. (1994).

Clapper Rail *Rallus longirostris*

DESCRIPTION: L = 33 cm (13 in); W = 51 cm (20 in); a large rail; streaked brown and tan above; buffy below with gray and white barring on flanks; dark crown; white preorbital strip; gray face; red eye; long, pinkish or yellowish bill; pale greenish legs.

HABITAT AND DISTRIBUTION: Common summer resident* along the immediate coast; uncommon to rare in winter: Saltmarshes.

DIET: Forages by walking or wading and searching visually for prey, mainly crabs and shrimp, although a wide variety of other aquatic life is also taken (e.g., insects, annelids, amphipods, molluscs, and bulrush seeds); may use olfactory cues in foraging as well.

REPRODUCTION: Monogamous; pairs form on breeding territory in April; bulky platform or domed nest, 27 cm dia. (11 in), is placed in grass tussocks or shrubs a few cm above the substrate in stands of tall (> 0.6 m) (24 in) marsh grass, and built mostly by the male of dried marsh-grass blades; clutch—5–8; incubation (both sexes)—20 days; fledging—young can fly at 10 weeks post-hatching, but may be independent of adults, foraging on their own by walking and swimming, by 6 weeks; mostly one brood per season, but some pairs may raise 2 broods successfully.

CONSERVATION STATUS: S3—MD; disturbance and destruction of coastal habitat threaten populations.

RANGE: Resident along coast from Connecticut to Belize, CA to southern Mexico; West Indies and much of coastal South America to southeastern Brazil and Peru.

Summer

Migration

Permanent

Winter

KEY REFERENCES: Eddleman and Conway (1998).

King Rail *Rallus elegans*

DESCRIPTION: L = 38 cm (15 in); W = 56 cm (22 in); a large rail; streaked brown and rusty above; tawny below with black and white barring on flanks; head and neck tawny with dark crown and darkish stripe through eye; long bill with dark upper and pinkish or yellowish lower mandible; pale reddish legs.

HABITAT AND DISTRIBUTION: Common (south) to rare (north) and local summer resident* (Apr–Oct) and rare winter visitor (Nov–Mar) along the immediate coast; rare to casual transient and summer resident* inland: Freshwater marshes.

DIET: Crustaceans (e.g., crayfish and fiddler crabs) and aquatic insects are the main prey taken, usually in daylight at shallow fresh or brackish

marshes with low, dense vegetation. Low tide is the preferred foraging period in tidal areas; fish, grasshoppers, frogs, and plant materials also eaten on occasion.

REPRODUCTION: Mostly monogamous, with pairs forming in April and nest construction in May; nests constructed mainly by the male of wet vegetation lined with dried grasses and placed in vegetation clumps over or near shallow water; average clutch size of 10–11 eggs; incubation (both sexes) lasts 22 days; young are semi-precocial, with day old chicks able to follow adults some distance; parents care for young for 7–9 weeks; young are able to fly at 9–10 weeks.

CONSERVATION STATUS: S1—PA, WV; S2—VA, DE; S3—MD, NJ; wetland destruction is the most serious threat.

RANGE: Breeds across the eastern half of the United States south to central Mexico; winters along coast from GA to TX and south to southern Mexico; resident in Cuba.

Summer

Migration

Permanent

Winter

KEY REFERENCES: Meanley (1992).

Virginia Rail *Rallus limicola*

DESCRIPTION: L = 26 cm (10 in); W = 36 cm (14 in); similar to the King Rail but about half the size; streaked brown and rusty above; tawny below with black and white barring on flanks; gray head with rusty crown; white throat; long reddish bill; pale legs.

HABITAT AND DISTRIBUTION: Common (along coast) to uncommon or rare (inland) or casual (WV) summer resident* (Mar–Oct); uncommon to rare in winter along the coast: Freshwater and brackish marshes.

DIET: Mainly crepuscular; forages by walking on floating vegetation, muddy shore, or in shallow water by probing for aquatic invertebrates (e.g., insects, molluscs, crustacea); small fish, reptiles, amphibians, and plant seeds are taken as well.

REPRODUCTION:
Monogamous; males arrive on breeding territory in late March or April, followed by females; partially or completely domed nest, 14–20 cm dia. (6–8 in), sometimes with side-entrance ramp, is placed in dense, emergent vegeta-

tion (e.g., cattail or bulrush), on or slightly above (< 15 cm) water, and built by both sexes of woven grasses; roof is made by pulling surrounding vegetation over the top; clutch—8–9; incubation (both sexes)—18–20 days; fledging—precocial young leave nest at 3–4 days, but cannot fly for 4–7 weeks; post-fledging parental care—30 days?; may have 2 broods per season in some populations.

CONSERVATION STATUS: S1—WV; S2—VA; S3—PA; destruction of wetland habitat.

RANGE: Breeds locally from southern Canada to southern South America; winters along coast and in subtropical

and tropical portions of breeding range.

KEY REFERENCES: Conway (1995).

Sora *Porzana carolina*

DESCRIPTION: L = 23 cm (9 in); W = 36 cm (14 in); a medium-sized rail; streaked brown and rusty above; grayish below with gray and white barring on flanks; gray head with rusty crown; black at base of bill; black throat and upper breast; short yellow bill; greenish legs. **Female** Amount of black on throat and breast reduced. **Immature** Browner overall, lacks black on throat and breast.

HABITAT AND DISTRIBUTION: Uncommon to rare and local summer resident* (Apr–Sep) in northern and southeastern PA, NJ, and DE, rare to casual elsewhere; uncommon to rare transient (Apr–May, Sep–Oct) throughout: Fresh and brackish water marshes.

DIET: Forages in emergent vegetation by walking on floating vegetation or debris, wading, or swimming, and raking with feet or picking aside debris to expose food items; main foods include seeds (e.g., sedges, grasses, bulrushes, rice), pondweed, duckweed, and invertebrates (e.g., aquatic insects, molluscs).

REPRODUCTION: Monogamous; pairs arrive on breeding territories in April or early May; open, partially domed, or domed platform nest, 12–20 cm dia. (5–8 in), is placed in sturdy emergent vegetation (e.g., cattails, sedges) just above water, and is built by both sexes of woven coarse grasses and plant stems; dome is made by pulling surrounding, supporting vegetation over the top and interweaving with grass blades and plant stems into a roof; usually a ramp is built of grasses to a side entrance; clutch—8–11; incubation (both sexes)—16–19 days; fledging—young are partially precocial, and are able to clamber off nest by 3–4 days post-hatching; parental care duration—up to 4 weeks post-hatching; 2 broods per season are possible.

CONSERVATION STATUS: S1—WV, VA, MD; S2—DE; S3—PA; populations have shown long-term (1966–1994) range-wide, evidently as a result of loss of freshwater wetland breeding habitat.

RANGE: Breeds from central Canada to southern United States; winters along coast and from

southern United States to northern South America and West Indies.

KEY REFERENCES: Melvin and Gibbs (1996).

Purple Gallinule *Porphyrula martinica*

DESCRIPTION: L = 33 cm (13 in); W = 59 cm (23 in); a large rail with short, cone-shaped bill and extremely long toes; iridescent green back; purple head and underparts; blue frontal shield of naked skin on forehead; red bill with yellow tip; yellow legs. **Immature** Buffy overall, darker on back.

HABITAT AND DISTRIBUTION: Rare summer visitor (Apr–Sep) in coastal VA; has bred*; rare to casual elsewhere: Freshwater or slightly brackish marshes, lakes, ponds, or impoundments with open pools and dense stands of emergent and floating vegetation (e.g., water lilies, lotus, spatterdock, pickerel weed, and water hyacinth).

DIET: The extremely long toes enable this rail to walk on lily pads and other floating vegetation without sinking; forages by walking on emergent vegetation and picking

seeds, flowers, and fruits; also feeds on aquatic invertebrates (e.g., insects, molluscs, annelids, spiders).

RANGE: Breeds from the eastern United States south to northern Argentina, West Indies; winters from southern Gulf states south through the breeding range.

KEY REFERENCES: West and Hess (2002).

Common Moorhen *Gallinula chloropus*

DESCRIPTION: L = 36 cm (14 in); W = 59 cm (23 in); a large rail with short, cone-shaped bill and extremely long toes; entirely sooty gray with white edging along wing and white undertail coverts; red frontal shield of naked skin on forehead; red bill with yellow tip; greenish-yellow legs. **Winter** Similar to summer but with olive bill, frontal shield, and legs.

HABITAT AND DISTRIBUTION: Uncommon to rare summer resident* (Apr–Oct) along the Coastal Plain and locally inland (e.g., northwestern PA); scattered summer records elsewhere; uncommon to rare in extreme southeastern VA in winter: Freshwater or brackish wetlands where emergent or shoreline vegetation is interspersed with pools containing floating or submerged plants.

DIET: Extremely long toes enable this rail to walk on lily pads and other floating vegetation without sinking; swims more than most other rails; gleans seeds and other parts of grasses, sedges, pondweeds, smartweeds, and other aquatic plants from water and leaf surfaces; also takes inverte-

brates, e.g., insects, molluscs, and crustacea; main foods often are sedge seeds and snails.

REPRODUCTION: Monogamous; adults arrive on breeding territory in April or early May; at least 3 types of nests are built depending on different uses: 1) egg nests for laying and hatching of eggs; 2) brood nests for roosting of hatched young; and 3) courtship nest for display purposes; nests are open cups (0.2–0.3 m dia. (8–12 in), placed in emergent or floating vegetation, often with a ramp, and built (both sexes) of blades and leaves picked from surrounding plants; clutch—5–9; incubation (both sexes)—19–22 days; fledging—young are semi-precocial, and able to forage for some food by 7 days of age; adults brood the young on the nest, or at special brood nests, up to 14 days post-hatching, and provide at least some food, sometimes assisted by young of previous broods, up to 45 days; fledging—40–50 days post-hatching; most young disperse from parents' territory at 72 days post-hatching; often produces 2 broods per season.

CONSERVATION STATUS: S1—WV, VA; S2—MD; S3—NJ, PA; moorhens are threatened with loss of wetland breeding habitats.

RANGE: Breeds locally in temperate and tropical regions of the world; winters mainly in subtropical and tropical zones.

▨	**Summer**
▧	**Migration**
▨	**Permanent**
■	**Winter**

KEY REFERENCES: Bannor and Kiviat (2002).

American Coot *Fulica americana*

DESCRIPTION: L = 41 cm (16 in); W = 66 cm (26 in); a large, black rail, more duck-like than rail-like in appearance and behavior; white bill with dark spot at tip; dark reddish-brown forehead shield (unfeathered); red eye; greenish legs and lobed toes. **Immature** Paler; pale gray bill, darker at tip.

HABITAT AND DISTRIBUTION: Common to uncommon transient (Apr–May, Oct–Nov) throughout; local summer resident* (May–Aug) along the

coast, northwestern PA, and scattered localities elsewhere; common to uncommon winter resident, mainly along the Coastal Plain but at open water elsewhere in the region: Ponds, lakes, marshes, bays, swamps, estuaries.

DIET: Swims in open water, tipping and diving for food instead of skulking through reeds like most rails; pumps head forward while swimming; feeds mostly on aquatic plants (90%), e.g., pondweed, sedges, algae, and wild and domestic grasses including rice.

REPRODUCTION: Monogamous; adults arrive on breeding territories in late April or early May; floating platform nest, 0.3 m dia. (12 in), is anchored to emergent vegetation (e.g., cattails, bulrushes, reeds, or marsh grass) and built mostly by the female of plant stems and blades gathered from the vicinity of the nest site; often more than one nest is built for use in courtship, and additional nests are built after hatching for brooding of the young; clutch—8–12; sometimes females place their eggs in nests of other pairs; incubation (both sexes)—23–26 days; fledging—young are precocial, and leave the nest 1–2 days post-hatching, return to brood nests at night up to 14 days of age, and are able to fly at 75 days, by which time they are independent of parents; one brood per season is usual.

CONSERVATION STATUS: S1—VA, WV, NJ, DE; S3—PA; disturbance and destruction of breeding habitat threatens populations.

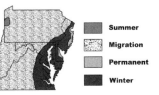

▨	**Summer**
▧	**Migration**
▨	**Permanent**
■	**Winter**

RANGE: Breeds from central Canada to Nicaragua and West Indies; winters along coast and from central United States to northern Colombia, West Indies.

KEY REFERENCES: Brisbin and Mowbray (2002).

ORDER CHARADRIIFORMES; FAMILY CHARADRIIDAE

Black-bellied Plover *Pluvialis squatarola*

DESCRIPTION: L = 31 cm (12 in); W = 64 cm (25 in); a Killdeer-sized plover; checked black and white on back; white crown, nape, and shoulder; black face, throat, and breast; white belly and undertail coverts. **Winter** Heavy black bill contrasts with whitish at base of bill; pale eyebrow; crown and back brownish mottled with white; breast is mottled with brown; belly white; black legs.

HABITAT AND DISTRIBUTION: Common to uncommon transient (Aug–Oct, Apr–May) and winter resident (Aug–Apr) along the coast; uncommon to rare in summer; rare or casual transient inland—most records are from the fall: Beaches, bays, mud flats, estuaries.

DIET: Forages both diurnally and nocturnally, focusing on low-tide periods; preferred foraging substrate is wet sand, where it hunts visually for insects (inland and breeding period) or marine invertebrates (e.g., annelids, crustacea, molluscs), using short runs followed by a quick peck to capture prey.

RANGE: Breeds in the high Arctic of Old and New

World; winters along temperate and tropical coasts of the world.

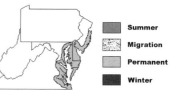

KEY REFERENCES: Paulson (1995).

American Golden-Plover *Pluvialis dominica*

DESCRIPTION: L = 28 cm (11 in); W = 59 cm (23 in); a Killdeer-sized plover; checked black, white, and gold on back and crown; white "headband" running from forehead, over eye, and down side of neck; black face and underparts. **Winter** Heavy black bill contrasts with white at base of bill; pale eyebrow; crown and back brown and white often flecked with gold; underparts white, speckled with brown on breast; gray legs.

HABITAT AND DISTRIBUTION: Uncommon fall transient (Aug–Oct) and rare spring transient (Apr–May) along the coast; rare to casual transient elsewhere: Prairie, pasture, plowed fields, grasslands, intertidal mud flats, shorelines, beaches.

DIET: Forages in open areas, usually with little vegetation; locates prey mainly by sight, using short runs followed by quick pecks; food items vary with availability, but include invertebrates (e.g., insects, annelids, spiders, crustacea), plant seeds, and, on occasion, fruits.

RANGE: Breeds in high Arctic of New World; winters

in South America.

KEY REFERENCES: Johnson and Connors (1996).

Wilson's Plover *Charadrius wilsonia*

DESCRIPTION: L = 20 cm (8 in); W = 41 cm (16 in); a smallish plover with a heavy, black bill; brown back and head; white forehead and eyeline; black lores; white chin and collar; black band across throat; white underparts; pinkish legs. **Female and Immature** Brown breast band.

HABITAT AND DISTRIBUTION: Uncommon to rare and local summer resident* (Apr–Sep) along the VA barrier islands; rare to casual elsewhere: Beaches, dunes, tidal flats, shorelines.

DIET: Forages during low tide by running and grasping prey from substrate surface, mainly crustacea, especially fiddler crabs.

REPRODUCTION: Monogamous; pair formation in May; nest, 8 cm dia. (3 in), is a scrape built by the male in debris above high-tide line along the beach; clutch—3; incubation (both sexes)—25 days; young are precocial and able to leave the nest 2–3 h post-hatching,

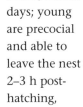

time to first flight and period of dependence on adults are unknown; single brood per season.

CONSERVATION STATUS: S1—VA, MD; beach disturbance (humans, predators) and development threaten nesting birds.

RANGE: Breeds along both coasts from northwestern Mexico in the west and VA in the east south to northern South America; West Indies; winters mainly in tropical portions of breeding range.

KEY REFERENCES: Corbat and Bergstrom (2000).

Semipalmated Plover *Charadrius semipalmatus*

DESCRIPTION: L = 18 cm (7 in); W = 38 cm (15 in); a small plover; black face with brown crown; white forehead; white postorbital stripe (sometimes not visible); orange eyering; orange bill, black at tip; white chin and collar; black breast band; white underparts; orange or yellow legs. **Winter** Brown breast band; dull, dark bill (may show some orange at base). **Immature** Similar to adult but eyering yellow, bill black at tip, brown at base, legs brown anteriorly, yellow posteriorly.

HABITAT AND DISTRIBUTION: Common to uncommon transient (Apr–May, Jul–Oct) throughout; rare in summer and winter along coast: Estuarine mud flats, saltmarshes, salt flats, beaches, bay shores, lake shores, flooded fields.

DIET: Forages mostly at intermediate and low tide levels, day or night, but more often during day or moonlit nights; searches visually for prey on wet, exposed benthos or in shallow water (< 2 cm deep) (1 in) by running, stopping, and tipping

whole body forward to peck or snatch prey (e.g., molluscs, annelids, crustacea, aquatic insects).

RANGE: Breeds in northern boreal and sub-Arctic regions of North America; winters mainly along temperate and tropical coasts from GA and CA to southern South America and the West Indies.

KEY REFERENCES: Nol and Blanken (1999).

Piping Plover *Charadrius melodus*

DESCRIPTION: L = 18 cm (7 in); W = 38 cm (15 in); a small plover; grayish-brown crown, back and wings; white forehead with distinct black band across forecrown; white chin, throat collar, and underparts; partial or complete black band across breast; orange bill tipped with black; orange legs; white rump. **Winter** Lacks black breast and crown bands; dark bill.

HABITAT AND DISTRIBUTION: Uncommon transient and summer resident* (Mar–Sep) along the immediate coast; rare to casual inland except along the Lake Erie shore where uncommon to rare as a transient and formerly a summer resident*. Reintroduction efforts are underway: Beaches.

DIET: Various marine and freshwater invertebrates (e.g., marine worms, insects, and crustacea) often taken near the water's edge with short runs followed by quick pecks at surface sand or detritus.

REPRODUCTION: Pair formation begins in mid-March

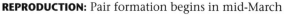

when birds arrive on breeding sites; usually monogamous, but serial polyandry also occurs; nest sites are simple scrapes in open sand, gravel or shell constructed by males several meters from water or shoreline; clutch size is usually four eggs; incubation 26–28 days with both sexes incubating; young are precocial and able to walk and peck within hours of hatching and able to fly at 27 days; family groups remain together for some time, apparently until southward departure on migration in some cases; only one brood is produced per year.

CONSERVATION STATUS: S1—VA, MD; "Threatened:—USFWS; the main threat to Piping Plover populations appears to be a scarcity of beach breeding habitat free from human disturbance or excessive predation.

RANGE: Breeds locally in southeastern and south central Canada, northeastern and north central United States; winters along coast from SC to Veracruz, also in West Indies.

KEY REFERENCES: Haig 1992.

Killdeer *Charadrius vociferus*

DESCRIPTION: L = 26 cm (10 in); W = 51 cm (20 in); a medium-sized plover; brown back and head; orange-buff rump; white forehead; white post-orbital stripe; orange eyering; dark bill; black lores; white chin and collar; 2 black bands across throat and upper breast; white underparts; pale legs.

HABITAT AND DISTRIBUTION: Common summer resident* (Mar–Oct) throughout; common to uncommon in winter (Nov–Feb) in the Coastal Plain and Piedmont, rare elsewhere: Open areas.

DIET: Forages on the ground by making short walks or runs, and then picking, probing, or chasing invertebrate prey (e.g., worms, insects, crustacea); sometimes uses "foot quivering" or "foot stamping" for reasons not well understood.

REPRODUCTION: Monogamous; pairs begin breeding in April; nest, 8–9 cm dia. (3 in), is a scrape placed in dirt, gravel, or sand of an open, bare area, e.g., a road shoulder, unpaved parking lot, or gravel roof; clutch—4; incubation (both sexes)—23–29 days; young are precocial and able to leave nest 2–3 h post-hatching, but cannot fly until about 30 days post-hatching; adults will feign a broken wing when young or nest are approached by a potential predator; parental care—perhaps up to 39 days post-hatching; double broods common in southern populations.

RANGE: Breeds from sub-Arctic Canada and AK south to central Mexico,

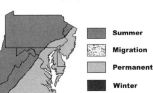

Summer
Migration
Permanent
Winter

Greater Antilles, western South America; winters along coast from Washington and MA, and inland from southern United States, south to northern and western South America, West Indies.

KEY REFERENCES: Jackson and Jackson (2000).

FAMILY HAEMATOPODIDAE

American Oystercatcher
Haematopus palliatus

DESCRIPTION: L = 48 cm (19 in); W = 89 cm (35 in); gull-sized bird; black hood; long, bright orange bill; red eyering; brown back; white breast and belly; pinkish legs; broad white bars on secondar-

ies and white base of tail show in flight.
  Immature Brown head; dull orange bill.
HABITAT AND DISTRIBUTION: Common summer resident* (Mar–Nov), mainly on the barrier islands; common to uncommon winter resident (Dec–Feb) from southern MD south: Beaches, bays.
DIET: Forages mostly in intertidal zones, mussel and oyster reefs, and shoals on oysters, mussels, and other molluscs by prying or hammering open the shells with its specially adapted bill and behaviors; also eats a variety of other marine invertebrates, including scavenging dead invertebrates, e.g., jellyfish and starfish, along the shoreline.

REPRODUCTION: Mostly monogamous; female arrives on breeding territory as much as three weeks before the male; pair formation occurs when the male arrives in March or early April; nest is a simple scrape lined with shell, pebbles, and other substrate detritus constructed in open beach 10–30 m (30–100 ft) beyond high tide line; clutch—2–3; incubation—26–28 days; young are precocial and able to walk within 24 h of hatching; fledging—35 days post-hatching; post-fledging parental care—young appear dependent on adults for at least 60 days after hatching and remain in family groups through the fall and perhaps into winter as well; single brood per season.
CONSERVATION STATUS: S1—DE; S3—MD, VA; beach-nesting shorebirds are threatened throughout the Mid-Atlantic region by nest predation and human disturbance of nesting sites.
RANGE: Breeds locally along coast from northwestern

Mexico in western North America and MA in eastern North

America south to southern South America, West Indies.

KEY REFERENCES: Nol and Humphrey (1994).

FAMILY RECURVIROSTRIDAE

Black-necked Stilt *Himantopus mexicanus*

DESCRIPTION: L = 36 cm (14 in); W = 69 cm (27 in); spindly shorebird with long neck, needle-like bill, and long legs; black head, nape, and back; white at base of bill, throat, and underparts; white patch behind eye; pink legs. **Female** Similar to male but paler. **Immature** Brown rather than black.

HABITAT AND DISTRIBUTION: Uncommon to rare summer resident* (Apr–Sep) in coastal DE, MD, and VA: Mud flats, lagoons, marshes, estuaries, ponds, flooded fields.

DIET: Forages by diurnal or nocturnal wading in mud or shallow water (< 13 cm) (5 in) and picking prey from surface, darting head underwater to grasp prey spotted visually, or scything (moving bill back and forth underwater) to locate and capture prey by touch; prey includes aquatic and terrestrial invertebrates (e.g., insects, crustacea, molluscs) and small vertebrates (e.g., fish, amphibians).

REPRODUCTION: Monogamous;

adults may arrive paired at breeding sites, often located on islands or islets, in April; nest, 13 cm dia. (5 in), is a scrape constructed by both pair members, more or less lined with vegetation and debris gathered from the vicinity of the site, sometimes built up a few cm above substrate and in loose association with other nesting stilts or avocets; clutch—4; incubation (both sexes)— 23–26 days; fledging—young are precocial, able to leave the nest 1–2 h post-hatching, but cannot fly until 27–31 days; post-fledging parental care— variable, but may last "well beyond" fledging; single brood per season.

CONSERVATION STATUS: S1—VA; S2—DE; stilt coastal breeding habitat is scarce in the region, and threatened by human disturbance, development, and predators.

RANGE: Breeds locally in the western United States and along coast in the

east from MA south locally along coasts of Middle and South America to northern Argentina; West Indies; winters from NC in the east and central CA in the west south through the breeding range.

KEY REFERENCES: Robinson et al. (1999).

American Avocet *Recurvirostra americana*

DESCRIPTION: L = 46 cm (18 in); W = 81 cm (32 in); long-legged, long-necked shorebird with long, upturned bill; black and white back and wings; white belly; rusty orange head; whitish at base of bill; gray legs. Winter Head and neck mostly whitish with little rusty tinge.

HABITAT AND DISTRIBUTION: Uncommon to rare and local resident summer resident (Apr–Oct) along the immediate coast; most numerous during fall migration (Sep); has bred in historical times in NJ; rare to casual in winter: Estuaries, ponds, lakes, mud flats, flooded pastures, bays.

DIET: Forages day or night, mostly at low tide,

by swinging extremely sensitive, submerged bill from side to side as it walks along in shallow water (15–20 cm) (6–8 in) ("scything") to capture invertebrate prey.

CONSERVATION STATUS: Historical records indicate that this species was once a normal part of the Mid-Atlantic shorebird breeding community, but market hunting caused its extirpation as a breeder in the region by the late 1800s.

RANGE: Breeds western and central Canada south through western United States to northern Mexico; winters southern United States to southern Mexico.

KEY REFERENCES: Robinson et al. (1997).

FAMILY SCOLOPACIDAE .

Greater Yellowlegs *Tringa melanoleuca*

DESCRIPTION: L = 36 cm (14 in); W = 64 cm (25 in); mottled grayish-brown above; head, breast, and

flanks white heavily streaked with grayish-brown; white belly and rump; long, yellow legs; bill long, often slightly upturned and darker at tip than at base. Winter Paler overall; streaking on breast and head is reduced.

HABITAT AND DISTRIBUTION: Common to uncommon transient (Apr–May, Aug–Sep) throughout; uncommon to rare resident (nonbreeding) along the coast: Mud flats, estuaries, marshes, ponds, lakes, lagoons, pools, flooded agricultural fields.

DIET: Forages visually by day, and uses sweeping motions to feel and capture prey by night; prey items include small fish, aquatic insects, crustacea, and other aquatic invertebrates taken by wading in shallow water, mostly alone or in small flocks.

RANGE: Breeds in northern Canada and AK; winters in coastal

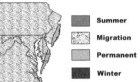

United States south to southern South America, West Indies.

KEY REFERENCES: Elphick and Tibbitts (1998).

Lesser Yellowlegs *Tringa flavipes*

DESCRIPTION: L = 28 cm (11 in); W = 51 cm (20 in); mottled grayish-brown and white above; head, breast and flanks white heavily streaked with grayish-brown and white; white belly and rump; yellow legs; white eyeline and eyering. **Winter** Paler overall; streaking on breast and head is reduced.

HABITAT AND DISTRIBUTION: Common to uncommon transient (Apr–May,

Aug–Sep) throughout, more numerous in fall; uncommon to rare summer resident (nonbreeding) in coastal areas, and winter resident (Oct–Mar) north to DE: Mud flats, marshes, pond or lake borders, flooded prairies, swales, estuaries.

DIET: Forages usually in shallow (3–6 cm) (1–2 in) water by walking rapidly and picking and probing for invertebrates (e.g., aquatic insects, molluscs, and crustacea), often with the neck stretched forward and the bill inserted at a 20–30% angle into the water.

RANGE: Breeds in northern Canada and AK; winters from the Atlantic (DE) and Pacific (southern CA) coasts of the United States south to southern South America; West Indies.

KEY REFERENCES: Tibbitts and Moskoff (1999).

Solitary Sandpiper *Tringa solitaria*

DESCRIPTION: L = 23 cm (9 in); W = 43 cm (17 in); dark gray back and wings with white spotting; heavily streaked with grayish-brown on head and breast; white eyering; long, greenish legs; tail dark in center, barred white on outer portions. **Winter** Streaking on head and breast reduced or faint.

HABITAT AND DISTRIBUTION: Common to uncommon transient (Aug–Sep, Apr–May) throughout: Mostly inland pools, ponds, lakes, or muddy shoreline.

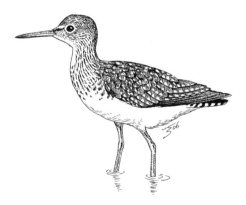

DIET: Bobs tail while foraging deliberately, and usually alone, in shallow, stagnant water or mud, picking or probing for aquatic insects, crustacea, molluscs, or, occasionally, frogs; sometimes vibrates foot to flush prey.

RANGE: Breeds across Arctic and boreal North America; winters south TX and Mexico south to southern South America; Greater Antilles.

KEY REFERENCES: Moskoff (1995).

Willet *Catoptrophorus semipalmatus*

DESCRIPTION: L = 38 cm (15 in); W = 69 cm (27 in); large, heavy bodied, long-billed shorebird; mottled grayish-brown above and below; whitish belly; broad white stripe on wing is conspicuous in flight. **Winter** Gray above, whitish below.

HABITAT AND DISTRIBUTION: Common summer resident* (Apr–Oct) along the immediate coast; rare in winter: Saltmarsh, estuaries, beaches, coastal ponds, tidal flats.

DIET: Forages diurnally and nocturnally at ebb, flood, and slack tides by picking prey (e.g., annelids, crustacea, molluscs, small fish) from substrate, "plowing," or tactile probing in wet sand or mud.

REPRODUCTION: Monogamous; pairs arrive on breeding territories in late April or early May; ground nest, 16 cm dia.) (6 in) is a scrape placed in beach dunes or bordering saltmarsh and built of stems and blades of neighboring vegetation lined with fine grasses; clutch—4; incubation (both sexes)—25–26 days; young are precocial, departing nest within a day or two after hatching, but are unable to fly until 4 weeks post-hatching; female usually leaves young within 2 weeks after hatching, but males normally continue care up to 4 weeks; single brood per season.

CONSERVATION STATUS: S2—MD; disturbance of coastal breeding colonies and destruction of coastal habitat threaten populations.

RANGE: Breeds in southwest-ern and south central Canada and northwestern

	Summer
	Migration
	Permanent
	Winter

and north central United States, along Atlantic and Gulf coasts (Nova Scotia to TX), West Indies; winters coastal North America (CA and VA) south to northern half of South America, West Indies.

KEY REFERENCES: Lowther et al. (2001).

Spotted Sandpiper *Actitis macularius*

DESCRIPTION: L = 20 cm (8 in); W = 33 cm (13 in); grayish-brown above, white with dark spots below; white eyebrow; bill orange with dark tip; legs pinkish. **Winter** Lacks spots; white below extending toward back at shoulder; grayish smear on side of breast; legs yellowish. Teeters almost continually while foraging; flight peculiar with stiff-winged bursts and brief glides.

HABITAT AND DISTRIBUTION: Common transient (Aug–Sep, May) and uncommon and local sum-mer resident* (Jun–Aug) throughout; rare to casual in winter along the coast: Streams, ponds, rivers, lakes, marshes, beaches.

DIET: Forages along waterline, searching visually for invertebrate prey (e.g., arthropods, annelids, mol-luscs, crustacea), then running and jabbing, prob-ing or "stitching" (rapid up and down head

movements) to grasp prey item from water or soil.

REPRODUCTION: Polyandrous; female arrives on breeding territory in early May followed 4–5 days later by the male; nest, 6–8 cm dia. (2–3 in), is a scrape excavated mostly by the female, placed within 100 m (330 ft) of water under or near low, dense herbaceous vegetation, and lined with grasses and debris; clutch—4; incubation (both sexes)—21 days; fledging—young are precocial, able to leave the nest by walking within a few hours of hatching, but cannot fly until 15–18 days post-hatching; post-fledging parental care—2 weeks; single brood per season.

CONSERVATION STATUS: S1—DE; S2—VA; S3—MD, WV; disturbance and destruction of wetland habi-tat threaten populations.

RANGE: Breeds throughout temperate and boreal North America; win-

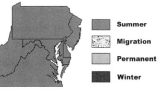

	Summer
	Migration
	Permanent
	Winter

ters from the southern United States south to northern Argentina, West Indies.

KEY REFERENCES: Oring et al. (1997).

Upland Sandpiper *Bartramia longicauda*

DESCRIPTION: L = 31 cm (12 in); W = 56 cm (22 in); the shape of this bird is unique—large, heavy body, long legs, long neck, small head and short bill. Plumage is light brown with dark brown streaks on back and wings; buff head, neck, and breast mottled with brown; large brown eye.

HABITAT AND DISTRIBUTION: Uncommon fall transient (Aug–Sep), rare spring transient (Apr–May), and rare and local summer resident* (May–Aug), mainly in PA, although there are scattered, recent breeding records from

elsewhere throughout the region: Prairie, pastures, plowed fields.

DIET: Forages in short grass (< 10 cm) (4 in), grazed pasture, or bare ground by picking invertebrates, mostly insects, from the ground or vegetation; also takes seeds and fruits seasonally.

REPRODUCTION: Monogamous; pairs arrive on breeding territory in April; ground nest, 11 cm dia. (4 in), is a scrape dug by both sexes in grassy areas, and lined with grasses, twigs, and leaves; sometimes a canopy over the nest is formed by pulling over surrounding vegetation; clutch—4; incubation (both sexes)—23–24 days; young are precocial, able to leave the nest within a day after hatching; able to fly 4–5 weeks post-hatching by which time they are independent of parents; single brood per season.

CONSERVATION STATUS: SH-WV, DE; S1—PA, NJ, VA, MD; native grassland breeding habitat is scarce and disappearing.

RANGE: Breeds across northeast and north central United States and in Great

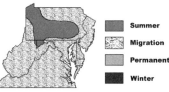

Plains of Canada; winters South America.

KEY REFERENCES: Houston and Bowen (2001).

Whimbrel *Numenius phaeopus*

DESCRIPTION: L = 46 cm (18 in); W = 84 cm (33 in); mottled gray and brown above; grayish below with dark flecks; crown striped with dark brown and gray.

HABITAT AND DISTRIBUTION: Common transient (Jul–Sep, Apr–May) along the immediate coast; rare in summer and winter: Tidal flats, marshes, meadows, fields, estuaries, oyster reefs.

DIET: Forages mostly during daylight hours by walking on mud, in shallow water, or through grasslands and probing for prey in soil or benthos or picking prey from substrate or vegetation; the long, decurved bill seems adapted principally for capture of marine invertebrates (e.g., fiddler crabs, annelids, molluscs) from burrow systems, but a wide variety of prey is taken inland on migration or during the breeding season including insects and fruits.

RANGE: Breeds in high Arctic of Old and New World; winters in subtropical and tropical regions.

KEY REFERENCES: Skeel and Mallory (1996).

Long-billed Curlew *Numenius americanus*

DESCRIPTION: L = 59 cm (23 in); W = 97 cm (38 in); a large, heavy bodied bird with very long, decurved bill; mottled brown and buff above; uniformly buffy below; rusty wing linings visible in flight.

HABITAT AND DISTRIBUTION: Rare to casual visitor along the coast of VA, mainly fall and winter: Estuaries, bay shores, grasslands, pastures, lawns, golf courses.

DIET: Forages on a variety of invertebrates, but the extremely long, decurved bill is best adapted for capturing shrimp and crabs from burrows on tidal flats in coastal wintering areas.

RANGE: Breeds in prairie regions of western United States and south-

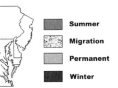

Summer
Migration
Permanent
Winter

western Canada; winters from central CA and southern portions of CA, AZ, TX, and LA south through southern Mexico.

KEY REFERENCES: Dugger and Dugger (2002).

Hudsonian Godwit *Limosa haemastica*

DESCRIPTION: L = 41 cm (16 in); W = 69 cm (27 in); mottled dark and light brown and white above; dark chestnut below; light eyebrow; long, upturned bill reddish at base, dark at tip; white rump; dark wings with white wing stripe visible in flight. **Winter** gray above, whitish below with white eyebrow.

HABITAT AND DISTRIBUTION: Rare fall transient (Aug–Oct) along the immediate coast: Estuaries, tidal flats, beaches, lakes, marshes, ponds, swales.

DIET: Forages by walking slowly in shallow water

over muddy substrate, alternately stepping, dipping, and probing deep into the mud in rapid up-and-down fashion, often submerging head below water surface; main prey is invertebrates (e.g., annelids, molluscs, crustacea, arthropods); plant tubers may be important at some inland stopover sites.

CONSERVATION STATUS: World populations estimated at < 50,000 birds; the species appears to depend on a few major stopover areas for successful migration.

RANGE: Breeds in AK, northwest Canada, and the Hudson Bay

Summer
Migration
Permanent
Winter

region; winters in southern South America.

KEY REFERENCES: Elphick and Klima (2002).

Marbled Godwit *Limosa fedoa*

DESCRIPTION: L = 46 cm (18 in); W = 79 cm (31 in); mottled buff and dark brown above; buffy below barred with brown; whitish line above eye; dark

line through eye; long, upturned bill pink at base, dark at tip; rusty wing linings. **Winter** Ventral barring faint or lacking.

HABITAT AND DISTRIBUTION: Rare transient (Aug–Oct, May), mainly in fall along the coast: Bay shores, mud flats, sand flats, estuaries, flooded grasslands.

DIET: Forages mostly at low tide in shallow water (5–13 cm) (2–5 in) by probing sand or mud substrate for polychaetes, molluscs, crustacea, and other invertebrates.

RANGE: Breeds northern Great Plains of Canada and extreme northern

Summer
Migration
Permanent
Winter

United States; winters along both coasts from CA and NC south to northern South America.

KEY REFERENCES: Gratto-Trevor (2000).

Ruddy Turnstone *Arenaria interpres*

DESCRIPTION: L = 26 cm (10 in); W = 48 cm (19 in); a plump bird with slightly upturned bill; rufous above, white below with black breast; black and white facial pattern; two white stripes on wings; white band across tail; orange legs. **Winter** Grayish above; white below with varying amounts of black on breast and face.

HABITAT AND DISTRIBUTION: Common summer resident (Apr–Sep--nonbreeding) along coast; common to uncommon transient (Jul–Sep, Apr–May) along the Coastal Plain; uncommon to rare win-

ter resident in coastal regions north to DE; rare to casual transient elsewhere in the region: Rocky and sandy beaches, tidal flats.

DIET: Forages by walking on sandy, rocky, or muddy substrate, pecking, probing, and turning over shells, seaweed, flotsam, and wrack for invertebrates (e.g., crustacea, molluscs, arthropods, annelids, and echinoderms).

RANGE: Breeds in high Arctic of Old and New World; winters in south temper-

Summer
Migration
Permanent
Winter

ate and tropical regions, with many nonbreeders remaining through the summer months.

KEY REFERENCES: Nettleship (2000).

Red Knot *Calidris canutus*

DESCRIPTION: L = 28 cm (11 in); W = 53 cm (21 in); a plump bird, mottled brown and russet above and on crown; rusty below. **Winter** Pale gray above; grayish breast; whitish belly; white eyebrow; white, finely barred rump.

HABITAT AND DISTRIBUTION: Common transient (Jul–Sep, May) along the immediate coast; rare at other seasons or elsewhere in the region: Tidal flats, bay shorelines, beaches.

DIET: Forages by walking rapidly in shallow (a few cm), receding tide water, and capturing prey (e.g., bivalves, gastropods, amphipods, mussel spat, horseshoe crab [*Limulus*] eggs) by pecking, probing, and plowing.

CONSERVATION STATUS: Knot populations that migrate through the Mid-Atlantic region (*C. canutus rufa*) have undergone marked declines in recent years, and have been proposed for listing as "Endangered" by the USFWS. Suggested reason for the decline is overharvest of horseshoe crab eggs, an important food for migrating knots, at major stopover sites for this population, like Delaware Bay.

RANGE: Breeds in Old and New World middle and high Arctic; winters in temperate and tropical coastal regions.

KEY REFERENCES: Harrington (2001).

Sanderling *Calidris alba*

DESCRIPTION: L = 20 cm (8 in); W = 41 cm (16 in); dappled brown, white, and black on back and crown; rusty flecked with white on face and breast; white belly; short black legs and bill; lacks hind toe (hallux); white wing stripe. **Winter** Mottled gray and white above; white below; white eyebrow; grayish shoulder patch extends to side of breast.

HABITAT AND DISTRIBUTION: Common transient and winter resident (Aug–May), and uncommon summer resident (Jun–Jul––nonbreeding) along the coast; rare to casual inland: Beaches, tidal flats, rocky shorelines.

DIET: Forages by running along sandy ocean beaches in front of advancing, or behind receding, wave waters, picking up crustacea, molluscs, and other aquatic invertebrates exposed by wave action; also probes in sand and mud for prey; often spends considerable time in chases and fights with other Sanderlings.

RANGE: Breeds in high Arctic of Old and New World; winters on temperate and tropical beaches.

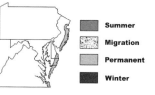

KEY REFERENCES: MacWhirter et al. (2002).

Semipalmated Sandpiper *Calidris pusilla*

DESCRIPTION: L = 15 cm (6 in); W = 31 cm (12 in); dark brown, russet and white above; finely streaked brown and white on head and breast; pale eyebrow with some rusty on crown and cheek; white belly; partial webbing between toes (hence "semipalmated"). **Winter** Grayish above with grayish wash on head and breast; white belly.

HABITAT AND DISTRIBUTION: Common to uncommon transient (Jul–Oct, Apr–May) throughout; common to uncommon in summer (Jun–Jul, nonbreeding) along the immediate coast: Mud flats, ponds, lakes.

DIET: Feeds by probing in soft mud for small arthropods, molluscs, annelids, and other invertebrates.

REPRODUCTION: Breeds in Arctic tundra during June and July.

CONSERVATION STATUS: Principal threats to populations are from wetland damage or loss.

RANGE: Breeds in high Arctic of North America; winters along both coasts of

Middle and South America to Paraguay (east) and northern Chile (west), West Indies.

KEY REFERENCES: Gratto-Trevor (1992).

Western Sandpiper *Calidris mauri*

DESCRIPTION: L = 15 cm (6 in); W = 31 cm (12 in); dark brown, russet, and white above; finely streaked brown and white on head and breast; pale eyebrow; partial webbing on toes; similar to Semipalmated Sandpiper but whiter on back, and with distinct rusty cheek and supraorbital patches. **Winter** Grayish above with grayish wash on head and breast; white belly.

HABITAT AND DISTRIBUTION: Common fall transient (Jul–Oct), uncommon winter resident (Nov–Mar), and rare spring transient (Apr–May) along the Coastal Plain; rare to casual transient inland: Beaches, mudflats, ponds, lakes, estuaries, swales.

DIET: Feeds by probing in wet mud, sand, or shallow water (2–3 cm) (1 in) or pecking and snatching prey from soil substrate; principal prey are arthropods (e.g., chironomid larvae and amphipods), molluscs, annelids, and other invertebrates.

RANGE: Breeds in high Arctic of northern AK and northeastern Siberia; winters along both coasts of United States from CA and NC south to northern South America, West Indies.

KEY REFERENCES: Wilson (1994).

Least Sandpiper *Calidris minutilla*

DESCRIPTION: L = 15 cm (6 in); W = 31 cm (12 in); dark brown and buff on back and wings; head and breast streaked with brown and white; relatively short, thin bill; white belly; yellowish legs (sometimes brown with caked mud). **Winter** Brownish-gray above; buffy wash on breast.

HABITAT AND DISTRIBUTION: Common to uncommon transient (Jul–Sep, Apr–May) throughout; uncommon to rare resident (nonbreeding) along the coast: Pond and lake margins, swales, mud flats.

DIET: Feeds by pecking and probing in mud for small arthropods (e.g., amphipods, chironomids), annelids, and other benthic and soil invertebrates. Coastal birds forage mostly during low-tide periods, day or night.

RANGE: Breeds in sub-Arctic tundra across northern

North America; winters from coastal (Oregon, NC) and southern United States south to northern half of South America, West Indies.

KEY REFERENCES: Cooper (1994).

White-rumped Sandpiper *Calidris fuscicollis*

DESCRIPTION: L = 20 cm (8 in); W = 38 cm (15 in); dark and tan on back and wings; finely streaked brown and white on head, breast, and flanks; white rump; dark legs; wings extend beyond tail. **Winter** As in breeding but gray above, white below with gray wash on breast and white eyebrow.

HABITAT AND DISTRIBUTION: Uncommon transient (Aug–Oct, May–Jun) and uncommon to rare in summer (Jun–Aug, nonbreeding) along the coast; rare to casual elsewhere: Swales, coastal mud flats, lake and pond borders.

DIET: Arthropods, annelids, molluscs, and other invertebrates taken by quick probing movements in open mud flats.

RANGE: Breeds in high Arctic of North America; winters in South America east of the Andes.

KEY REFERENCES: Parmelee (1992).

Baird's Sandpiper *Calidris bairdii*

DESCRIPTION: L = 20 cm (8 in); W = 38 cm (15 in); a medium-sized sandpiper, dark brown and buff on back and wings; buffy head and breast flecked with brown; broad, pale line above eye; folded wings extend beyond tail.

HABITAT AND DISTRIBUTION: Rare fall transient (Aug–Oct), mainly along the coast: Ponds, lakes, swales, wet grasslands, marshes, tidal flats, estuaries, beaches; prefers freshwater wetlands.

DIET: Forages mostly by walking in exposed sand, soil, or mud, and picking prey, mainly insects and crustacea, from substrate.

RANGE: Breeds in high Arctic of North America and northeastern Siberia; winters in western and southern South America.

KEY REFERENCES: Moskoff and Montgomerie (2002).

Pectoral Sandpiper *Calidris melanotos*

DESCRIPTION: L = 23 cm (9 in); W = 43 cm (17 in); crown dark with chestnut streaks; light eyeline; dark lores; chestnut ear patch; dark brown, tan, and white on back and wings; head and breast whitish, densely flecked with brown; belly white, the dividing line between mottled brown breast and white belly distinct; dark bill; yellow legs. **Female and Winter Male** Paler; lack chestnut on crown and ear.

HABITAT AND DISTRIBUTION: Common to uncommon transient (Apr–May, Aug–Oct) throughout; more numerous in fall; rare to casual in winter along the coast: Wet grasslands, ponds, lakes, swales, agricultural fields, marshes.

DIET: Forages, often in low vegetation, bordering or in water, by picking and probing substrate for arthropods or, occasionally, seeds.

RANGE: Breeds in the high Arctic of northern North America; winters in southern half of South America.

Summer
Migration
Permanent
Winter

KEY REFERENCES: Holmes and Pitelka (1998).

Purple Sandpiper *Calidris maritima*

DESCRIPTION: L = 23 cm (9 in); W = 38 cm (15 in); dark brown back and wings; head finely streaked with brown and white; belly white; breast and flanks whitish spotted with brown; bill yellow-orange at base, dark at tip; yellow-orange legs; white eyering. **Winter** Back marked with gray and buff; head and breast slate-gray with gray streaking on flanks.

HABITAT AND DISTRIBUTION: Uncommon to rare and local winter resident (Oct–May) along the immediate coast: Rocky shores, cobble beaches, jetties, breakwalls, ledges, tidal flats, muddy pools, sandy beaches.

DIET: Forages by picking, jabbing, and probing in crevices, barnacles, mussels, and other debris for invertebrates (e.g., molluscs, especially *Littorina*, crustacea, annelids).

RANGE: Breeds in Arctic of northeastern North America, Greenland,

Summer
Migration
Permanent
Winter

Iceland, northern Scandinavia, and Siberia; winters along Atlantic Coast of North America; also shores of North and Baltic seas in Old World.

KEY REFERENCES: Payne and Pierce (2002).

Dunlin *Calidris alpina*

DESCRIPTION: L = 23 cm (9 in); W = 41 cm (16 in); rusty and dark brown on back and wings; white streaked with black on neck and breast; crown streaked with russet and black; large black smudge on belly; long bill droops at tip. **Winter** As in breeding but with dark gray back; head and breast pale gray with some streaking; white eyebrow; white belly.

HABITAT AND DISTRIBUTION: Common (along coast) to uncommon or rare (inland) transient (Oct, Apr–May); common in winter (Nov–Mar) along the coast, uncommon to rare in summer (Jun–Sep, nonbreeding): Estuaries, bay shores, flooded fields, ponds, lakes, beaches, mud flats.

DIET: Forages diurnally and nocturnally using tactile probing in benthos of shallow water or in wet soil to locate and capture invertebrate prey (e.g., bivalves, amphipods, polychaetes, and crustacea).

RANGE: Breeds in Arctic of Old and New World; winters in temperate and northern tropical regions.

KEY REFERENCES: Warnock and Gill (1996).

Curlew Sandpiper *Calidris ferruginea*

DESCRIPTION: L = 20 cm (8 in); W = 41 cm (16 in); rusty above and below with long, decurved bill. **Winter** Dark gray back; head and breast pale gray; white eyebrow; white belly.

HABITAT AND DISTRIBUTION: Rare transient (Jul–Oct, May) along the immediate coast: Estuaries, lagoons, lakes, muddy and sandy beaches, mud flats, tidal pools, sewage farms, saltmarsh.

DIET: Forages by walking in shallow water and probing muddy or sandy bottom for invertebrates, e.g., molluscs, annelids, and crustacea.

RANGE: Breeds in Old World Arctic; winters in Old World temperate and tropical regions.

KEY REFERENCES: Hammerson and Cannings 2004.

Stilt Sandpiper *Calidris himantopus*

DESCRIPTION: L = 23 cm (9 in); W = 43 cm (17 in); a trim, long-necked, long-legged sandpiper, dark brown and buff on back and wings; crown streaked with dark brown and white; whitish eyebrow; chestnut pre- and postorbital stripe; buff streaked with brown on neck; buff barred with brown on breast and belly; long, yellow-green legs; white rump; long, straight bill (twice length of head), slightly expanded at tip. **Winter** Gray and white mottling on back; gray head with prominent eyebrow; gray throat and breast; whitish belly with gray flecks.

HABITAT AND DISTRIBUTION: Uncommon fall transient (Jul–Oct) and rare spring transient (Apr–May) along the coast; rare to casual elsewhere: Ponds, pools, marshes, flooded pastures.

DIET: Forages by wading, often breast deep, and dipping, probing, or stitching in muddy bottom for benthic invertebrates and, less commonly, seeds.

RANGE: Breeds in Arctic of north central Canada; winters in central South America.

KEY REFERENCES: Klima and Jehl (1998).

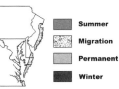

Buff-breasted Sandpiper *Tryngites subruficollis*

DESCRIPTION: L = 20 cm (8 in); W = 43 cm (17 in); dark brown feathering edged in buff or whitish on back; buffy head and underparts, spotted with brown on breast; streaked crown; large, dark eye; pale legs; wing linings appear white in flight.

HABITAT AND DISTRIBUTION: Rare fall transient along the immediate coast (Aug–Oct), casual in spring and elsewhere in the region: Open areas, e.g., grassland, plowed and stubble fields, short-grass prairie, overgrazed pastures; often travels in small flocks that twist and turn erratically in flight over freshly plowed fields.

DIET: Forages by walking on the ground and picking prey from low vegetation or the soil surface; main foods are insects, spiders, crustacea, and, occasionally, seeds.

CONSERVATION STATUS: Buff-breasted Sandpipers were much more plentiful during the 1800s until populations were decimated by hunting.

RANGE: Breeds in Arctic of northern AK and north-western Canada; winters in Paraguay, Uruguay, and northern Argentina.

KEY REFERENCES: Lanctot and Laredo (1994).

Ruff *Philomachus pugnax*

DESCRIPTION: Male: L = 31 cm (12 in); W = 61 cm (24 in); Female: L = 26 cm (10 in); W = 53 cm (21 in); dark brown, tan, and white on back; spectacular collar of ruffled feathers around neck and breast—black, chestnut, or white; black below; bill and legs yellowish; rump white except for dark central line. **Female** (called a Reeve) Dark gray above and below; white belly; white rump with dark central line; yellow legs and base of bill. **Winter** Dark gray and white on back; grayish head and breast; whitish belly; bill yellowish at base; legs yellowish.

HABITAT AND DISTRIBUTION: Rare transient (Aug–Oct, Apr–Jun) along the immediate coast; casual elsewhere and at other seasons: Estuaries, mud flats, saltmarsh, pond margins.

DIET: Forages by walking in shallow water and picking and probing for molluscs, crustacea, fish, and annelids.

RANGE: Breeds in Old World Arctic; winters in Old World temperate and tropical regions.

KEY REFERENCES: Hammerson (2005b).

Short-billed Dowitcher
Limnodromus griseus

DESCRIPTION: L = 28 cm (11 in); W = 48 cm (19 in); a rather squat, long-billed sandpiper (bill is 3 times length of head); mottled black, white, and buff on back, wings, and crown; cinnamon buff on neck and underparts barred and spotted with dark brown; belly white; light line above eye with dark line through eye; white tail barred with black; white on rump extending in a wedge up back; bill dark; legs greenish. **Winter** Grayish crown, nape, and back; white line above eye and dark line through eye; grayish neck and breast fading to whitish on belly; flanks and undertail coverts spotted and barred with dark brown. **Juvenile** Patterned like adult but paler brown above and on head, and with buffy neck and breast.

HABITAT AND DISTRIBUTION: Common (coast) to rare or casual (inland) transient (Jul–Oct, Apr–May); common in summer, uncommon to rare in winter along the coast: Tidal flats, estuaries, beaches, saltmarshes, sewage ponds, flooded fields; prefers saline habitats in contrast to the Long-billed Dowitcher, which prefers freshwater wetlands.

DIET: Often forages with a rapid up-and-down motion, breast-deep in water, usually in small flocks, probing for invertebrates, e.g., annelids, arthropods, crustacea, and molluscs.

CONSERVATION STATUS: Populations of this species that migrate through the Mid-Atlantic region have undergone marked declines in recent decades, perhaps due to loss of stopover habitat or food sources.

RANGE: Breeds across central Canada and southern AK; winters coastal

United States (CA, SC) south to northern South America, West Indies.

KEY REFERENCES: Jehl et al. (2001).

Long-billed Dowitcher *Limnodromus scolopaceus*

DESCRIPTION: L = 28 cm (11 in); W = 48 cm (19 in); a squat, long-billed sandpiper (bill is 3 times length of head); mottled black, white, and buff on back, wings, and crown; cinnamon buff on neck and underparts barred and spotted with dark brown; belly white; white line above eye with dark line through eye; white tail barred with black; white on rump extending in a wedge up back; bill dark; legs greenish. **Winter** Grayish crown, nape, and back; white line above eye and dark line through eye; grayish neck and breast fading to whitish on belly; flanks and undertail coverts spotted and barred with dark brown. **Juvenile** Patterned like adult in winter.

HABITAT AND DISTRIBUTION: Uncommon fall transient (Aug–Oct), rare winter resident (Nov–Mar) and spring transient (Apr–May) along the coast; casual inland: Mud flats, wet meadows, shortgrass marshes, flooded pastures; prefers freshwater wetlands, in contrast to the Short-billed

Dowitcher, which prefers coastal marine wetlands.

DIET: Forages by probing in shallow water with a rapid up-and-down movement, feeding mostly on annelids, arthropods, crustacea, and molluscs.

RANGE: Breeds in coastal AK, northwestern Canada, and northeastern Siberia; win-

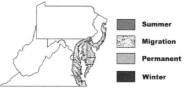

- Summer
- Migration
- Permanent
- Winter

ters along both coasts from British Columbia and VA south to Yucatan and El Salvador.

KEY REFERENCES: Takekawa and Warnock (2000).

Wilson's Snipe *Gallinago delicata*

DESCRIPTION: L = 28 cm (11 in); W = 43 cm (17 in); dark brown back with white stripes; crown striped with dark brown and buff; neck and breast spotted and scalloped with grayish-brown, buff, and white; whitish belly; tan rump; tail banded with rust and black; very long bill (3 times length of head). Normally solitary, retiring, and wary, flushed birds give an explosive "zhrrt" and fly off in erratic swoops and dips. Courting males fly high above the breeding marsh, producing a winnowing sound caused by air flow over modified tail feathers. The display is often given at night, and the source is hard to locate, perhaps serving as the origin of the term "snipe hunt" for something non existent or impossible to find.

HABITAT AND DISTRIBUTION: Common to uncommon transient (Sep–May) throughout; uncommon to rare winter resident (Oct–Mar) along the Coastal Plain, Piedmont, and lowland, open marshes else-

where in the region; uncommon to rare and local summer resident* (Apr–Sep) in northern PA, WV's Canaan Valley, and northwestern NJ: Marshes, bogs, swamps, flooded pastures, wet ditches.

DIET: Forages by probing in wet organic soil, mud, or shallow water; detects prey by touch with sensitive bill, and is able to grasp prey by opening the distal end of the bill; major prey items include soil insects, annelids, crustacea, and molluscs; plant material is also ingested, but does not appear to serve as a significant nutritional source.

REPRODUCTION: Monogamous, but extrapair copulations are common; male arrives on breeding territory in April, 10–14 days before the female; open platform nest, 10–17 cm dia. (4–7 in), is placed on the ground in a marsh, bog, or swamp in low, dense vegetation (e.g., sedge, marsh grass, or sphagnum moss), and built by the female, who excavates a small scrape and weaves coarse grass stems into a bowl, which is then lined with fine grasses; clutch—2–4; incubation (mostly by female)—18–20 days; fledging—young are precocial, able to leave the nest within 1 h of hatching, and able to fly at 14–20 days post-hatching; post-fledging parental care—young are probably independent at 20 days when able to fly significant distances; single brood per season.

CONSERVATION STATUS: S1—WV; S3—PA, NJ; disturbance and destruction of wetland habitat threaten populations.

RANGE: Breeds in north temperate and boreal regions of North America from

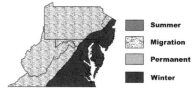

- Summer
- Migration
- Permanent
- Winter

AK and northern Canada south to the northern United States; winters in temperate and tropical regions from the northern United States to northern South America and the Greater Antilles. Formerly, this species was considered conspecific with Old World and South American breeding populations of snipes, which are, in any event, very closely related.

KEY REFERENCES: Tuck (1972), Mueller (1999).

American Woodcock *Scolopax minor*

DESCRIPTION: L = 28 cm (11 in); W = 43 cm (17 in); a fat, short-legged, long-billed bird with no apparent neck and hardly any tail; dark and grayish-brown above; gray-brown neck and breast becoming rusty on belly and flanks; dark brown crown with thin, transverse gray stripes; large, brown eye; overall impression of flushed bird is cinnamon buff because of underparts and wing linings.

HABITAT AND DISTRIBUTION: Common to uncommon transient and summer resident* (Feb–Nov) throughout; uncommon to rare and local winter resident (Dec–Jan) along the coast: Moist woodlands, swamp borders.

DIET: Forages both diurnally and nocturnally with timing of activity periods varying by season; feeds mainly by probing moist loam or sandy loam soils for earthworms (Oligochaeta) and other invertebrates; other prey includes amphibians and plant parts.

REPRODUCTION: Polygynous with no pair bond; males establish singing/aerial display territories in March; nest is a shallow bowl shaped in surface detritus on the ground by the female; egglaying begins in March and April in the Mid-Atlantic; clutch—4; incubation—21 days; fledging—young are precocial, and leave nest within a few hours of hatching, but they cannot fly until 14–18 days post-hatching; post-hatching female care—31–38 days; single brood per season.

RANGE: Breeds across the eastern half of the United States and southeastern Canada; winters from the southern half of the breeding range into south TX and southern FL.

KEY REFERENCES: Keppie and Whiting (1994).

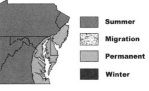

Wilson's Phalarope *Phalaropus tricolor*

DESCRIPTION: L = 23 cm (9 in); W = 43 cm (17 in); **Female** A trim, thin-billed, long-necked bird; chestnut and gray on the back; gray crown and nape; black extending from eye down side of neck; white chin; chestnut throat; white breast and belly. **Male** Similar but paler. **Winter** Grayish-brown above, white below; gray head with white eyebrow; white chin; rusty throat.

HABITAT AND DISTRIBUTION: Uncommon fall transient (Aug–Sep) and rare spring transient (May) along the coast: Ponds, lakes, mud flats, flooded pastures.

DIET: Unlike other shorebirds, phalaropes often spin while foraging; aquatic invertebrates are the main prey items (e.g., Diptera, brine shrimp, copepods) captured mostly by spinning, chasing, and pecking prey from water surface.

RANGE: Breeds in western United States, southwestern Canada, and in Great Lakes region; winters western and southern South America.

KEY REFERENCES: Colwell and Jehl (1994).

Red-necked Phalarope *Phalaropus lobatus*

DESCRIPTION: L = 20 cm (8 in); W = 36 cm (14 in); **Female** Black with rusty striping on back and flanks; white below; neck and nape orange; head black; chin white. Male Similar in pattern but paler. **Winter** Black back with white striping; white below; white forehead; gray-black crown; white eyebrow; black postorbital stripe.

HABITAT AND DISTRIBUTION: Uncommon to rare transient (Sep–Oct, May), mainly offshore: Mainly pelagic in winter, but also found occasionally at marshes, ponds, or lakes.

DIET: Unlike other shorebirds, phalaropes often spin while foraging, pecking small invertebrate prey from water surface. The spinning behavior draws prey to the surface.

RANGE: Breeds in Arctic regions of the Old and New World; winters mostly at sea in southern Pacific and Indian oceans.

Summary
Migration
Permanent
Winter

KEY REFERENCES: Rubega et al. (2000).

Red Phalarope *Phalaropus fulicarius*

DESCRIPTION: L = 23 cm (9 in); W = 43 cm (17 in); **Female** Black back striped with buff; neck and underparts rusty orange; black cap; white cheek; relatively short heavy bill is yellow at base, black

at tip. **Male** Similar in pattern but paler. **Winter** Uniform pearl gray on back; white below; white forehead and front half of crown; gray rear half of crown and nape line; dark postorbital stripe; bill blackish.

HABITAT AND DISTRIBUTION: Rare transient and winter visitor, mainly offshore (Sep–May): Mainly pelagic in winter, but also found occasionally on exposed tidal bars, beaches and mud flats.

DIET: Unlike other shorebirds, phalaropes often spin while foraging; feeds on zooplankton and other marine invertebrates during migration and winter.

RANGE: Breeds in Old and New World Arctic; winters at sea in tropical and subtropical oceans.

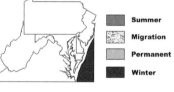

Summer
Migration
Permanent
Winter

KEY REFERENCES: Tracy et al. (2002).

FAMILY LARIDAE

Parasitic Jaeger *Stercorarius parasiticus*

DESCRIPTION: L = 48 cm (19 in); W = 107 cm (42 in); dark back and crown; whitish-gray collar and breast; white belly; pointed, protruding central tail feathers; wings dark but whitish at base of primaries. **Dark Phase** Completely dark brown. **Immature** Dark brown above; barred reddish-brown or brown and white below.

HABITAT AND DISTRIBUTION: Uncommon transient (Sep–Nov, May) and rare summer and winter visitor, mainly offshore but occasionally in lower Chesapeake Bay: Pelagic, usually < 10 km (6 mi) from shore; in-shore marine, estuaries, bays, beaches.

DIET: Parasitic Jaegers obtain much of their food during migration by stealing prey from gulls, terns, and other seabirds; also forage by flying low along shorelines and coastal areas, attempting to surprise and capture waterfowl, songbirds, shorebirds, and small mammals; pairs sometimes hunt cooperatively to herd and trap prey; eggs and young of beach-nesting species are common prey items as well.

RANGE: Breeds in Old and New World Arctic and sub-Arctic regions; winters in temperate and tropical oceans.

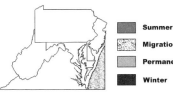

KEY REFERENCES: Wiley and Lee (1999).

Laughing Gull *Larus atricilla*

DESCRIPTION: L = 43 cm (17 in); W = 102 cm (40 in); black head; gray back and wings with black wingtips; white collar, underparts, and tail; scarlet bill and black legs; partial white eyering. **Winter** As in breeding but head whitish with dark gray ear patch; black bill. **Immature** Brownish above and on breast; white belly and tail with black terminal bar.

HABITAT AND DISTRIBUTION: Common transient and summer resident* (Apr–Nov) along the coast; rare to casual in winter mainly along the VA coast: Beaches, bays, estuaries, lakes, garbage dumps.

DIET: Feeds by walking, wading, or swimming, and picking items from substrates; main foods are marine invertebrates (e.g., molluscs, crustacea), small fish, beach carrion, organic garbage; often begs for food from picnickers at the beach.

REPRODUCTION: Monogamous; pair formation occurs at stopover sites during migration; pairs arrive at breeding colonies in April; colony sites are located mostly on small natural or dredge spoil coastal islands; open cup nest, 0.3 m dia. (12 in), is placed on the ground in or near low, dense vegetation, and built (both sexes) of stems and leaves of marsh vegetation; clutch—3; incubation (both sexes)—23–27 days; fledging—young are semi-precocial, and able to walk around in vicinity of nest at 5–7 days, but are unable to fly until 35–40 days post-hatching; post-fledging parental care—2–3 weeks; single brood per season.

RANGE: Breeds along both coasts in North America from New Brunswick on the east and northern Mexico on the west, south to northern South America, West Indies; winters along coast from the southern U.S. to central South America.

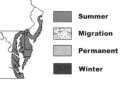

KEY REFERENCES: Burger (1996).

Franklin's Gull *Larus pipixcan*

DESCRIPTION: L = 38 cm (15 in); W = 92 cm (36 in); black head; gray back and wings; white bar bordering black wingtips spotted terminally with white; white collar, underparts, and tail; underparts variously tinged with rose; scarlet bill and legs; partial white eyering. **Winter** As in breeding but head with a partial, dark hood; black bill. **Immature** Similar to winter adult but with black terminal tail band.

HABITAT AND DISTRIBUTION: Rare summer and fall visitor to the VA coast (Jun–Oct); casual elsewhere: Estuaries, bays, pastures, flooded fields, lakes, lagoons, open garbage dumps.

DIET: Forages, often in flocks, on fish and marine invertebrates in coastal waters by picking prey from the water surface while swimming; on land in pastures, agricultural fields, or along shorelines, it takes a wide variety of invertebrates, small vertebrates, organic garbage, and carrion, as well as some plant material (e.g., grain).

RANGE: Breeds north central United States and south central Canada; winters along Pacific coast of Middle and South America.

KEY REFERENCES: Burger and Gochfeld (1994).

Little Gull *Larus minutus*

DESCRIPTION: L = 28 cm (11 in); W = 56 cm (22 in); a small gull; black head; gray back; wings dark below and gray above with white border; white collar, underparts, and tail; black bill; red legs. **Winter** light forehead, incomplete gray hood, black ear patch. **Immature** Black tail band and black stripe running diagonally across gray upper wing; pinkish legs.

HABITAT AND DISTRIBUTION: Rare visitor (Dec–Apr) to the immediate coast, most numerous in spring (Apr): Marshes, bays, rivers, estuaries, sewage treatment ponds.

DIET: Forages by flying and hovering low over open water and picking, dipping, or plunging for prey at or below the surface; main prey items are small fish and aquatic invertebrates.

RANGE: Breeds in central and western Asia, northern Europe, and locally in the Great Lakes region of North America. This species was believed to be restricted to the Old World as a breeder until 1962 when it was discovered breeding on Lake Ontario. Since then, other small North American breeding colonies have been discovered, mostly in the Great Lakes region. Winters in coastal regions of the Baltic, Mediterranean, Black, and Caspian seas, and rarely in North America in the Great Lakes area

and along the Atlantic, Pacific, and Gulf coasts of the United States.

KEY REFERENCES: Ewins and Weseloh (1999).

Black-headed Gull *Larus ridibundus*

DESCRIPTION: L = 41 cm (16 in); W = 102 cm (40 in); white with gray back and wings; front two-thirds of head with dark brown hood; red bill and legs; white outer primaries with black tips.
Winter white head with black ear patch.
Immature Like winter adult but with black tail band and brown stripe running diagonally across  gray wing; bill flesh-colored with dark tip; legs flesh-colored; incomplete brown hood in first breeding plumage.

HABITAT AND DISTRIBUTION: Rare visitor (Nov–Apr) to the immediate coast, most numerous in spring (Apr): Marshes, bays, estuaries, beaches, mud flats.

DIET: Forages mainly on marine invertebrates (e.g., shrimp, annelids, molluscs), carrion, and aquatic insects taken by walking along shoreline or on mud flats, and picking or probing for prey; also feeds on fish taken from water surface.

RANGE: Breeds in northern regions of the Old World, casual in Newfoundland; winters in temperate and northern tropical  regions of the Old World and along the Atlantic Coast of North America from Labrador to VA.

KEY REFERENCES: Nova Scotia Museum of Natural History (2005).

Bonaparte's Gull *Larus philadelphia*

DESCRIPTION: L = 36 cm (14 in); W = 81 cm (32 in); a small gull; black head; gray back; wings gray with white outer primaries tipped in black; white collar, underparts, and tail; black bill; red legs.
Winter White head with black ear patch.
Immature Like winter adult but with black tail band and brown stripe running diagonally across gray wing; flesh-colored legs; black bill.

HABITAT AND DISTRIBUTION: Common winter resident (Oct–Apr) along the Atlantic coast and Lake Erie; uncommon to rare transient (Oct–Nov, Apr–May) elsewhere, more numerous in spring; rare to casual in summer (May–Sep): Bays, estuaries, tidal flats, in-shore marine waters, lakes, rivers, marshes, impoundments.

DIET: Feeds on fish, zooplankton, arthropods, and other aquatic invertebrates by diving from the air, taking food from water surface in flight, swimming and surface diving, or picking up food while walking on the ground; often congregates in flocks at recently exposed tidal flats, tidal rips, upwellings, and similar situations where surface fish or aquatic invertebrates are concentrated.

RANGE: Breeds across north central and northwestern North America in 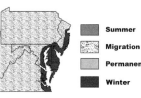 Canada and AK; winters south along both coasts, from Washington on the west and Nova Scotia on the east, to central Mexico; also the Great Lakes, Bahamas, and Greater Antilles.

KEY REFERENCES: Burger and Gochfeld (2002).

Ring-billed Gull *Larus delawarensis*

DESCRIPTION: L = 41 cm (16 in); W = 122 cm (48 in); white body; gray back; gray wings with black outer primaries tipped with white; bill orange-yellow with black subterminal ring; legs pale yellow. **Immature** Like adult but mottled with brown or gray; bill pale with black tip; whitish tail witish-black terminal band.

HABITAT AND DISTRIBUTION: Common to uncommon transient (Aug–Oct, Mar–Apr) throughout; uncommon to rare resident (nonbreeding) along the Coastal Plain; uncommon winter resident (Nov–Feb) in the Piedmont and at large bodies of open water elsewhere in the region.

DIET: Feeds mainly on fish and aquatic insects when foraging in water; also feeds by walking along beaches scavenging animal detritus or in fields picking up crop seeds, earthworms, insects, small mammals, and birds (eggs and nestlings).

RANGE: Breeds in the Great Plains region and scattered areas elsewhere in the northern U. S. and Canada; winters across most of United States except northern plains

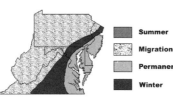

Summer
Migration
Permanent
Winter

and mountains south to southern Mexico and the Greater Antilles.

KEY REFERENCES: Ryder (1993).

Herring Gull *Larus argentatus*

DESCRIPTION: L = 61 cm (24 in); W = 145 cm (57 in); white body; gray back; gray wings with black outer primaries tipped with white; bill yellow with red spot on lower mandible; feet pinkish; yellow eye. **Winter** Head and breast smudged with brown. **Immature** Mottled dark brown in first winter; bill black; feet pinkish; tail dark; in second and third winter, grayish above, whitish below variously mottled with brown; tail with broad, dark, terminal band; bill pinkish with black tip; legs pinkish.

HABITAT AND DISTRIBUTION: Common to uncommon transient (Oct–Nov, Feb–Apr) throughout; common resident* along the Coastal Plain; common to uncommon and local winter resident (Oct–Apr) in the Piedmont; uncommon and local winter resident inland elsewhere at larger rivers and lakes: Beaches, bays, marine, lakes, rivers.

DIET: Along the coast, Herring Gulls forage by walking along shoreline and picking prey from beach surface, picking or dipping prey from water surface while swimming, or by shallow dives from the air to take prey from just beneath the water surface; inland they use similar behaviors to take prey from lakes, lake shores, plowed fields, and open garbage dumps; feeds on a wide variety of vertebrate (e.g., fish and young of conspecifics and other seabirds) and invertebrate prey (e.g., littoral marine invertebrates along the coast or aquatic insects inland); also eats carrion and organic garbage opportunistically. Casts pellets of indigestible material.

REPRODUCTION: Monogamous, mating for life; breeds in colonies, often located on islands with little or no vegetation in concert with other seabirds; male establishes small nesting territory in March or April; nest, 30 cm dia. (12 in), is a shallow, bowl-shaped scrape lined with grass, leaves, and feathers built by both sexes; clutch—

3; incubation (both sexes)—30 days; fledging—young are semi-precocial and able to walk within one or two days, but are flightless until 40–50 days post-hatching; post-fledging parental care—6–9 weeks, and perhaps up to four months; single brood per season.

CONSERVATION STATUS: S3—DE; disturbance of coastal breeding colonies and destruction of coastal habitat threaten populations.

RANGE: Breeds in boreal and north temperate regions of both Old and New World; winters along coasts of far north southward into temperate and north tropical regions.

KEY REFERENCES: Pierotti and Good (1994).

Iceland Gull *Larus glaucoides*

DESCRIPTION: L = 59 cm (23 in); W = 107 cm (42 in); white body; gray back and upper wing coverts; white primaries and terminal edge of secondaries; yellow eye; bill pinkish or yellowish with red spot on lower mandible; feet pinkish; long-winged—wingtips extend beyond tail in sitting bird. **Winter** Head and breast smudged with brown; wingtips with some gray. **First Winter** White with light brown markings; bill black; legs pink-

ish; tail white marked with brown. **Second and Third Winter** Paler than adult; some brown smudging; bill pinkish with black tip.

HABITAT AND DISTRIBUTION: Rare transient and winter visitor (Dec–Apr), mainly along the coast and offshore: Mostly maritime, but occasionally in coastal waters at tidal and intertidal zones and rips, beaches.

DIET: Forages by flying low over the water and picking prey from the surface or by swimming and dipping for prey; feeds mostly on small fish, but also takes molluscs, echinoids, ascideans, crustacea, annelids, and other small marine invertebrates.

RANGE: Breeds on islands of the far north in the North Atlantic; winters along northern coasts of the North Atlantic and Baltic Sea.

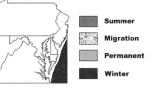

KEY REFERENCES: Snell (2002).

Lesser Black-backed Gull *Larus fuscus*

DESCRIPTION: L = 53 cm (21 in); W = 114 cm (45 in); white body; dark gray back and wings; outer primaries black sparsely tipped with white, appearing an almost uniform gray below; bill yellow

with red spot on lower mandible; eyes and legs yellow. **Winter** Head and breast streaked with brown. **First Winter** White heavily marked with brown; bill black; feet pale pinkish. **Second Winter** Body white marked with brown; back and wings dark gray marked with brown; bill pinkish with black tip; legs pale yellow or pink.

HABITAT AND DISTRIBUTION: Rare transient and winter visitor (Oct–Apr), mainly along the coast and offshore: Beaches, marine.

DIET: Omnivorous—feeds on a variety of coastal invertebrate and vertebrate prey and carrion.

RANGE: Breeds in Iceland and coastal northern Europe; winters along

coasts of Europe and eastern North America.

KEY REFERENCES: Paludan (1952).

Glaucous Gull *Larus hyperboreus*

DESCRIPTION: L = 69 cm (27 in); W = 152 cm (60 in); a very large gull; white body; pale gray back; yellow eyering and eye; wings pale gray with outer primaries broadly tipped with white above, mostly white from below; bill yellow with red spot on lower mandible; feet pinkish; wingtips barely extend beyond tail in sitting bird. **Winter** Head and breast smudged with brown. **First Winter** Mostly white or with light brown markings; bill pinkish with black tip; legs pinkish; tail

white marked with brown. **Second and Third Winter** Back grayish; bill yellowish with dark tip.

HABITAT AND DISTRIBUTION: Rare winter resident (Dec–Mar) along immediate coast and offshore: Coastal regions, lakes, agricultural fields, landfills.

DIET: Forages opportunistically on fish, plankton, crustacea, carrion, garbage, eggs, and nestling birds.

RANGE: Breeds in circumpolar regions of Old and New World; winters in coastal

regions of northern Eurasia and North America.

KEY REFERENCES: Gilchrist (2001).

Great Black-backed Gull *Larus marinus*

DESCRIPTION: L = 76 cm (30 in); W = 168 cm (66 in); a large gull; white body; black back; yellow eye; wings appear black above with white trailing edge, outer primaries tipped with white; bill very large, yellow with red spot on lower mandible; legs pinkish. **First Winter** White checked with brown above, paler below; rump paler; bill black; legs pale pink. **Second and Third Winter** Body white streaked with brown; black mottled with brown on back and wings; bill pinkish or yellowish with black tip; legs pale pinkish.

HABITAT AND DISTRIBUTION: Common resident* along the Atlantic coast and common winter resi-

dent (Sep–Apr) on the Lake Erie coast; scarce else-where in the region: Seacoasts and large lakes.

DIET: Forages by walking along shoreline, swimming, or making shallow dives for prey, which include fish, invertebrates, small vertebrates, carrion, and refuse.

REPRODUCTION: Monogamous; pairs form on male breeding territory in colonies in April; most colonies are located on higher, grassy areas of islands or in saltmarshes; nest, 20–56 cm dia. (8–22 in), is usually a scrape excavated by both pair members and lined with wet vegetation and feathers, often screened from the prevailing wind by a shrub, driftwood, or other detritus; clutch—3; incubation (both sexes)—26–28 days; fledg-ing—young are precocial, able to walk from the nest 24 h post-hatching, but cannot fly until 45 days post-hatching; post-fledging parental care—5–6 weeks; single brood per season.

CONSERVATION STATUS: S1—DE; disturbance of coastal breeding colonies and destruction of coastal habitat threaten populations.

RANGE: Breeds in coastal northeastern North America, east-ern Great

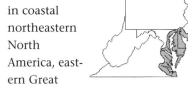

Lakes, and northeastern Eurasia; winters in breed-ing range and south to the southeastern United States and southern Europe.

KEY REFERENCES: Good (1998).

Black-legged Kittiwake *Rissa tridactyla*

DESCRIPTION: L = 43 cm (17 in); W = 92 cm (36 in); a small gull; white body; gray back; white, slightly forked tail; dark eye; wings gray above, white below tipped with black; bill yellow; legs dark. **Winter** Nape and back of crown gray with dark ear patch. **First Winter** White marked with black half "collar" on nape; gray back; dark diago-nal stripe on upper wing; black ear patch; tail white with black terminal band; bill and legs black.

HABITAT AND DISTRIBUTION: Uncommon to rare and irregular winter visitor (Nov–Mar), mainly off-shore: Pelagic, coasts.

DIET: Feeds on fish and marine invertebrates cap-tured by dipping or seizing from water surface while in flight or swimming, or making shallow dives (< 6 m) (20 ft) into the water to capture prey below the surface.

RANGE: Breeds circumpolar in Old and New World; winters in northern and temperate

seas of the Northern Hemisphere.

KEY REFERENCES: Baird (1994).

Gull-billed Tern *Sterna nilotica*

DESCRIPTION: L = 38 cm (15 in); W = 92 cm (36 in); a medium-sized tern with a heavy, black bill; black cap and nape; white face, neck, and underparts; gray back and wings; tail with shallow fork; black legs. **Winter** Crown white finely streaked with black.

HABITAT AND DISTRIBUTION: Uncommon to rare summer resident* (Apr–Aug) along the coast: Marshes, wet fields, grasslands, agricultural fields, bays.

DIET: This bird normally does not dive for prey like most terns. It swoops and sails over open grassland and marshland, hawking for flying insects, or plucking vertebrates (e.g., fish, reptiles, small mammals and birds, frogs) and invertebrates (e.g., crabs, insects) from the ground, vegetation, or water surface; casts pellets of undigested material.

REPRODUCTION: Monogamous; birds arrive at nesting colonies in late April or May; colonies are usually 5–50 nests, including nesting pairs of other terns or skimmers, located in open beach habitat with a sand, shell, or wrack substrate on coastal islands; the nest is a shallow scrape built by both pair members, lined with bits of shell or detritus, often located on a slight rise or hummock; clutch—3; incubation (both sexes) 22–23 days; fledging—young are semi-precocial, able to walk from nest a few days after hatching, but are unable to fly until 28–35 days post-hatching; post-fledging parental care—up to 2 weeks after first flight; single brood per season.

CONSERVATION STATUS: SH—DE; S1—NJ, MD; S2—VA; Gull-billed Terns, like most beach-nesting shorebirds, are threatened by predation and human disturbance.

RANGE: Breeds locally in temperate and tropical regions of the world; winters in subtropics and tropics.

▨	Summer
▨	Migration
▨	Permanent
▧	Winter

KEY REFERENCES: Parnell et al. (1995).

Caspian Tern *Sterna caspia*

DESCRIPTION: L = 53 cm (21 in); W = 132 cm (52 in); a large tern with large, heavy, blood-orange bill; black cap and nape; white throat, and underparts; gray back and wings; tail moderately forked; black legs. **Winter** Black cap streaked with white.

HABITAT AND DISTRIBUTION: Common transient (Aug–Oct, Apr–May) along the coast; uncommon to rare and local summer resident (Apr–Aug) and scarce breeder*, mainly on coastal islands; rare to casual transient inland: Beaches, marsh, bays, marine, estuaries, lakes.

DIET: Forages by flying 3–30 m (10–100 ft) over water, head down, searching for fish, and then folding wings to plunge into water and capture prey, mostly small fish; casts pellets of indigestible material.

REPRODUCTION: Monogamous; birds arrive paired at breeding colonies, usually located on a coastal island, in April; nest is a scrape (19–20 cm dia. (8 in), placed in an open area of sand or shell with little or no vegetation; both pair members excavate the scrape, and line it with bits of vegetation and beach debris; clutch—1–3; incubation (both sexes)—25–28 days; fledging—young are semi-precocial, and able to leave the nest within 1–2 days post-hatching; post-fledging parental care—young remain with adults for several months while learning where and how to fish; single brood per season.

CONSERVATION STATUS: S1—VA; Caspian Terns, like other beach-nesting species, are threatened by human disturbance, development, and predation at breeding colony sites.

RANGE: Breeds locally inland and along the coast in temperate, tropical, and

- Summer
- Migration
- Permanent
- Winter

boreal areas of the world; winters in south temperate and tropical regions.

KEY REFERENCES: Cuthbert and Wires (1999).

Royal Tern *Sterna maxima*

DESCRIPTION: L = 51 cm (20 in); W = 114 cm (45 in); a large tern with yellow or yellow-orange bill; black cap with short, ragged crest; black nape; white face, neck, and underparts; gray back and wings; tail forked; black legs. **Winter** white forehead and crown; black nape.

HABITAT AND DISTRIBUTION: Common summer resident* (Apr–Oct) along the coast: Bays, inlets, tidal creeks, lagoons, and inshore along beaches.

DIET: Forages over shallow (10–20 m) (30–70 ft) waters both diurnally and nocturnally mainly for

small fish and shrimp, which are taken by diving from 5–10 m (16–33 ft) above the water surface.

REPRODUCTION: Monogamous; pairs arrive at breeding colonies in April; nest is a scrape on the beach excavated by both sexes in bare sand and shell substrates above high tide line; clutch—1; incubation (both sexes)—30–31 days; fledging—young are born semi-precocial, leaving the nest to join groups of mixed ages of other young ("creches") at 2–3 days of age; adults recognize their own young in these groups, and feed them until fledging at 30–31 days post-hatching; association with adults by young continues after fledging, with family groups remaining together during migration and at wintering sites where some food-begging by offspring persists; single brood per season.

CONSERVATION STATUS: S1—NJ, MD; S2—VA; disturbance of beach nesting sites is the main threat to Royal Terns.

RANGE: Breeds locally in temperate and tropical regions of

- Summer
- Migration
- Permanent
- Winter

Western Hemisphere and west Africa; winters in warmer portions of breeding range.

KEY REFERENCES: Buckley and Buckley (2002).

Sandwich Tern *Sterna sandvicensis*

DESCRIPTION: L = 38 cm (15 in); W = 86 cm (34 in); a trim, medium-sized tern with long, pointed black bill, tipped with yellow; black cap and short, ragged crest; black nape; white throat, neck, and underparts; gray back and wings; tail deeply forked; black legs. **Winter** White forehead; crown white, finely streaked with black; black nape.

HABITAT AND DISTRIBUTION: Uncommon summer resident* (Apr–Sep), mainly on VA coastal islands, although found along bays and inlets during the postbreeding period (Jul–Sep): In-shore marine, saltmarsh, estuaries, bays; often associated with Royal or Roseate terns at breeding colonies and while foraging.

DIET: Forages by flying, 5–7 m (16–23 ft) above water surface, head down, searching for prey; when prey are sighted, the bird hovers, rises slightly, and dives vertically to plunge into water and capture prey, usually small fish, squid, or crustacea; generally feeds in flocks with other tern species; successful dives quickly attract other individuals to foraging sites, e.g., areas where small fish are chased to the surface by larger predatory fish; casts pellets of indigestible material.

REPRODUCTION: Monogamous; birds arrive at island breeding colonies in April already paired for the most part; the small breeding territory (0.5 m dia. (20 in), is usually located on a site bare of vegetation on a coastal island, often within the immediate vicinity of large numbers of conspecifics, Royal Terns, or other tern or gull species; nest, 7–10 cm dia. (3–4 in), is a scrape excavated by both pair members in sand or shell substrate and lined with bits of beach debris; clutch—1–2; incubation (both sexes)—23–25 days; fledging—young are semi-precocial and can walk from nest within 1–2 days, and are able to fly at 25–28 days post-hatching; post-fledging parental care—several months; single brood per season.

CONSERVATION STATUS: S1—VA; breeding colonies are threatened by human disturbance, coastal development, and predation.

RANGE: Breeds locally along Atlantic, Gulf, and Caribbean

Summer
Migration
Permanent
Winter

coasts of North and South America from VA to Argentina, West Indies; winters through subtropical and tropical portions of breeding range and on the Pacific Coast of Middle and South America from southern Mexico to Peru.

KEY REFERENCES: Shealer (1999).

Roseate Tern *Sterna dougallii*

DESCRIPTION: L = 41 cm (16 in); W = 71 cm (28 in); a medium-sized tern; black bill reddish-orange at base; black cap and nape; white face, throat; underparts white tinged with pink; gray back and wings; black outer primaries; tail very long, white, and deeply forked, extends well beyond folded wings in sitting bird; red legs. **Winter** white forehead and crown; back of crown and nape black; bill black. **First Breeding** Like adult but with white forehead.

HABITAT AND DISTRIBUTION: Rare to casual transient and summer visitor (May–Jul) along the coast and offshore; has bred*: Beaches, marine, bays. Flies with distinctive rapid, shallow wing beats.

DIET: Forages over shoals, submerged sandbars, tide rips, reefs, and other marine sites where schools of small fish are brought near the water surface; captures prey by diving from 1–6 m (3–20 ft) above the surface.

CONSERVATION STATUS: SH—NJ, VA; "Endangered"—USFWS; like many other beach-nesting shorebirds, the Roseate Tern is threatened by human

disturbance, coastal development, and predation on breeding colonies.

RANGE: Breeds locally along coasts of temperate and tropical regions of the world (in United States from ME to NC); winters mainly in tropical portions of breeding range.

KEY REFERENCES: Gochfeld et al. (1998).

Common Tern *Sterna hirundo*

DESCRIPTION: L = 36 cm (14 in); W = 79 cm (31 in); a medium-sized tern; red bill, black at tip; black cap and nape; white face, neck, and underparts; gray back and wings; entire outer primary, tips, and basal portions of other primaries are black (gray from below); forked tail with gray outer edgings, white inner edgings; red legs. **Winter** White forehead and crown; back of crown and nape black; bill black; sitting bird shows a dark bar at the shoulder (actually upper wing coverts). **First Breeding** Similar to adult winter.

HABITAT AND DISTRIBUTION: Common transient and summer resident* (Apr–Sep) along the immediate coast; rare to casual transient (Aug–Oct, Apr–May) inland: In-shore coastal waters, bays, inlets, shoals, saltmarsh creeks, lakes, rivers.

DIET: Forages by flying 2–3 m (7–10 ft) above the water and diving to capture fish, usually from water surface.

REPRODUCTION: Monogamous; pairs arrive on breeding colonies in April; Mid-Atlantic colony sites are usually on sparsely vegetated coastal islands; nest, 18 cm dia. (7 in), is a scrape excavated by both sexes in loose sand or shell, and lined with beach debris; clutch 2–3; incubation (both sexes)—20–33 days; young are semi-precocial, able to leave nest 2–3 days post-hatching, and are able to fly at 22–29 days; post-fledging parental care—at least 20–30 days; young require time to learn successful foraging; single brood per season.

CONSERVATION STATUS: S1—DE; S3—VA, NJ; SX—PA; coastal colony sites threatened by disturbance.

RANGE: Breeds locally in boreal and temperate regions of Old and New

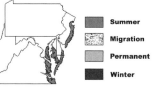

World (mostly Canada, eastern United States along the coast and Great Lakes, and West Indies); winters in south temperate and tropical regions.

KEY REFERENCES: Nisbet (2002).

Arctic Tern *Sterna paradisaea*

DESCRIPTION: L = 41 cm (16 in); W = 79 cm (31 in); a medium-sized tern; red bill; black cap and nape; white face; grayish white neck, breast, and belly; white undertail coverts; gray back; wings gray above with dark outer edge of outer primary, white below with primaries narrowly tipped in black; tail deeply forked; red legs. **Winter** White

forehead and crown; back of crown and nape black; bill black.

HABITAT AND DISTRIBUTION: Rare offshore spring transient (May), rare to casual along coast: Mostly marine; occasionally along coast at beaches, sand flats, and spits, more often at night than during the day.

DIET: Forages by flying, bill down, < 6 m (20 ft) above water surface, briefly hovering, then plunge diving or picking prey from at or near the water surface; feeds mostly on small fish (< 15 cm) (6 in) and crustacea.

RANGE: Breeds in the Arctic; winters in the Antarctic; migrates mostly off-shore.

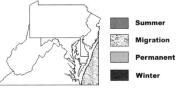

KEY REFERENCES: Hatch (2002).

Forster's Tern *Sterna forsteri*

DESCRIPTION: L = 38 cm (15 in); W = 79 cm (31 in); a medium-sized tern; orange or yellow bill with black tip; black cap and nape; white face and underparts; gray back and wings; forked tail, white on outer edgings, gray on inner; orange legs. **Winter** White head with blackish ear patch; bill black.

HABITAT AND DISTRIBUTION: Common transient and summer resident* (Apr–Nov) along the coast; rare to casual transient inland; uncommon in winter, mainly in southeastern VA: Marshes, estuaries, bays, beaches.

DIET: Forages by flying 6–8 m (20–26 ft) above shallow water (< 1 m) (3 ft) with bill pointing downward and then plunging when prey is spotted, or hovering and then plunging; main food items are small fish.

REPRODUCTION: Monogamous; pairs arrive at breeding colonies in May; colony sites in the Mid-Atlantic region are often on coastal islands; ground nest, 18 cm dia. (7 in), is a shallow scrape dug in sand or shell substrate; clutch—2–3; incubation (both sexes)—23–28 days; young are semi-precocial, walking off nest at 4–5 days post-hatching; able to fly at 4–5 weeks old; association with adults post-fledging is not known; single brood per season.

CONSERVATION STATUS: S1—DE; S3—VA; disturbance of coastal breeding colonies and destruction of coastal habitat threaten populations.

RANGE: Breeds locally across the northern United States, south central and south-western Canada, and Pacific, Atlantic, and Gulf coasts; winters coastally from CA and MD along the Coastal Plain of the southeastern United States south to Yucatan and El Salvador; also Greater Antilles.

KEY REFERENCES: McNicholl et al. (2001).

Least Tern *Sterna antillarum*

DESCRIPTION: L = 23 cm (9 in); W = 51 cm (20 in); a small tern; yellow bill with black tip; black cap and nape with white forehead; black band from bill to crown; white underparts; gray back and wings; tail deeply forked; yellow legs. **Winter** White head with blackish postorbital stripe and nape; bill brown; legs dull yellow.

HABITAT AND DISTRIBUTION: Common transient and summer resident* (May–Sep) along the coast: Bays, beaches, estuaries, lagoons, tidal marshes, rivers, lakes, ponds.

DIET: Forages principally for small fish (< 9 cm) (3.5 in) captured by diving into the water from 1–10 m (3–33 ft) in the air; invertebrates (e.g., crustacea, insects) taken on occasion as well.

REPRODUCTION: Monogamous; pairs form after arrival at breeding colonies in May; nest, 7–10 cm

dia. (3–4 in), is a simple scrape excavated by the male in substrate of sand, shell, pebbles, or soil, usually in a bare, open area perhaps slightly elevated above high-water level; colony sites include beaches, spoil islands (formed by dredged material deposits), gravel roofs, parking lots, and similar sites; clutch—2–3; incubation (both sexes)—19–25 days; fledging—young are precocial, able to leave the nest 2 days post-hatching, but cannot fly until 18 days or so later; post-fledging parental care—apparently none; single brood per season.

CONSERVATION STATUS: S1—NJ, DE; S2—VA, MD; "Endangered"—USFWS (inland populations); Least Tern nesting colonies are especially vulnerable to human recreational and development activities, in addition to a number of predators, e.g., foxes, raccoons, skunks, opossums, and crows, and flooding common to bare areas near water.

RANGE: Breeds along both coasts from central CA (west) and ME (east) south to southern Mexico and the West Indies; inland populations breed along rivers of Mississippi drainage; winters along coast of northern South America.

Summer
Migration
Permanent
Winter

KEY REFERENCES: Thompson et al. (1997).

Black Tern *Chlidonias niger*

DESCRIPTION: L = 26 cm (10 in); W = 61 cm (24 in); a small tern; dark gray throughout except white undertail coverts; slightly forked tail. **Winter** Dark gray above, white below with dark smudge at shoulder; white forehead; white crown streaked with gray; gray nape; white collar.

HABITAT AND DISTRIBUTION: Common fall transient (Jul–Sep) and uncommon to rare spring transient (May) along the coast. Uncommon to rare transient (Jul–Sep, Apr–May) inland, more numerous in spring in WV. Summer resident* (May–Jul) in western PA (Crawford and Erie counties) and as a nonbreeder at Chincoteague: Marshes, bays, estuaries, lakes, ponds, flooded fields.

DIET: Forages by flying head down 1–3 m (3–10 ft) over water, vegetation, or bare ground, scanning for prey, then dropping or swooping to pick prey from the substrate, or if necessary, give chase in zigzag aerial pursuit; food items vary by season, habitat, and availability, and include small fish, flying insects (e.g., Odonata, Lepidoptera, Orthoptera), and other invertebrates (e.g., crayfish, spiders, molluscs).

REPRODUCTION: Monogamous; pairs may form at communal feeding and roosting sites; nesting begins in May with selection of a pond or lakeshore site with shallow (0.5–1.2 m) (20–47 in),

still water, emergent vegetation (e.g., cattails or bulrushes), and mats of floating dead vegetation somewhat distant from shore; the birds nest semicolonially; both members of the pair build the open platform nest on floating, dead vegetation to which they add additional vegetation to form a mound in which they shape a small cup for egg placement; clutch—2–3; incubation (both sexes)—20–22 days; fledging—20–24 days post-hatching; post-fledging parental care—two weeks or more; one brood per season.

CONSERVATION STATUS: S1—PA; Black Tern populations have declined sharply since the 1960s, presumably because of loss of suitable wetland breeding habitat.

RANGE: Breeds in temperate and boreal regions of Northern Hemisphere; winters in tropics.

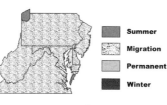

Summer
Migration
Permanent
Winter

KEY REFERENCES: Dunn and Agro (1995).

Black Skimmer *Rynchops niger*

DESCRIPTION: L = 46 cm (18 in); W = 114 cm (45 in); black above; white below; long, straight razor-shaped bill, red with black tip; lower mandible is longer than the upper. **Immature** Like adult but mottled brown and white above.

HABITAT AND DISTRIBUTION: Common transient and summer resident* (Apr–Oct) along the coast; rare in winter; rare to casual inland: Beaches, bays, estuaries.

DIET: Flies over the water with the lower mandible dipped below the surface to capture prey, mainly small fish; foraging is dependent to some extent on tide cycles (low tide preferred, when prey are concentrated in tidal pools), so the bird feeds both diurnally and nocturnally.

REPRODUCTION: Monogamous; birds arrive back from migration in late April and early May; skimmers nest colonially; colony sites are often located on small coastal islands of sand or shell with sparse vegetation; pairs select nesting sites at colonies in mid-June; both pair members build the open scrape nest, kicking out a bowl-shaped depression in the substrate; clutch—3–4; incubation—21–25 days; fledging—young are semi-precocial and can walk from nest within 2–3 days post-hatching, but are not able to fly until 30 days post-hatching; post-fledging parental care—several weeks; single brood per season.

CONSERVATION STATUS: S1—NJ, MD, DE; S2—VA; skimmers, like most beach-nesting shorebirds, are threatened by predation and human disturbance.

RANGE: Breeds from southern CA (west) and NY (east) south to southern

Summer
Migration
Permanent
Winter

South America along coasts and on major river systems; winters in south temperate and tropical portions of breeding range, West Indies.

KEY REFERENCES: Gochfeld and Burger (1994).

ORDER COLUMBIFORMES; FAMILY COLUMBIDAE

Rock Pigeon *Columba livia*

DESCRIPTION: L = 33 cm (13 in); W = 59 cm (23 in); the Rock Pigeon or domestic pigeon is highly variable in color, including various mixtures of brown, white, gray, and dark blue; the "average" bird shows gray above and below with head and neck iridescent purplish-green; white rump; dark terminal band on tail.

HABITAT AND DISTRIBUTION: Common permanent resident* throughout: Cities, towns, agricultural areas.

DIET: Plant seeds, fruits, other vegetable matter (bread, french fries, popcorn, etc.), occasionally invertebrates.

REPRODUCTION: Monogamous, normally pairing for life; pairs breed throughout the year, but mostly Feb–Sep; open platform nest is constructed by the female of twigs, roots, straw, and other plant material brought to her by the male; the nest is placed on nearly any flat surface protected from weather and predators, e.g., ledges, lofts, attics, eaves, crags, or cliffs; clutch size—2; incubation—18 days; fledging—25–32 days post-hatching during summer, as much as 45 days post-hatching in winter; post-fledging dependence on adults is not known; multiple broods per year.

RANGE: Resident of Eurasia and North Africa; introduced into the Western

Hemisphere where resident nearly throughout at farms, towns, and cities.

KEY REFERENCES: Johnston (1992).

Mourning Dove *Zenaida macroura*

DESCRIPTION: L = 31 cm (12 in); W = 46 cm (18 in); tan above, orange-buff below; brownish wings spotted with dark brown; tail long, pointed, and edged in white; gray cap; black whisker; purplish-bronze iridescence on side of neck; blue eyering; brown eye and bill; pink feet.

HABITAT AND DISTRIBUTION: Common summer resident* (Apr–Oct) throughout; common winter resident in lowlands and foothills while uncommon to rare or absent in highlands: Woodlands, pastures, old fields, second growth, scrub, residential areas.

DIET: Feeds mostly on the ground in open areas on herbaceous plant seeds and grains, often in flocks.

REPRODUCTION: Monogamous; pair formation in April and May in the Mid-Atlantic region; open platform nest is placed on horizontal supporting structure of a tree or shrub branch or branches, vines, shelves, or occasionally on the ground; average nest height is 5 m (16 ft); flimsy nest is built by the female with sticks, twigs, pine needles, and grasses brought by the male; clutch—2; incubation—14 days; fledging—15 days post-hatching; young are fed crop milk exclusively for 1–4 days post-hatching by both the male and the female, then fed crop milk mixed with increasing proportions of regurgitated seeds until fledging; post-fledging parental care—12 days; multiple broods per season.

RANGE: Breeds from central and southern Canada south through

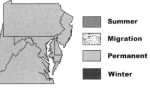

United States to highlands of central Mexico, Costa Rica, Panama, Bahamas, Greater Antilles; winters in temperate and tropical portions of breeding range.

KEY REFERENCES: Blockstein (1989), Baskett et al. (1993), Mirarchi and Baskett (1994).

Common Ground-Dove *Columbina passerina*

DESCRIPTION: L = 18 cm (7 in); W = 28 cm (11 in); a very small dove; grayish-brown above with dark brown spotting on wing; gray tinged with rose below with scaly brown markings on breast; gray pileum; red eye; pink bill; flesh-colored legs; rusty wing patches show in flight; black outer tail feathers with white edgings. **Female** Like male but grayish below (not pinkish), and gray crown is less distinct.

HABITAT AND DISTRIBUTION: Rare summer visitor (Apr–Oct) in extreme southeastern VA, specifically Mackay Island National Wildlife Refuge: Open or scrubby areas, farm land, sandy or grassy shrublands, coastal dunes.

DIET: Forages by walking on the ground, thrusting head forward with each step, and picking up weed and grass seeds from the ground; fruits, snails, and insects are also taken depending on seasonal or site availability.

RANGE:
Southern United States south to northern half of South America, West Indies.

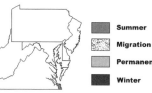

Summer
Migration
Permanent
Winter

KEY REFERENCES: Bowman (2002).

ORDER CUCULIFORMES; FAMILY CUCULIDAE

Black-billed Cuckoo *Coccyzus erythropthalmus*

DESCRIPTION: L = 31 cm (12 in); W = 41 cm (16 in); a thin, streamlined bird with long, graduated tail; red eyering; black bill; brown above; white below; dark brown tail with white tips.

HABITAT AND DISTRIBUTION: Common (north and highlands) to uncommon or rare (south and lowlands) transient and summer resident* (May–Sep): Deciduous and mixed forest, thickets, overgrown fields with shrubby second growth, swamps, brushy marshes.

DIET: Forages in trees and shrubs using slow, deliberate, reptilian movements followed by quick runs, flights, or snatches to capture large insects, e.g., cicadas, katydids, caterpillars, and grasshoppers.

REPRODUCTION: Monogamous?; pairs arrive on breeding territory in May, and commence breeding relatively late (Jun–Jul egg dates); the loose platform or bowl nest, 13–15 cm dia. (5–6 in), is placed in horizontal branches in thickets of shrubs or trees, usually 1–2 m above ground, and built (both sexes?) of twigs, grasses, and plant fibers lined with leaves, pine needles, plant fibers, down, grasses, and bark strips; clutch—2–3; incubation (both sexes)—10–11 days; fledging—young clamber off the nest at 6–7 days post-hatching but cannot fly until about 3 weeks of age; post-fledging parental care—?; single brood per season; Black-billed Cuckoos occasionally lay their eggs in the nests of other cuckoo pairs as well as those of at least 11 other species of birds (e.g.,

Yellow-billed Cuckoo, Chipping Sparrow, American Robin, and Gray Catbird).

CONSERVATION STATUS: S1—DE; S3—WV; short-term (1980–1994) and long-term (1966–1994) declines have been recorded throughout the cuckoo's range, for reasons that are not well understood; perhaps due to loss of winter habitat in the Neotropics.

RANGE: Breeds in the eastern two-thirds of the United States and southern  Canada; winters in the northern half of South America.

KEY REFERENCES: Hughes (2001).

Yellow-billed Cuckoo *Coccyzus americanus*

DESCRIPTION: L = 31 cm (12 in); W = 43 cm (17 in); a thin, streamlined bird with long, graduated tail; dark upper bill mandible with yellow lower mandible and eyering; brown above; white below; dark brown tail with white tips; rusty wing patches visible in flight.

HABITAT AND DISTRIBUTION: Common transient and summer resident* (May–Sep) throughout; more numerous in lowlands: Open deciduous and mixed woodlands, second growth, old fields, shrubby thickets.

DIET: Forages by perching in trees and shrubs, scanning carefully surrounding vegetation, and moving with slow, almost reptilian deliberation to capture prey, mostly large insects (e.g., caterpillars, cicadas, katydids, and beetles); feeds on fruit to some extent during nonbreeding periods.

REPRODUCTION: Monogamous; breeding evidently begins in May, although there is little information; open platform nest, 8 cm dia. (3 in), is placed on the horizontal branch or fork of a tree or shrub, 1–6 m (3–20 ft) above ground, and built by both pair members of twigs, leaves, bark strips, and plant stems; clutch—2–3; incubation (both sexes)—11 days; fledging—the young are able to clamber off the nest 7–9 days post-hatching but cannot fly until 20 days; post-fledging parental care—?; there is some evidence of cooperative breeding with adult nonpair members observed feeding young; the Yellow-billed Cuckoo is at least an occasional social parasite, known to lay its eggs in the nests of at least 11 other species, most commonly that of the American Robin.

CONSERVATION STATUS: Western populations of the species have been decimated over the past century, evidently due to loss of riparian breeding habitat and pesticides; significant short- (1980–1994) and long-term (1966–1994) declines have also been recorded for eastern populations for reasons that are not well understood, but may be related to reversion of old field habitats to forest or conversion to intensive agriculture.

RANGE: Breeds in southeastern Canada and nearly throughout the United  States into northern Mexico and West Indies; winters in north and central South America; some evidence of breeding populations in the tropics.

KEY REFERENCES: Hughes (1999).

ORDER STRIGIFORMES; FAMILY TYTONIDAE

Barn Owl *Tyto alba*

DESCRIPTION: L = 41 cm (16 in); W = 114 cm (45 in); tawny and gray above; white sparsely spotted with brown below; white, monkeylike face, and large, dark eyes; long legs.

HABITAT AND DISTRIBUTION: Uncommon to rare permanent resident*. Easily overlooked because of nocturnal habits. Recent breeding bird surveys found the bird to be rare in northern and western PA and rare or absent over much of WV: Old fields, pasture, grasslands, farmland; roosts in barns, caves, abandoned mine shafts.

DIET: Almost strictly nocturnal with extraordinary ability to detect and capture prey in low light. Feeds mainly on small mammals, but also birds, reptiles, amphibians, and, rarely, even fish and invertebrates; prey are usually swallowed whole, with fur, feathers, and bones regurgitated later as pellets. Voles (*Microtus*) are the main prey item in many parts of the range.

REPRODUCTION: Monogamous, often pairing for life; 4–7 eggs laid in Feb–Mar in cavity nest sites—cliffs, caves, trees, lofts, barns, sheds, nest boxes; incubation lasts 30 days; young fledge at 60 days but are cared for by adults for 3–5 weeks after first flight; can produce more than one brood per year.

CONSERVATION STATUS: S1—WV; S3—PA, VA, MD, DE, NJ; poorly known because of nocturnal habits; principal limiting factor seems to be nest sites and reversion of open habitats to forest.

RANGE: Resident in temperate and tropical regions nearly throughout the world.

KEY REFERENCES: Marti (1992).

Summer

Migration

Permanent

Winter

FAMILY STRIGIDAE

Eastern Screech-Owl *Megascops asio*

DESCRIPTION: L = 23 cm (9 in); W = 56 cm (22 in); **Red Phase** A small, long-eared, yellow-eyed owl with a pale bill; rufous above; streaked with rust, brown, and white below; rusty facial disk with white eyebrows. **Gray Phase** Similar to red phase but gray rather than rusty.

HABITAT AND DISTRIBUTION: Common resident* throughout: Woodlands, residential areas.

DIET: Forages by sallying from low (2–3 m) (6–10 ft) perches and grasping prey in talons; diet includes small birds, mammals, reptiles, amphibians, large insects (e.g., Lepidoptera, Orthoptera, Coleoptera), crustacea, and other invertebrates in varying amounts depending on availability.

REPRODUCTION: Mostly monogamous, mating for life; breeding begins in March; nest is a natural cavity, abandoned woodpecker hole, or nest box, 0–20 m (0–66 ft) above ground; clutch—3–5; incubation (female, fed by male)—27–34 days; fledging—28 days post-hatching; post-fledging parental care—8–10 weeks; single brood per season.

RANGE: Resident from southeastern Canada and the

eastern half of the United States to northeastern Mexico.

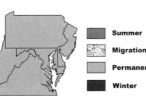

KEY REFERENCES: Gehlbach (1995).

Great Horned Owl *Bubo virginianus*

DESCRIPTION: L = 56 cm (22 in); W = 132 cm (52 in); a large owl; mottled brown, gray, buff, and white above; grayish-white below barred with brown; yellow eyes; rusty or grayish facial disc; white throat (not always visible); long ear tufts.

HABITAT AND DISTRIBUTION: Uncommon permanent resident* throughout: Woodlands.

DIET: Forages mainly at night by sallying or dropping on to prey from a perch; prey items include small to medium-sized mammals (90%) (e.g., rabbits, opossums, skunks, rodents), birds, other vertebrates, or even invertebrates on occasion.

REPRODUCTION: Monogamous, may mate for life; usually the pair uses an open platform nest in a tree built by another species (e.g., Red-shouldered Hawk or Red-tailed Hawk); clutch—2, with laying beginning in February or early March; incubation (female—male brings food to the nest)—33 days; fledging—7 weeks post-hatching; post-

fledging parental care—several weeks; single brood per season.

RANGE: New World except polar regions.

KEY REFERENCES: Houston et al. (1998).

Snowy Owl *Bubo scandiacus*

DESCRIPTION: L = 61 cm (24 in); W = 152 cm (60 in); white with various amounts of gray and black barring; yellow eyes. **Immature** Much more buff and brown spotting and barring than adults.

HABITAT AND DISTRIBUTION: Rare to casual and irregular winter visitor (Dec–Feb). This species appears in the Mid-Atlantic states in certain years, and is absent in others, at 5–10 year intervals, depending on food availability within its normal, boreal range: Tundra, fields, pastures, meadows, marshes.

DIET: Mostly lemmings (*Lemmus, Synaptomys, Dicostonyx*) when available; a wide variety of small to medium-sized mammalian and avian prey when lemming populations in the tundra crash, including waterfowl, hares (*Lepus*) and rabbits (*Sylvilagus*), and shorebirds; occasionally eats fish.

RANGE: Resident in northern polar regions of Old and New World.

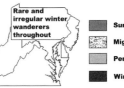

KEY REFERENCES: Parmelee (1992).

Barred Owl *Strix varia*

DESCRIPTION: L = 46 cm (18 in); W = 107 cm (42 in); a large, dark-eyed owl; brown mottled with white above; white barred with brown on throat; breast and belly whitish streaked with brown; tail barred gray and dark brown; yellowish bill.

HABITAT AND DISTRIBUTION: Uncommon resident* throughout: Mature deciduous and mixed forest.

DIET: Forages nocturnally from a perch in forest mid-story; main prey is small mammals (mice, rabbits, squirrels), but large insects, birds, amphibians, reptiles, and fish are also taken.

REPRODUCTION: Monogamous, perhaps mating for life; usually nests in a tree cavity, with egglaying in March or April; clutch—2–3; incubation (female)—28–33 days; fledging—young leave nest at 4–5 weeks post-hatching, but can't fly until 10–12 weeks; post-fledging parental care—several weeks.

CONSERVATION STATUS: S2—DE; S3—NJ; conversion of bottom land habitat to agriculture and residential.

RANGE: Resident from southern and western Canada, and the eastern half of

	Summer
	Migration
	Permanent
	Winter

the United States; locally in northwestern United States and in the central plateau of Mexico.

KEY REFERENCES: Mazur and James (2000).

Long-eared Owl *Asio otus*

DESCRIPTION: L = 36 cm (14 in); W = 100 cm (39 in); medium-sized, trim, yellow-eyed owl with long ear tufts; mottled brown and white above; whitish streaked with brown below; rusty facial disks.

HABITAT AND DISTRIBUTION: Uncommon to rare and local resident* in PA, northwestern NJ, WV and the highlands of VA; rare winter visitor (Nov–Mar) elsewhere in the region: Coniferous and mixed woodlands, especially dense cedar (*Thuja*) thickets in winter, when several birds can be found in communal roosts.

DIET: Forages mostly at night by flying low (0.5–2m) (2–7 ft) over open areas, coursing back and forth with a series of rapid wingbeats followed by short glides; small mammals (e.g., voles, deer mice, and young rabbits) are the main prey items.

REPRODUCTION: Monogamous with some polygyny reported; pairs may form at communal roosts in winter; nest sites are selected shortly after the pair arrives at the breeding territory in March or early April; normally the abandoned open platform nest of another species (e.g., American Crow, Cooper's Hawk, or Red-shouldered Hawk) is used; clutch—5–6; incubation (female)—26–28 days, male feeds female; fledging—young climb off nest 21 days or so post-hatching and are able to fly at 35 days post-hatching; post-fledging parental care—5–6 weeks; single brood per season.

CONSERVATION STATUS: S1—WV, VA; SH—DE, MD; S2—NJ, PA; small breeding population size in limited and decreasing coniferous and mixed forest habitat.

RANGE: Breeds in temperate and boreal regions of the Northern

Hemisphere; winters in breeding range, but also migrates to temperate

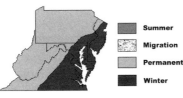

Summer
Migration
Permanent
Winter

areas and mountains of northern tropical regions.

KEY REFERENCES: Marks et al. (1994).

Short-eared Owl *Asio flammeus*

DESCRIPTION: L = 38 cm (15 in); W = 107 cm (42 in); a medium-sized owl; brown streaked with buff above; buff streaked with brown below; round facial disk with short ear tufts and yellow eyes; buffy patch shows on upper wing in flight; black patch at wrist visible on wing lining.

HABITAT AND DISTRIBUTION: Uncommon to rare winter resident (Nov–Mar) throughout, more numerous in coastal areas; rare to casual summer resident* (Apr–Oct) in NJ and PA (strip mine areas of Clarion and Jefferson counties and grasslands of the Philadelphia airport): Grasslands, marshes, estuaries, agricultural fields.

DIET: Diurnal and nocturnal in foraging habits, depending on prey density and nestling status; hunts by coursing low over open areas—reminiscent of Northern Harrier; small mammals, especially voles (*Microtus*), are the main prey item although birds, particularly shorebirds in coastal areas, can be important.

REPRODUCTION: Monogamous, with pairs forming at communal winter roosts in February and March;

open nest, 25 cm dia. (10 in), is built by female in April or May, and placed in open grassy areas (grass—0.5 m in height) (20 in), often on a knoll, ridge, or hummock; female scrapes out a nest bowl in the ground, and lines it with grasses and down feathers; clutch—5–6; incubation—21–29 days; fledging—nestlings walk away from nest at about 18 days, but do not fly until 31–36 days post-hatching; parental care post-fledging—?; single brood per season.

CONSERVATION STATUS: S1—PA, WV, VA, MD; SH—DE, NJ; loss of grassland habitat is presumed to be the main cause of the sharp population decline observed in the eastern U.S. over the past 3 decades.

RANGE: Breeds in tundra, boreal, and north temperate areas of Northern

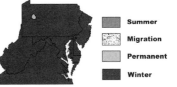

Summer
Migration
Permanent
Winter

Hemisphere, also in Hawaiian Islands; winters in southern portions of breeding range south to northern tropical regions.

KEY REFERENCES: Holt and Leasure (1993).

Northern Saw-whet Owl *Aegolius acadicus*

DESCRIPTION: L = 20 cm (8 in); W = 51 cm (20 in); a small owl; dark brown above with white spotting; white below with broad chestnut stripes; facial disk with gray at base of beak, gray eyebrows, the rest finely streaked brown and white; yellow eyes. **Immature** Dark brown back spotted with white; solid chestnut below; chestnut facial disk with white forehead and eyebrows.

HABITAT AND DISTRIBUTION: Rare resident* in highland spruce-fir forest of the region; rare and irregular winter resident (Oct–Mar) elsewhere; has been reported in summer in cedar swamps of the NJ Pine Barrens: Coniferous forest, especially spruce-fir; dense thickets of pine and cedar (*Thuja*) in winter. A tame owl, easily approached.

DIET: Nocturnal, hunting mostly for small mammals,

e.g., *Peromyscus*, *Microtus*, and *Sorex*, by sallying from low perches in forest openings; also insects and small birds.

REPRODUCTION: Usually monogamous, but some evidence of polygyny in populations where prey are abundant; males establish breeding territories in February, pairs form in March; nests are established in old woodpecker cavities or nest boxes in April or May; eggs are laid directly on the floor of the cavity; clutch—5–6; incubation—27 days; fledging—30 days post-hatching; parental care— at least four weeks post-fledging; one brood per season.

CONSERVATION STATUS: S1—VA, MD, NJ; S2—WV; S3—PA; loss of highland coniferous and mixed forest habitat.

RANGE: From central Canada south to northern United States and in western mountains south to southern Mexico; winters in breeding range south and east to southern United States.

■	Summer
▒	Migration
▤	Permanent
■	Winter

KEY REFERENCES: Cannings (1993).

ORDER CAPRIMULGIFORMES; FAMILY CAPRIMULGIDAE

Common Nighthawk *Chordeiles minor*

DESCRIPTION: L = 23 cm (9 in); W = 59 cm (23 in); mottled dark brown, gray and white above; whitish below with black bars; white throat; white band across primaries and tail; tail slightly forked. **Female** Buffy rather than white on throat.

HABITAT AND DISTRIBUTION: Common to uncommon transient and summer resident* (May–Sep) throughout the region, although spotty in distribution (requires bare ground, beaches, or gravel roofs as nest sites): A variety of open and semi-open situations in both urban and rural environments.

DIET: Crepuscular; normally forages high above the ground, often giving its "beezzt" call while hawking erratically for flying insects (e.g., Hymenoptera, Homoptera, Coleoptera, Lepidoptera, Ephemeroptera, Diptera).

REPRODUCTION: Monogamous?; birds arrive on breeding territory in May; male has a spectacular dive display in which the bird plummets toward the ground, swerving up at the last moment and making a whirring sound with the wings; eggs are placed on bare ground at open sites like over-grazed pasture, short-grass prairie, beaches, gravel roofs, parking lots, dirt roads, or road shoulders; clutch—2; incubation (mostly female)—18–20 days; fledging—18 days post-hatching; post-fledging parental care—?; single brood per season.

CONSERVATION STATUS: S2—DE; S3—PA, WV, NJ, MD; breeding populations of nighthawks have undergone significant short- (1980–1994) and long-term (1966–1994) declines for reasons that are not clear; loss of breeding sites, e.g., gravel roofs, and poisoning resulting from spraying of pesticides for mosquitoes have been suggested, but breeding range is so broad and summer feeding and nesting habitat so generally available that these explanations seem inadequate; declines could be due to loss of wintering ground habitat in the Neotropics.

RANGE: Breeds locally from central and southern Canada south through  United States south to Panama; winters in South America.

KEY REFERENCES: Poulin et al. (1996).

Chuck-will's-widow *Caprimulgus carolinensis*

DESCRIPTION: L = 31 cm (12 in); W = 64 cm (25 in); a large nightjar; tawny, buff, and dark brown above; chestnut and buff barred with dark brown below; white throat; long, rounded tail with white inner webbing on outer 3 feathers; rounded wings. **Female** Has buffy tips rather than white inner webbing on outer tail feathers.

HABITAT AND DISTRIBUTION: Uncommon summer resident* (May–Sep) of the Coastal Plain and southern VA Piedmont; has been recorded in recent years as a rare and irregular summer resident in southwestern PA, WV, and the mountains of VA: Open deciduous, mixed and pine woodlands; also hedgerows, second growth, and wood margins.

DIET: Forages mostly at dusk, flying low in woodland

openings, gaps, and margins as well as over grasslands; main prey is flying insects, but small birds and bats are occasionally taken.

REPRODUCTION: Monogamous?; pairs arrive on breeding territory by April or early May; no nest is built; eggs are laid on the ground in leaf litter; clutch—2; incubation (female)—20 days; fledging—17 days post-hatching; post-fledging parental care—14 days; single brood per season.

CONSERVATION STATUS: S1—WV; S3—DE; long-term (1966–1994) population declines in Mid-Atlantic region; declines may be due to winter habitat loss or sampling error.

RANGE: Breeds across the eastern half of the United States, mainly in the south- 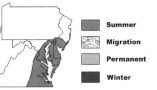 east; permanent resident in south FL and parts of the West Indies; winters from eastern Mexico south through Central America to Colombia; also winters in south FL and the West Indies.

KEY REFERENCES: Straight and Cooper (2000).

Whip-poor-will *Caprimulgus vociferus*

DESCRIPTION: L = 26 cm (10 in); W = 48 cm (19 in); dark brown mottled with buff above; grayish barred with dark brown below; dark throat and breast with whitish patch; rounded tail with white outer 3 feathers; rounded wings. **Female** Has buffy tips rather than white on outer tail feathers.

HABITAT AND DISTRIBUTION: Common to uncommon summer resident* (Apr–Sep) throughout: Open deciduous and mixed forest; also second growth forest and woodlots.

DIET: Forages by sallying from perches in trees, mostly at dusk and dawn, but will continue through the night in bright moonlight; main prey items are flying insects (e.g., moths, beetles, ants, flies, grasshoppers, and mosquitoes).

REPRODUCTION: Monogamous; pairs arrive on

breeding territories in April; eggs are laid in leaf litter, usually under an overhanging shrub; clutch—2; incubation (both sexes)—19–21 days; fledging—20 days post-hatching; post-fledging parental care—at least 14 days; single brood per season.

CONSERVATION STATUS: S3—WV, MD; declines have occurred for reasons unknown, perhaps loss of winter habitat in the tropics.

RANGE: Breeds across south-eastern and south central Canada and the eastern

Summer

Migration

Permanent

Winter

United States, also in the southwestern United States through the highlands of Mexico and Central America to Honduras; winters from northern Mexico to Panama, Cuba, also rarely along Gulf and Atlantic coasts of southeastern United States.

KEY REFERENCES: Cink (2002).

ORDER APODIFORMES; FAMILY APODIDAE

Chimney Swift *Chaetura pelagica*

DESCRIPTION: L = 13 cm (5 in); W = 31 cm (12 in); dark throughout; short square tail; long narrow wing.

HABITAT AND DISTRIBUTION: Common transient and summer resident* (May–Sep) throughout: Widely distributed over most habitat types wherever appropriate nesting and roosting sites are available (chimneys, cliffs, caves, crevices, hollow

trees); often seen over towns and cities where chimneys are available for roosting.

DIET: Forages high in the air, 20–150 m above ground (70–500 ft) for flying insects (e.g., Diptera, Homoptera, Hymenoptera, Ephemeroptera), capturing them on the wing with a rapid, fluttering flight.

REPRODUCTION: Monogamous, typically with one or more helpers to assist in raising of offspring; pairs arrive at breeding sites in May; shelf nest, 5–7.5 cm (2–3 in) wide x 10 cm (4 in) long, is placed on a protected, vertical surface, formerly cliff faces or the inside wall of a hollow tree, but now usually a chimney; both sexes build the nest, which is constructed of twigs glued together and to the chimney wall with their sticky saliva to form a half saucer-shaped platform; clutch—4; incubation (both sexes)—19 days; fledging—young crawl off from the nest to hang on wall at about 19 days, but do not fly until 28–30 days post-hatching; post-fledging parental care—7 days?; mostly one brood per season.

CONSERVATION STATUS: Short-term (1980–1994) and long-term (1966–1994) declines have been reported from throughout the breeding range for reasons that are not understood. Reduced availability of chimney nesting sites has been suggested as a possible reason, but there are no data to support this contention.

RANGE: Breeds throughout eastern North America; winters mainly in the upper

Summer

Migration

Permanent

Winter

Amazon basin of Peru, Ecuador, Chile, and Brazil.

KEY REFERENCES: Cink and Collins (2002).

FAMILY TROCHILIDAE

Ruby-throated Hummingbird
Archilochus colubris

DESCRIPTION: L = 10 cm (4 in); W = 10 cm (4 in); green above; whitish below; iridescent red throat **Female and Immature** Green above, whitish below.

HABITAT AND DISTRIBUTION: Common transient and summer resident* (May–Sep) throughout:

Woodlands, second growth, brushy pastures and fields, gardens.

DIET: Forages on nectar when available, often from red or orange flowers; also hawks flying insects and gleans arthropods from vegetation; young are fed by regurgitation and also by insertion of small whole insects into gaping mouth; follows Yellow-bellied Sapsuckers, and uses sap from sap wells where the breeding ranges of these two species overlap.

REPRODUCTION: Polygynous?; male associates with female only during courtship and egglaying; males arrive on breeding territory in May, 5–7 days before female; open cup nest, 4–5 cm dia. (2 in), is placed 5–7 m (16–23 ft) up in a tree, saddled on a horizontal or down-sloping branch, and is built by the female alone of plant down, spider silk, lichens, and bud scales; clutch—2; incubation (female)—12–14 days; fledging—18–22 days post-hatching; post-fledging care by female—4–7 days; 2 broods per season in some populations.

RANGE: Breeds in southern Canada and the eastern half of the United States;

Summer
Migration
Permanent
Winter

winters from southern Mexico to Costa Rica; also south FL and Cuba.

KEY REFERENCES: Robinson et al. (1996), Rappole (1999).

ORDER CORACIIFORMES; FAMILY ALCEDINIDAE

Belted Kingfisher *Ceryle alcyon*

DESCRIPTION: L = 33 cm (13 in); W = 56 cm (22 in); blue-gray above with ragged crest; white collar, throat, and belly; blue-gray breast band. **Female** Like male but with chestnut band across belly in addition to blue-gray breast band.

HABITAT AND DISTRIBUTION: Common summer resident* (Mar–Oct) throughout; withdraws from highlands and northern parts of region in winter: Rivers, lakes, ponds, bays.

DIET: Usually forages for small fish by perching above water, scanning visually for prey, then diving into the water, grasping the prey in its bill; other prey items taken commonly include crayfish, aquatic insects, molluscs, amphibians, reptiles, young birds, and small mammals.

REPRODUCTION: Monogamous; pairs form on breeding territory in April; nest construction is done by both sexes and takes 4–7 days; nest consists of a burrow, 1–2 m (3–7 ft) in length, dug into a high bank of sandy clay soil; nest chamber is 20–30 cm (8–12 in) in diameter and 15–18 cm (6–7 in) in height; eggs are laid in May or early June; clutch—6–7; incubation—22 days; fledging—

27–29 days post-hatching; post-fledging parental care—at least one week; one brood per season.

CONSERVATION STATUS: Short (1980–1994) and long (1966–1994) term declines in populations of the Mid-Atlantic region; kingfishers are very sensitive to stream health, requiring clear, unpolluted water for foraging; declines likely are related to loss of this habitat.

RANGE: Breeds throughout most of temperate and boreal North America excluding arid southwest; winters from southern portion of breeding range south through Mexico and Central America to northern South America; West Indies; Bermuda.

KEY REFERENCES: Hamas (1994).

ORDER PICIFORMES; FAMILY PICIDAE

Red-headed Woodpecker *Melanerpes erythrocephalus*

DESCRIPTION: L = 26 cm (10 in); W = 46 cm (18 in); red head; black upper back; white lower back; white below; white bill. **Immature** Head brownish; upper back brown.

HABITAT AND DISTRIBUTION: Uncommon to rare and local resident* throughout; withdraws in winter from highland areas: Open beech and oak woodlands, riparian stands, bottom lands, park lands, and savanna.

DIET: Forages by pecking, chiseling, probing, and gleaning from

bark and dead trunks and limbs for arthropods; sallies flycatcher-fashion for flying insects; feeds seasonally on nuts, acorns, seeds, and fruits, including crops; stores nuts for later use; occasionally forages on the ground for invertebrates.

REPRODUCTION: Monogamous; pairs begin nest-hole excavation in April or May; cavity is excavated in dead trunk or snag; clutch—5; incubation (both sexes)—12–14 days; fledging—23–31 days post-hatching; post-fledging parental care—3 weeks; double broods normal in southern populations.

CONSERVATION STATUS: S1—DE; S2—WV, NJ; long-term declines (1966–1994) reported throughout the bird's range for reasons not understood.

RANGE: Eastern North America from the central plains and southern Canada to TX and FL.

KEY REFERENCES: Smith et al. (2000).

Red-bellied Woodpecker *Melanerpes carolinus*

DESCRIPTION: L = 26 cm (10 in); W = 41 cm (16 in); nape and crown red; barred black and white on back and tail; white rump; dirty white below with reddish tinge on belly. **Female** Gray crown with red nape.

HABITAT AND DISTRIBUTION: Common resident* nearly throughout except in highlands; scarce or absent from parts of northern PA: Riparian forests, coniferous, deciduous and mixed woodlands, bottomland forest.

DIET: Forages mostly on dead limbs, snags,

and stumps for arthropods; also feeds on mast (acorns, fruit, nuts) and seeds in season; also takes small or young vertebrates opportunistically.

REPRODUCTION: Monogamous; pairs form in winter or spring; nest hole excavation begins in March or April; nest cavity is excavated in a dead trunk or snag by the male initially with the female assisting in later stages; clutch—4; incubation (both sexes)—12 days; fledging—24–27 days post-hatching; post-fledging parental care—5 weeks; single brood per season normally.

RANGE: Eastern United States.

KEY REFERENCES: Shackleford et al. (2000).

Yellow-bellied Sapsucker
Sphyrapicus varius

DESCRIPTION: L = 20 cm (8 in); W = 38 cm (15 in); black and white above; creamy yellow below with dark flecks on sides; breast black; throat, forehead, and crown red; black and white facial pattern. **Female** White throat. **Immature** Black wings with white patch, white rump, and checked black and white tail of adults; but barred brownish and cream on head, back, breast, and belly.

HABITAT AND DISTRIBUTION: Uncommon summer resident* (May–Sep) in northern PA; rare and local summer resident in the mountains (> 900 m) (3000 ft) in southern PA, VA, and Pocahontas, Randolph, and Barbour counties in eastern WV; uncommon transient (Sep–Oct, Apr–May) elsewhere; winter resident (Sep–May) in the Piedmont and Coastal Plain of MD, VA, and DE: Deciduous and mixed woodlands, bottomland and riparian forest, second growth, orchards, park lands, wooded residential areas.

DIET: Forages by poking holes or slits in tree trunks ("sap wells"), and harvesting the sap that flows into them; also captures insects, especially for nestlings, by gleaning, probing, scaling, and drilling in bark and dead snags; feeds on buds and fruits seasonally.

REPRODUCTION: Monogamous; males arrive on breeding territory in April, followed a week or so later by females; nest cavity is excavated by the male, often in an aspen (*Populus*) infected with fungus; clutch—4–6; incubation (both sexes)—11–13 days; fledging—23–29 days post-hatching; post-fledging parental care—at least 7–10 days; single brood per season.

CONSERVATION STATUS: SH—MD; S1—WV, VA; loss of highland forest breeding habitat.

RANGE: Breeds in temperate and boreal, wooded regions of Canada and

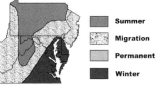

the United States; winters from the southern United States south to Panama and the West Indies.

KEY REFERENCES: Walters et al. (2002).

Downy Woodpecker *Picoides pubescens*

DESCRIPTION: L = 18 cm (7 in); W = 31 cm (12 in); black and white above; white below; red cap on back portion of crown. **Female** Black cap.

HABITAT AND DISTRIBUTION: Common resident* throughout: Forests, second growth, and park lands.

DIET: Differential foraging habitat by sex; males tend to forage on snags, small diameter twigs, outer branches, and woody weed stems, while females tend to forage on larger branches and trunks; main food items are arthropods, especially insect eggs and larvae in bark and rotted wood, exposed by picking and probing with the sharp bill, and extracted using the long, sticky, barbed tongue; also eats fruits, nuts, and acorns seasonally.

REPRODUCTION: Monogamous; pairs arrive on breeding territories in April; both sexes excavate the cavity nest, usually built on the underside of a dead snag that leans from the vertical; clutch—4–5; incubation (both sexes)—12 days; fledging—18–21 days post-hatching; post-fledging parental care—3 weeks; single brood per season.

RANGE: Temperate and boreal North America south to southern CA, AZ, NM, and TX.

KEY REFERENCES: Jackson and Ouellet (2002).

Hairy Woodpecker *Picoides villosus*

DESCRIPTION: L = 23 cm (9 in); W = 38 cm (15 in); black and white above; white below; red cap on back portion of crown. **Female** Black cap.

HABITAT AND DISTRIBUTION: Uncommon resident* throughout: Coniferous, mixed, deciduous, and riparian woodlands.

DIET: Forages mainly on trunks of both dead and live trees, pecking, gleaning, and excavating for invertebrates (e.g., Coleoptera, Hymenoptera, Lepidoptera, Hemiptera).

REPRODUCTION: Monogamous; some pairs remain together year-round while others form in early spring; nest hole is usually placed in a living, fungus-infected tree, and excavation (both sexes) is

begun in late March or early April, requiring 2–3 weeks; clutch—3–4; incubation (both sexes)—11–12 days; fledging—28–30 days; post-fledging parental care—"several weeks"; single brood per season.

RANGE: Temperate and boreal North America south through the mountains of western Mexico and Central America to Panama. Bahamas.

KEY REFERENCES: Jackson et al. (2002).

Red-cockaded Woodpecker *Picoides borealis*

DESCRIPTION: L = 20 cm (8 in); W = 38 cm (15 in); barred black and white above; white below flecked with black along sides; white cheek bordered below by black stripe; red spot behind eye. **Female** No red spot behind eye.

HABITAT AND DISTRIBUTION: Rare resident* of extreme southeastern VA: Pine savanna—especially mature longleaf and loblolly pine stands.

DIET: Forages, often in family groups, by flicking bark from the trunk or branches of pines to capture arthropod adults,

eggs, and larvae (e.g., beetles, centipedes, roaches); occasionally eats seeds and fruits.

REPRODUCTION: Monogamous, often mating for life; family groups called "clans," consisting of the breeding pair and 2 or 3 adult offspring, raise the young produced in a given year; nest is a cavity constructed by the breeding male, usually placed 10–13 m up in a mature pine (e.g., longleaf or loblolly); bark is removed around the opening of the cavity, evidently to increase sap flow and discourage nest predators (e.g., rat snakes); eggs are laid in late April or early May; clutch—3–4; incubation—10–11 days; fledging—26–29 days; post-fledging group care—2–5 months; normally one brood is produced per year.

CONSERVATION STATUS: S1—VA; "Endangered"—USFWS; mature pine savanna required as foraging and breeding habitat for Red-cockaded Woodpeckers has been reduced significantly throughout its range.

RANGE: Southeastern United States.

KEY REFERENCES: Jackson (1994).

Northern Flicker *Colaptes auratus*

DESCRIPTION: L = 31 cm (12 in); W = 51 cm (20 in); barred brown and black above; tan with black spots below; black breast; white rump. Eastern forms have yellow underwings and tail linings, gray cap and nape with red occiput, tan face and throat with black mustache. **Female** Lacks mustache.

HABITAT AND DISTRIBUTION: Common transient and summer resident* (Mar–Oct) throughout; uncommon (south, Piedmont, Coastal Plain) to rare (north, mountains) winter resident (Nov–Feb): Forests, parklands, orchards, woodlots, and residential areas.

DIET: Forages mostly on the ground, probing with bill and long, sticky tongue for ants (usually > 60% of

diet) and other insects; sometimes takes fruits and seeds, depending on availability.

REPRODUCTION: Monogamous; pair formation begins with arrival of birds at breeding territory in late March or April; nest cavity is excavated by both pair members in dead or diseased tree trunks or large branches; clutch—6–8; incubation (both sexes)—11 days; fledging—24–27 days post-hatching; post-fledging parental care—?; single brood per season.

CONSERVATION STATUS: Flickers have shown long-term (1966–1994) declines in breeding populations for reasons that are not understood.

RANGE: Nearly throughout temperate and boreal North America

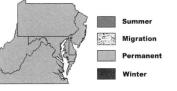

south in highlands of Mexico to Oaxaca. Leaves northern portion of breeding range in winter.

KEY REFERENCES: Moore (1995).

Pileated Woodpecker *Dryocopus pileatus*

DESCRIPTION: L = 43 cm (17 in); W = 69 cm (27 in); black, crow-sized bird; black and white facial pattern; red mustache and crest; white wing lining. **Female** Black mustache.

HABITAT AND DISTRIBUTION: Uncommon resident* throughout: Deciduous and mixed forest.

DIET: Forages by excavating large, oblong cavities in rotted snags, stumps, and logs for carpenter ants, beetle larvae, and other insect inhabitants of dead wood, using the long, sticky, barbed tongue to extract and capture prey; also gleans and pecks from bark surfaces and occasionally scales bark to expose prey; some fruits and mast taken when available.

REPRODUCTION: Monogamous, perhaps with the same mate for life; breeding begins with excavation of the cavity (both sexes) in March or April, which takes 3–6 weeks; nest holes constructed 15–20 m (50–70 ft) up in trees that are living (though often infected with fungus); eggs are laid in May or early June; clutch—4; incubation (both sexes)—18 days?; fledging—24–31 days post-hatching; post-fledging parental care—"several months;" single brood per year.

CONSERVATION STATUS: S3—DE; destruction of floodplain forest.

RANGE: Much of temperate and boreal North America excluding Great Plains and Rocky Mountain regions.

▨	**Summer**
▦	**Migration**
▨	**Permanent**
▨	**Winter**

KEY REFERENCES: Bull and Jackson (1995).

ORDER PASSERIFORMES; FAMILY TYRANNIDAE

Olive-sided Flycatcher *Contopus cooperi*

DESCRIPTION: L = 20 cm (8 in); W = 33 cm (13 in); olive above; white below with olive on sides of breast and belly; white tufts on lower back sometimes difficult to see.

HABITAT AND DISTRIBUTION: Uncommon to rare transient (Sep, May) nearly throughout; casual in western WV. Rare and local summer resident* (May–Sep) in the mountains of PA, eastern WV (Randolph and Pocahontas counties), and VA (Mount Rogers): Breeds in hemlock and spruce-fir forests where it often perches on lone, tall, dead snags; found in a variety of forested habitats during migration.

DIET: Forages by sallying from perches in forest openings and canopies for flying insects.

REPRODUCTION: Mostly monogamous; male arrives on breeding territory in late May or early June followed in 1–2 weeks by female; open cup nest, 12 cm dia. (5 in) is placed, often in a twig cluster, on the horizontal branch of a conifer (spruce, fir, hemlock) and built (female) of twigs, rootlets, and lichens lined with fine grasses and pine needles; clutch—3; incubation (female)—15–16 days; fledging—19–21 days post-hatching; post-fledging parental care—10–17 days; single brood per season.

CONSERVATION STATUS: S1—WV; SH—MD, VA; short-term (1980–1994) and long-term (1966–1994) declines have been recorded in the Mid-Atlantic region and range-wide, perhaps due to winter habitat loss, although highland spruce-fir forest in VA and MD is threatened by various factors.

RANGE: Breeds in boreal forest across the northern tier of North America and through mountains in west south to TX and northern Baja California; winters in mountains of South America from Colombia to Peru.

▨	**Summer**
▦	**Migration**
▨	**Permanent**
▨	**Winter**

KEY REFERENCES: Altman and Sallabanks (2000).

Eastern Wood-Pewee *Contopus virens*

DESCRIPTION: L = 15 cm (6 in); W = 26 cm (10 in); dark olive above; greenish wash on breast; belly whitish; white wing bars; slight crest.

HABITAT AND DISTRIBUTION: Common transient and summer resident* (May–Sep) throughout: Deciduous and mixed woodlands.

DIET: Forages by sallying for flying insects (e.g., Diptera, Homoptera, Lepidoptera, Hymenoptera) from open perches usually in woodland subcanopy (8–15 m up) (26–49 ft); sometimes eats fruits.

REPRODUCTION:

Monogamous; pairs arrive on breeding territory in May; open cup nest, 8–9 cm dia. (3–4 in), is placed 5–20 m (16–66 ft) up in a tree on a horizontal branch, usually 1–3 m (3–10 ft) from the trunk; nest is built of grass, plant fibers, bark strips, leaves, twigs, moss, and rootlets, covered often with lichens and spider silk, and lined with fine grass, moss, and hair; clutch—3; incubation (female)- 12–13 days?; fledging—16–18 days? post-hatching; single brood per season?

CONSERVATION STATUS: Breeding populations have declined sharply since 1966 for reasons that are not known.

RANGE: Breeds in the eastern United States and southeastern Canada; winters in northern South America.

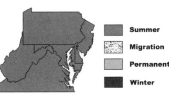

| | Summer |
| Migration |
| Permanent |
| Winter |

KEY REFERENCES: McCarty (1996).

Yellow-bellied Flycatcher *Empidonax flaviventris*

DESCRIPTION: L = 13 cm (5 in); W = 20 cm (8 in); greenish above; yellowish below; white wing bars and eyering; black legs.

HABITAT AND DISTRIBUTION: Rare and local summer resident* (May–Sep) in northern PA; scattered breeding records for WV (Cranberry Back Country Wilderness, Pocahontas County), and VA (Mount Rogers); rare transient (Sep, May) elsewhere: Breeds in boreal bogs in hemlock and spruce-fir forests; on migration in a variety of forest and shrublands.

DIET: Forages by hawking from low perches in forest understory for flying insects or hover-gleaning insects, especially from leaf undersurfaces; also takes some small fruits seasonally.

REPRODUCTION: Monogamous; males arrive on breeding territory in May; open cup nest, 8 cm dia. (3 in), is placed on the ground, usually in a depression in sphagnum moss or similar soft substrate, and built (female) of rootlets, moss, plant fibers, twigs, pine needles, and sedge stalks lined with fine rootlets, fine sedge strips, and hair-like fungus; clutch—3–4; incubation (female)—15 days?; fledging—13 days post-hatching?; post-fledging parental care—> 9 days, poorly known; Appalachian populations may be double-brooded.

CONSERVATION STATUS: S1— PA, WV, VA; bog breeding habitat is limited and threatened.

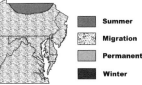

| | Summer |
| Migration |
| Permanent |
| Winter |

RANGE: Breeds along the extreme northern tier of the eastern United States and in Canada south of the Arctic west to British Columbia; winters from central Mexico south to Panama.

KEY REFERENCES: Gross and Lowther (2001).

Acadian Flycatcher *Empidonax virescens*

DESCRIPTION: L = 15 cm (6 in); W = 23 cm (9 in); olive green above; whitish below; yellowish flanks; white wing bars; whitish eyering; bluish legs.

HABITAT AND DISTRIBUTION: Common to uncommon transient and summer resident* (May–Sep) throughout; somewhat scarcer as a breeder in northern PA and northern NJ: Mature deciduous and mixed forest, especially bottomlands, swamps, and riparian forest.

DIET: Forages 2–12 m (7–40 ft) up in forest midstory by sallying from perches to glean insect prey from under surfaces of leaves or to capture flying insects; also takes small fruits seasonally.

REPRODUCTION: Mostly monogamous; males arrive on breeding territory in May followed 4–7 days later by females; open cup nest, 8 cm dia. (3 in), is placed in the fork near the end of a horizontal branch of an understory tree, shrub, or sapling, often over an open area, and built by the female of caterpillar and spider silk, bark strips, plant fibers, catkins, twigs, and grape tendrils, and lined with fine grasses and plant fibers; clutch—3; incubation (female)—13–15 days; fledging—12–18 days post-hatching; post-fledging parental care—3 weeks; some double-brooding.

RANGE: Breeds across the eastern United States; winters from Nicaragua south through Central America to northern South America.

KEY REFERENCES: Whitehead and Taylor (2002).

Alder Flycatcher *Empidonax alnorum*

DESCRIPTION: L = 15 cm (6 in); W = 23 cm (9 in); brownish-olive above, whitish below; greenish on flanks; white wing bars and eyering.

HABITAT AND DISTRIBUTION: Uncommon summer resident* (May–Sep) in northern and western PA and NJ; rare and local summer resident* (May–Sep) in the highlands of WV, western MD, and VA; rare transient (Sep, May) elsewhere: Swamp thickets, bogs, riparian forest.

DIET: Forages by hawking or sallying for flying insects from perches relatively low (2–5 m) (7–23 ft) in forest or scrub habitats; also hover-gleans or gleans arthropods from foliage; sometimes eats fruit during the nonbreeding period.

REPRODUCTION: Monogamous; breeding biology is poorly known; bulky open cup nest, 8 cm dia. (3 in), is located low (1–2 m) (3–7 ft) in a shrub or sapling, and built by the female in late May or early June of plant down, grass, moss, twigs, plant stems, and bark strips and lined with grass and

pine needles; clutch—3–4; incubation (both sexes)—11 days?; fledging—14 days post-hatching; post-fledging parental care—?; single brood per season; Alder Flycatchers are among the latest songbirds to arrive on their northern breeding territories and among the earliest to depart on southward migration because most of the pre-Basic molt occurs on the wintering sites.

CONSERVATION STATUS: S1—VA; S2—MD, NJ; S3—PA, WV; breeding habitat for Alder Flycatchers is limited to a few highland boreal bog sites in VA, WV, and MD, making the species particularly vulnerable to extirpation in these states.

RANGE: Breeds across the northern tier of the continent in Canada and

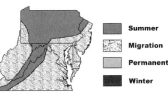

Summer
Migration
Permanent
Winter

AK and south through the Appalachians to NC; winters in northern South America.

KEY REFERENCES: Lowther (1999).

(3–10 ft) for flying insects or hover-gleaning from plant surfaces; also takes some small fruits seasonally.

REPRODUCTION: Mostly monogamous; male arrives on breeding territory in May; open cup nest, 8 cm dia. (3 in), is placed in the crotch of a bush or shrub and built (female) of grasses, bark shreds, and plant fibers lined with fine grasses, hair, feathers, and rootlets; clutch—3–4; incubation (female)—12–15 days; fledging—14–15 days post-hatching; post-fledging parental care—14 days; single brood per season.

CONSERVATION STATUS: S3—DE; loss of floodplain forest breeding habitat.

RANGE: Breeds through much of north and central

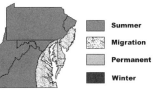

Summer
Migration
Permanent
Winter

United States; winters in northern South America.

KEY REFERENCES: Sedgwick (2000).

Willow Flycatcher *Empidonax traillii*

DESCRIPTION: L = 15 cm (6 in); W = 23 cm (9 in); brownish-olive above; whitish below turning to a creamy yellow on belly; greenish on flanks; white wing bars and eyering.

HABITAT AND DISTRIBUTION: Uncommon transient and summer resident* (May–Sep) in the Piedmont and highlands; uncommon to rare transient and rare summer resident* (May–Sep) in the Coastal Plain: Willow thickets, shrubby fields, riparian thickets.

DIET: Forages by sallying from low perches (1–3 m)

Least Flycatcher *Empidonax minimus*

DESCRIPTION: L = 13 cm (5 in); W = 20 cm (8 in); brownish above; white below; wing bars and eyering white. Bobs tail.

HABITAT AND DISTRIBUTION: Uncommon transient and summer resident* (May–Sep) in PA and the Piedmont and Appalachian highlands elsewhere; uncommon to rare transient and rare summer resident* (May–Sep) in the Coastal Plain: Open woodlands.

DIET: Forages by hawking and sallying for flying insect prey (e.g., Hymenoptera,

Coleoptera, Diptera) from perches 1–15 m (3–50 ft) above ground; also gleans or hover-gleans prey from leaf surfaces; occasionally feeds on other arthropods (e.g., spiders) and small fruits.

REPRODUCTION: Mostly monogamous with some serial polygyny; male arrives on breeding territory in early May; female arrives a few days later; open cup nest, 7–9 cm dia. (3–4 in), is usually placed in the upright crotch or fork of a sapling or tree at a height of 3–15 m (10–50 ft); nest is built by female of grasses, bark fibers, plant down, leaves, lichens, moss, spider silk, and hair lined with plant down, fine grasses, and hair; clutch—4; incubation—13–15 days; fledging—12–17 days post-hatching; post-fledging parental care—?; single brood per season is normal, double brooding has been documented.

CONSERVATION STATUS: S3—NJ, MD; SH—DE; Least Flycatchers have shown both short (1980–1994) and long (1966–1994) term declines in breeding populations in the Mid-Atlantic region for reasons that are not understood.

RANGE: Breeds in northern United States and southern and central Canada west 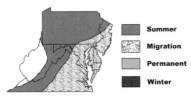 to British Columbia, south in Appalachians to north GA; winters from central Mexico to Panama.

KEY REFERENCES: Briskie (1994).

Eastern Phoebe *Sayornis phoebe*

DESCRIPTION: L = 18 cm (7 in); W = 28 cm (11 in); dark brown above; whitish or yellowish below. Dark cap often has crest-like appearance. Bobs tail.

HABITAT AND DISTRIBUTION: Common transient and summer resident* (Mar–Oct) throughout; rare winter resident (Nov–Feb), mainly along the coast: Riparian forest; deciduous and mixed forest edge; good nest site localities seem to determine breeding habitat—cliffs, eaves, bridges.

DIET: Forages by sallying from perches (< 10 m above ground) (30 ft) for flying insects (e.g., Hymenoptera, Coleoptera, Orthoptera); also feeds on other arthropods, fruits, and even small fish on occasion.

REPRODUCTION: Mostly monogamous but with some evidence of polygyny; pairs form with arrival on breeding territories in March or early April; bulky, open cup nest, 11–12 cm dia. (4–5 in), is placed on a covered ledge (e.g., building eaves, cliffs, culverts, bridges); nest is built by the female of mud, moss, lichens, and leaves, and lined with fine grass and hair; clutch—5; incubation—16 days; fledging—16 days post-hatching; post-fledging parental care—2 weeks?; 2 broods per season are common.

RANGE: Breeds from eastern British Columbia across central and southern  Canada south through the eastern United States (except southern Coastal Plain); winters in the southeastern United States south through eastern Mexico to Oaxaca and Veracruz.

KEY REFERENCES: Weeks (1994).

Great Crested Flycatcher *Myiarchus crinitus*

DESCRIPTION: L = 23 cm (9 in); W = 33 cm (13 in); brown above; throat and breast gray; belly yellow; rufous wings and tail; two white wingbars; slight crest.

HABITAT AND DISTRIBUTION: Common transient and summer resident* (May–Sep) throughout: Deciduous and mixed forest.

DIET: Forages by hawking or sallying for flying insects (e.g., Lepidoptera, Coleoptera, Orthoptera); eats some fruits during migration and wintering periods.

REPRODUCTION: Monogamous; males arrive on breeding territory in late April or early May, females 7–12 days later; cavity nest is placed in a natural hole, abandoned woodpecker hole, or nest box; female builds nest of leaves, feathers, roots, bark strips, plant stems, hair, rootlets, grass, and twigs; most nests contain a shed snake skin; clutch—5; incubation (female)—13–15 days; fledging—13–15 days post-hatching; post-fledging parental care—3 weeks; single brood per season.

RANGE: Breeds in the eastern United States, central and southeastern

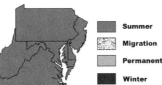

Canada; winters in southern FL, Cuba, and southern Mexico through Middle America to Colombia and Venezuela.

KEY REFERENCES: Lanyon (1997).

Western Kingbird *Tyrannus verticalis*

DESCRIPTION: L = 23 cm (9 in); W = 38 cm (15 in); pearl-gray head and breast; dark pre- and postorbital stripe; back and belly yellowish; wings dark brown; tail brown with white outer tail feathers; red crown patch usually concealed.

HABITAT AND DISTRIBUTION: Rare fall transient (Sep–Oct) along the coast: Grasslands, savanna, pastures, old fields.

DIET: Forages mainly from low perches in open areas by sallying, hawking, or hover-gleaning insects (e.g., Hymenoptera, Hemiptera, Orthoptera. Lepidoptera); also eats fruits on occasion.

RANGE: Breeds in western North America from southern British Columbia and Manitoba to northern Mexico; winters from southern Mexico to Costa Rica.

KEY REFERENCES: Gamble and Bergin (1996).

Eastern Kingbird *Tyrannus tyrannus*

DESCRIPTION: L = 23 cm (9 in); W = 38 cm (15 in); blackish above; white below; terminal white band on tail; red crest is usually invisible.

HABITAT AND DISTRIBUTION: Common transient and summer resident* (May–Aug) throughout: Wood margins, open farmland, hedgerows, savanna, marshes.

DIET: Forages by hawking flying insects from low perches (1–4 m) (3–13 ft), e.g., shrubs or fence wire; also feeds on fruit when available, especially during migration and on wintering grounds.

REPRODUCTION: Monogamous, but with some extra-pair copulations; male is aggressive to intruders on breeding territory (hence the "king" moniker); divebombs any large bird that happens to cross

his airspace, and harasses mammals, including humans, that trespass; males arrive on breeding territory in May, followed shortly thereafter by females; open, elliptical, cup nest, 12x13 cm) (5 x 5 in) is placed2–8 m (7–26 ft) above ground in horizontal branches of a tree or shrub, often a hawthorn (*Crataegus*), and is built by the female of twigs, roots, plant stems, and bark strips, lined with plant down, rootlets, and hair; clutch—3–4; incubation (female)—14–17 days; fledging—16–17 days post-hatching; post-fledging parental care—5–6 weeks; single brood per season.

RANGE: Breeds from central Canada south through eastern and central United

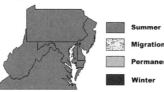

	Summer
	Migration
	Permanent
	Winter

States to eastern TX and FL; winters in central and northern South America.

KEY REFERENCES: Morton (1971), Murphy (1996).

FAMILY LANIIDAE.

Loggerhead Shrike *Lanius ludovicianus*

DESCRIPTION: L = 23 cm (9 in); W = 33 cm (13 in); gray above, paler below; black mask, wings, and tail; white outer tail feathers and wing patch.

HABITAT AND DISTRIBUTION: Uncommon to rare and local resident* in VA west of the Coastal Plain, MD (Washington and Frederick counties), south central PA, and eastern and southern WV; rare to

casual visitor* elsewhere in the region: Old fields, grassland, farmland, savanna, hedgerows; prefers open, barren areas where perches (such as fence-posts or telephone lines) are available for sallying.

DIET: Hunts by scanning from exposed perches, then sallying to capture and kill prey with its sharp, powerful bill; once captured, small prey (e.g., insects) are often eaten on the spot, while larger prey (e.g., large insects, small mammals, birds, reptiles, amphibians) are grasped with the feet, carried to a sharp thorn, barbed wire, or shrub crotch, and impaled or wedged; impaling or wedging of prey is thought to serve two purposes: 1) provide a convenient means for stripping edible parts from larger prey, and 2) storing prey for later consumption.

REPRODUCTION: Mostly monogamous; males may remain on or near breeding territories throughout the year in the Mid-Atlantic; nesting begins in later March or April; open cup nest, 15 cm dia. (6 in), is placed low (1–3 m up) (3–10 ft) in the crotch or a thorny tree or shrub, and is built (mostly female) of sticks, twigs, and plant strips woven together with bark strips and rootlets and lined with grasses, hair, moss, feathers, and lichens; clutch—5; incubation (female)—15–17 days; fledging—18 days post-hatching; post-fledging parental care—3–4 weeks?; 2 broods per season are common in sedentary populations.

CONSERVATION STATUS: S1—WV, PA, MD; S2—VA; SH—DE; many breeding populations of Loggerhead Shrikes have declined significantly over the past half century; suggested reasons include reversion of old field habitat to forest, conversion of grasslands to intensive farming, and pesticides.

RANGE: Breeds locally over most of the United States and southern Canada south  through highlands of Mexico to Oaxaca; winters in all but north portions of breeding range.

KEY REFERENCES: Yosef (1996).

Northern Shrike *Lanius excubitor*

DESCRIPTION: L = 26 cm (10 in); W = 36 cm (14 in); pearly gray above, faintly barred below; black mask, primaries, and tail; white wing patch and outer tail feathers; heavy, hooked bill; often pumps tail while sitting on an exposed perch.

HABITAT AND DISTRIBUTION: Rare and irregular winter visitor (Nov–Feb) to NJ; casual elsewhere; more numerous in some years than others: Open woodlands and farmland, coastal marshes, savanna, forest edge, old fields, hedgerows.

DIET: Forages by locating prey (e.g., large insects, small birds, rodents, and lizards) from an exposed perch, then flies to capture it in the air or on the ground, or flies, hovers, and then drops to the ground to capture prey; vertebrates are killed by biting the neck; larger prey are often carried to a spine, crotch, or wire barb to facilitate dismem-

berment for consumption; casts pellets.

RANGE: Breeds across Arctic and sub-Arctic regions of the Old and New World; winters in north temperate and southern boreal regions.

KEY REFERENCES: Cade and Atkinson (2002).

FAMILY VIREONIDAE

White-eyed Vireo *Vireo griseus*

DESCRIPTION: L = 13 cm (5 in); W = 20 cm (8 in); greenish-gray above, whitish washed with yellow below; white eye, yellowish eyering, and forehead ("spectacles"); white wingbars. **Immature** Has brown eye until Aug–Sep.

HABITAT AND DISTRIBUTION: Common to uncommon summer resident* (May–Sep) except in highlands, northern PA, and northern NJ where scarce: Riparian woodland, thickets, swamp borders, overgrown fields.

DIET: Forages deliberately using short hops or flights to pick or glean arthropods (e.g., Lepidoptera, Diptera, Coleoptera, spiders) or fruits from twigs and leaves. Regurgitates larger seeds.

REPRODUCTION: Monogamous; male arrives on breeding territory in April, followed by female; pendulous, open cup nest, 8 cm dia. (3 in), is placed low (1–2 m) (3–7 ft) in a shrub or tree; both sexes build the nest, which is attached and held together by spider and caterpillar silk, and constructed of leaves, bark shreds, rootlets, paper bits, plant down, and lichens lined with fine grass, rootlets, and hair; clutch—3–4; incubation (both sexes)—13–15 days; fledging—9–11 days post-hatching; post-fledging parental care—23 days; 2 broods per season are common.

CONSERVATION STATUS: Declines in some breeding populations have occurred, perhaps due to loss of shrub and thicket habitat resulting from reversion to forest or conversion to intensive agriculture.

RANGE: Breeds in the eastern United States from southern Minnesota to MA southward to FL and eastern Mexico to Veracruz. Winters from the southeastern United States south through eastern Mexico and Central America to northern Nicaragua; Bahamas and Greater Antilles.

KEY REFERENCES: Hopp et al. (1995).

Yellow-throated Vireo *Vireo flavifrons*

DESCRIPTION: L = 15 cm (6 in); W = 26 cm (10 in); greenish above with gray rump; yellow throat, breast, and "spectacles" (lores, forehead, eyering); white belly and wingbars.

HABITAT AND DISTRIBUTION: Common to uncommon transient and summer resident* (May–Sep) nearly throughout (scarce in swamplands of southern NJ and highlands of central PA and

southeastern WV): Deciduous and mixed forest, riparian woodland.

DIET: Forages usually in forest canopy or subcanopy, carefully scanning bark and leaf surfaces, then flitting, hopping, or flying to prey proximity and gleaning or picking prey (e.g., Lepidoptera, Hemiptera, Coleoptera) from substrate; takes fruit when available.

REPRODUCTION: Monogamous; males arrive on breeding territory in May, followed by female; pensile nest, 7–9 cm dia. (3–4 in), is suspended from the horizontal fork of a branch usually high in a deciduous tree, and built by both sexes of bark strips, grass, rootlets, plant down, and pine needles, bound together with caterpillar and spider silk, and lined with fine grass, bark strips, and rootlets; clutch—3–4; incubation (both sexes)—13 days; fledging—13 days post-hatching; post-fledging parental care—at least 4 weeks, perhaps longer; single brood per season.

CONSERVATION STATUS: S3—DE; loss of floodplain forest.

RANGE: Breeds in the eastern United States. Winters from southern Mexico, south through Central and northern South America; Bahamas; Greater Antilles.

KEY REFERENCES: Rodewald and James (1996).

Blue-headed Vireo *Vireo solitarius*

DESCRIPTION: L = 15 cm (6 in); W = 26 cm (10 in); greenish on the back with a gray rump; whitish below with yellow flanks; 2 prominent, white wingbars; gray head and white "spectacles" (lores, forehead, eyering).

HABITAT AND DISTRIBUTION: Common transient and summer resident* (Apr–Oct) in highlands and northern PA; uncommon to rare transient (Sep–Oct, Apr) elsewhere; rare winter visitor (Nov–Mar), mainly along the coast: Coniferous and mixed woodlands.

DIET: Forages deliberately, usually in midcanopy, examining leaves, twigs, and branches for arthropods; captures prey by gleaning, picking, or sallying; fruits form a variable portion of the diet during the nonbreeding period, when individuals often join mixed-species foraging flocks.

REPRODUCTION: Monogamous; male arrives on breeding territory in April followed one to several days later by the female; pensile nest, 7–9 cm dia. (3–4 in), is suspended from the fork of horizontal branches of a sapling 2–5 m (7–16 ft) above ground and built by both pair members of bark strips, lichens, arthropod silk, moss, hornet nest bits, leaves, hair, and feathers lined with fine grasses, rootlets, and conifer needles; clutch—4; incubation (both sexes)—13–14 days; fledging—12–13 days post-hatching; post-fledging parental care—probably 2–4 weeks; single brood per season.

CONSERVATION STATUS: S3—NJ, MD; loss of mature highland coniferous and mixed forest.

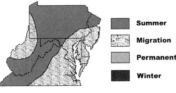

RANGE: Breeds across central and southern Canada, northern and western United States. Winters from the southern United States through Mexico and Central America to Costa Rica; Cuba.

KEY REFERENCES: James (1998), Morton et al. 1998.

Warbling Vireo *Vireo gilvus*

DESCRIPTION: L = 15 cm (6 in); W = 23 cm (9 in); grayish-green above, pale yellow below; white eye stripe; buffy cheek; no wingbars.

HABITAT AND DISTRIBUTION: Uncommon transient and local summer resident* (May–Aug) in Piedmont and highlands; rare transient and rare to casual summer resident* along the Coastal Plain: Mature deciduous woodland; riparian and bottomland forest; often frequents tall elms, cottonwoods, and sycamores.

DIET: Forages mostly in canopy and subcanopy, gleaning arthropods from leaf surfaces, often by hovering; some fruits seasonally.

REPRODUCTION: Monogamous; pairs arrive on breeding territory in May; pendant nest, 7–8 cm dia. (3 in), is placed in the fork of a horizontal branch of a deciduous tree, and built (female) of arthropod silk, bark strips, plant down, and fibers, hair, twigs, lichen, and rootlets lined with fine grasses, feathers, plant fibers, rootlets, pine needles, and leaves; clutch—3–4; incubation (both sexes)—12–14 days; fledging—13–14 days post-hatching; post-fledging parental care—2 weeks; single brood per season in eastern populations.

CONSERVATION STATUS: S2—DE; loss of floodplain forest.

RANGE: Breeds across North America south of the Arctic region to central Mexico. Winters from Guatemala to Panama.

KEY REFERENCES: Gardali and Ballard (2000).

Philadelphia Vireo *Vireo philadelphicus*

DESCRIPTION: L = 13 cm (5 in); W = 20 cm (8 in); grayish-green above, variably yellowish below; white eye stripe and dark eyeline; no wingbars.

HABITAT AND DISTRIBUTION: Uncommon fall transient (Sep) and rare spring transient (May) throughout: Deciduous and mixed forest thickets, shrublands, and second growth.

DIET: Forages by deliberately gleaning, with much peering and poking, in outer clumps of leaves and twigs, even hanging occasionally like a chickadee; main prey are insects and spiders; some fruit when available.

RANGE: Breeds in eastern and central boreal North America in Canada and extreme northern United States (MN to ME). Winters Guatemala to Panama.

KEY REFERENCES: Moskoff and Robinson (1996).

Red-eyed Vireo *Vireo olivaceus*

DESCRIPTION: L = 15 cm (6 in); W = 26 cm (10 in); olive above, whitish below; gray cap; white line above eye; dark line through eye; red eye (brown in juvenile).

HABITAT AND DISTRIBUTION: Common transient and summer resident* (May–Sep) throughout: Deciduous and mixed woodlands.

DIET: Forages deliberately by gleaning arthropods from leaves in canopy and subcanopy; also feeds seasonally on fruits.

REPRODUCTION: Monogamous; male arrives on breeding territory in May followed within a few days by the female; pendant nest, 7–8 cm dia. (3 in), is placed in the fork of a horizontal branch of a tree or sapling and built (female) of bark shreds, spider silk, wasp-nest paper, grasses, twigs, plant fibers, and pine needles lined with fine grasses, pine needles, hair, and plant fibers; clutch—3; incubation (mostly female)—11–13 days; fledging—10–12 days post-hatching; post-fledging parental care—15–16 days; occasional second broods in southern populations.

RANGE: Breeds over much of North America (except western United States, AK, and northern Canada). Subspecies breed in Central and South America. United States subspecies winters in northern South America.

KEY REFERENCES: Cimprich et al. (2000).

FAMILY CORVIDAE

Blue Jay *Cyanocitta cristata*

DESCRIPTION: L = 28 cm (11 in); W = 41 cm (16 in); blue above; whitish below; blue crest; black nape

and lores; white wing bar and wing spots; black necklace; blue tail barred with black and tipped with white. Jays have the curious habit of imitating the calls of various hawks, e.g., the Broad-winged Hawk, apparently to startle a potential predator or facilitate kleptoparasitism of prey from another bird.

HABITAT AND DISTRIBUTION: Common resident* throughout: Deciduous, coniferous, and mixed forest, parklands, residential areas.

DIET: Forages by picking and gleaning large insects (e.g., Odonata, Orthoptera, Homoptera) and other invertebrates, nuts (e.g., acorns), fruits, and small vertebrates from the ground or from trees and shrubs; caches acorns and other nut or seed types for later use; often forages in small, single-species flocks.

REPRODUCTION: Monogamous; pair bond is maintained year-round; breeding begins in late March or early April; bulky open cup nest, 17–21 cm dia. (7–8 in), is placed 1–30 m (3–100 ft) above ground, usually on a horizontal branch of a tree, and built by both sexes of sticks, twigs, bark bits, moss, lichen, leaves, and grass lined with rootlets; clutch—4–6; incubation (female)—17–18 days; fledging—17–21 days post-hatching; post-fledging parental care—3–5 weeks?; some evidence of assistance with feeding

Summer
Migration
Permanent
Winter

nestlings by extrapair individuals, but this practice may be exceptional.

RANGE: Temperate and boreal North America east of the Rockies.

KEY REFERENCES: Tarvin and Woolfenden (1999).

American Crow *Corvus brachyrhynchos*

DESCRIPTION: L = 46 cm (18 in); W = 92 cm (36 in); black throughout.

HABITAT AND DISTRIBUTION: Common resident* throughout: Woodlands, grasslands, farmlands; riparian and bottomland forest and wooded residential areas used as migration or winter roosts.

DIET: Forages opportunistically for carrion and refuse from roadsides, parks, urban areas, and landfills; feeds on waste grain and invertebrates at agricultural fields; preys on eggs and young of birds; caches food items for later use.

REPRODUCTION: Monogamous, sometimes with one or more helpers assisting with raising of offspring; pairs begin nesting in March; open platform nest is placed in a crotch near the top of a tree, often a conifer, and built by both sexes of sticks, bark, plant stems, stalks, and husks, roots, and leaves, lined with bark strips, plant fibers, hair, moss, rootlets, grass, and leaves; clutch—4–5; incubation (female)—16–18 days; fledging—5 weeks post-hatching; post-fledging parental care—young birds can remain in family groups for weeks or months after fledging; single brood per season.

CONSERVATION STATUS: The American Crow seems to have been hit harder by appearance of West Nile virus in the Western Hemisphere than any other bird species, with some local populations suffering declines of 40% or more (Marra et al. 2004).

RANGE:

Temperate and boreal North America; migratory in

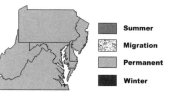

Summer

Migration

Permanent

Winter

northern and western montane portions of range.

KEY REFERENCES: Verbeek and Caffrey (2002).

Fish Crow *Corvus ossifragus*

DESCRIPTION: L = 38 cm (15 in); W = 84 cm (33 in); black throughout.

HABITAT AND DISTRIBUTION: Common (Coastal Plain) to uncommon or rare (Piedmont) resident*, withdraws toward the coast and southward in winter: Floodplain forests, bayous, coastal waterways, estuaries, beaches, marshes, farm lands, urban and suburban areas.

DIET: Forages opportunistically for carrion and refuse from roadsides, beaches, parks, urban areas, and landfills; preys on eggs and young of colonial birds; caches food items for later use.

REPRODUCTION: Monogamous with occasional helper at nest; pairs apparently form at winter roosts, and leave to establish breeding territories in April; open platform nest, 0.5 m dia. (20 in), is usually placed on horizontal branches against the main trunk, high in a tree, often a conifer, and built by both sexes of sticks, and lined with bark

strips, Spanish moss, hair, pine needles, and plant fibers; clutch—4–5; incubation (female)—18–19 days; fledging—32–40 days post-hatching; post-fledging parental care—at least 1 month; single brood per season.

CONSERVATION STATUS: S3—WV; limited habitat at periphery of range.

RANGE: Coastal Plain and parts of the Piedmont of the eastern United States

Summer

Migration

Permanent

Winter

from ME to eastern TX.

KEY REFERENCES: McGowan (2001).

Common Raven *Corvus corax*

DESCRIPTION: L = 61 cm (24 in); W = 122 cm (48 in); black throughout; large heavy bill; shaggy appearance in facial region; wedge-shaped tail.

HABITAT AND DISTRIBUTION: Uncommon resident* of highlands in the Appalachian and Blue Ridge mountains throughout: Rugged crags, cliffs, canyons as well as a variety of forested and open lands.

DIET: Forages, often in groups, for carrion, especially roadkills or large predator kills (e.g., deer); also feeds opportunistically on a variety of plant (e.g., fruits, seeds, crops) and animal (e.g., eggs, nestlings, insects, small mammals) prey; casts pellets; steals food on occasion from other predators (kleptoparasitism); caches food for later consumption.

REPRODUCTION: Monogamous; year-round pair bond; breeding begins in February; open platform nest, 0.4–1.5 m dia. (16–60 in), is located on a cliff ledge, tree crown, or comparable artificial site (e.g., a tower or telephone pole), and built by both pair members of sticks lined with a central cup woven of twigs, hair, mud, grass, bark shreds; nests are reused year after year and often have "an unpleasant odor" from rotted prey remains; clutch—5; incubation (female)—20–25 days; fledging—4–7 weeks post-hatching; post-fledging

parental care—young may remain with adults for several months after leaving the nest, foraging and roosting together; single brood per season.

CONSERVATION STATUS: S2—MD; S1—NJ; populations were extirpated from much of the Mid-Atlantic region by the early twentieth century, but have rebounded, at least in many highland areas; Ravens are subject to control measures in other parts of the range, due to supposed effects on domestic livestock (e.g., plucking the eyes from newborn lambs) and crops.

RANGE: Arctic and boreal regions of the Northern Hemisphere, south through

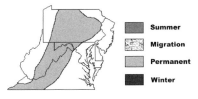

Summer
Migration
Permanent
Winter

mountainous regions of the Western Hemisphere to Nicaragua, and in the Eastern Hemisphere to North Africa, Iran, the Himalayas, Manchuria, and Japan.

KEY REFERENCES: Boarman and Heinrich (1999).

FAMILY ALAUDIDAE

Horned Lark *Eremophila alpestris*

DESCRIPTION: L = 18 cm (7 in); W = 33 cm (13 in); brown above; white below with black bib; face and throat whitish or yellowish with black forehead and eyeline; black horns raised when singing. **Female** Similar pattern but paler. **Immature** Nondescript brownish above; whitish below with light streaking; has dark tail with white edging and long hind claw like adult.

HABITAT AND DISTRIBUTION: Locally common to uncommon resident*; scarce in winter in highlands: Overgrazed pasture, airports, golf courses, plowed fields, surface mines, roadsides, sand flats.

DIET: Mostly granivorous, but feeds young on invertebrates; forages by walking on ground in open areas and gleaning plant seeds and invertebrates, or occasionally chasing flushed prey.

REPRODUCTION: Monogamous; pair formation begins in late January or February with first clutch laid in March or April; open cup nest, 8–10 cm dia. (3–4 in), is placed in a hole, cavity, or depression in the bare ground excavated by the female, and built of woven grass, shredded plant stalks, stems, and roots, and lined with plant down, hair, feathers, and rootlets; clutch—3; incubation (female)—11–12 days; fledging— young leave the nest at 10 days post-hatching, but cannot fly until several days later; post-fledging parental care—3–4 weeks; two broods per season are common.

CONSERVATION STATUS: S2—WV; S3—NJ; clearing of land during the settlement period resulted in a significant expansion of Horned Lark breeding range; subsequent reforestation resulting from abandonment of farm land has resulted in a loss of breeding habitat.

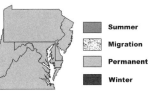

Summer
Migration
Permanent
Winter

RANGE: Breeds in North America south to southern Mexico; winters in southern portion of breeding range.

KEY REFERENCES: Beason (1995).

FAMILY HIRUNDINIDAE

Purple Martin *Progne subis*

DESCRIPTION: L = 20 cm (8 in); W = 41 cm (16 in); iridescent midnight blue throughout. **Female and Immature** Dark, iridescent blue above; dirty white, occasionally mottled with blue, below; grayish collar.

HABITAT AND DISTRIBUTION: Locally common transient and summer resident* (Mar–Sep) nearly throughout; scarcer in highlands: Forests and open areas—limited by nesting sites, which are almost entirely artificial for eastern populations.

DIET: Forages by capturing flying insects on the wing from 0–150 m (0–500 ft) above ground, often flying higher than other swallow species.

REPRODUCTION: Mostly monogamous although there is extensive evidence of extrapair fertilizations, especially for mates of yearling males (47%); males arrive at breeding colony site in March with females arriving up to 4 weeks later; almost all nests for eastern martins are placed in artificial martin houses with multiple nest holes; prior to European settlement, eastern martins likely were not colonial, using natural cavities in trees or abandoned woodpecker holes, as is still done by western martins; nest is built inside the

hole, mostly by the female, of twigs, grass, leaves, and mud, lined with green leaves; clutch—4–5; incubation (mostly female)—15–18 days; fledging—28–29 days post-hatching; post-fledging parental care—7–10 days; single brood per season.

CONSERVATION STATUS: Purple Martins have undergone short-term (1980–1994) declines in the Mid-Atlantic region for reasons that are not understood.

RANGE: Breeds in open areas over most of temperate North America

south to the highlands of southern Mexico; winters in South America.

KEY REFERENCES: Morton et al. (1990), Brown (1997).

Tree Swallow *Tachycineta bicolor*

DESCRIPTION: L = 15 cm (6 in); W = 33 cm (13 in); iridescent blue above; pure white below with slightly forked tail. **Female** Similar in pattern to male, but duller. **Immature** Grayish-brown above; whitish below.

HABITAT AND DISTRIBUTION: Common transient (Aug–Oct, Apr), and common to uncommon and local summer resident* (May–Aug) throughout: Lakes, ponds, marshes, open fields—any open area during migration. Limited in summer by breeding sites (holes in hollow trees or nest boxes).

DIET: Mostly flying insects; berries and other plant material when arthropod prey is not available.

REPRODUCTION: Males arrive and begin defense of nest site (box or hole in tree) mid-March to early April; females arrive up to 7 days later; monogamous with some polygyny (5%) and extrapair copulations (up to 50% of offspring produced in some populations); nest is built in late April and early May almost entirely by the female, although the male may bring material; the nest is placed in a cavity 2–3 m above the ground or water, and built of grasses, rootlets, mosses, pine needles, and often lined with feathers from other species (e.g., waterfowl, domestic fowl); clutch size—5–6; incubation—14–15 days; young altricial, fledging—18–22 days post-hatching; young are fed by adults for up to 3 days post-fledging. Usually only one brood is produced per season.

CONSERVATION STATUS: Tree Swallows have shown short-term declines (1980–1994) in the Eastern Region.

RANGE: Breeds over most of boreal and north temperate North America; winters from southern United States south to Costa Rica and Greater Antilles.

KEY REFERENCES: Robertson et al. (1992).

Northern Rough-winged Swallow
Stelgidopteryx serripennis

DESCRIPTION: L = 15 cm (6 in); W = 31 cm (12 in); brown above; grayish-brown throat and breast becoming whitish on belly.

HABITAT AND DISTRIBUTION: Common to uncommon transient and summer resident* (Apr–Sep) throughout: Lakes, rivers, ponds, streams, most open areas during migration.

DIET: Forages by capturing flying insects on the wing, often low and over water; occasionally gleans insects in flight from water surface.

REPRODUCTION: Monogamous; pairs arrive at nesting sites in April; nest is usually placed in a hole, burrow, or crevice in a stream bank, although nearly any horizontal tube in a wall (e.g., drainpipes) can be used; usually the pair

selects an existing hole, although excavation of their own burrow has been reported; nest is built close to the mouth of the burrow by the female of twigs, plant strips and stems, grass, and leaves lined with rootlets, fine grass, and often some green plant material; pairs usually nest singly or in small colonies; clutch—5–7; incubation (female)—16 days; fledging—17–21 days post-hatching; post-fledging parental care—?; single brood per season.

RANGE: Breeds locally over most of temperate North America; winters from southern TX and northern Mexico south to Panama.

KEY REFERENCES: DeJong (1996).

Bank Swallow *Riparia riparia*

DESCRIPTION: L = 15 cm (6 in); W = 31 cm (12 in); brown above; white below with brown band across breast and down to belly (like a "T.")

HABITAT AND DISTRIBUTION: Common to uncommon transient (Jul–Aug, May) throughout; uncommon summer resident* (May–Aug) in PA, MD (mostly in Chesapeake Bay area), and the Coastal Plain of VA; uncommon to rare and local summer resident elsewhere in the region. Lakes, rivers, ponds, most open areas during migration. Nests in holes in stream, river, coast, and quarry banks, normally in colonies.

DIET: Forages by flying over open areas, average height above ground or water = 15 m (50 ft) to capture flying insects.

REPRODUCTION: Monogamous; pair formation occurs as birds arrive at breeding colonies in April; burrow nest is placed in a vertical bank, usually in the vicinity of other nesting Bank Swallows (colonies 5–2,000 pairs), excavated by both pair members. The end of the 0.5 m (20 in) long burrow is lined with a nest mat composed of grass, feathers, rootlets, plant stems, and leaves; clutch—3–5; incubation (mostly by female)—13–16 days; fledging—18–21 days post-hatching; post-fledging parental care—10–12 days; single brood per season.

CONSERVATION STATUS: S2—WV, DE; S3—VA, MD; limited stream-bank breeding colony sites.

RANGE: Cosmopolitan, breeding over much of Northern Hemisphere continents; winters in the tropics of Asia, Africa, and the New World.

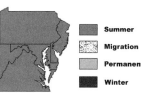

	Summer
	Migration
	Permanent
	Winter

KEY REFERENCES: Garrison (1999).

Cliff Swallow *Petrochelidon pyrrhonota*

DESCRIPTION: L = 15 cm (6 in); W = 31 cm (12 in); dark above; whitish below with dark orange or blackish throat; orange cheek; pale forehead; orange rump; square tail. **Immature** Similar to adults but duller.

HABITAT AND DISTRIBUTION: Uncommon to rare and local transient and summer resident* (Apr–Sep) nearly throughout; scarce or absent as a breeder on the Coastal Plain and most of NJ, where found only at a few colonies: Most open areas, generally near water. This species seems to be limited by nesting locations during the breeding season (farm buildings, culverts, bridges, cliffs—near potential mud source).

DIET: Forages, often in flocks, by capturing small flying insects (e.g., Hymenoptera, Hemiptera, Coleoptera, Diptera) on the wing.

REPRODUCTION: Monogamous, although copulation outside the pair bond is common; breeds in colonies; pairs form at colony sites with arrival of the birds in April or May; both sexes construct the domed, often gourd-shaped, mud nest, side entrance), lined with grass, and usually placed at the juncture of a vertical wall with a horizontal overhang (e.g., cliff, bridge abutment, barn eaves); clutch—3–4; incubation (both sexes)—13–15 days; fledging—21–26 days post-hatching; post-fledging parental care—3–5 days; single brood per season.

CONSERVATION STATUS: S1—DE; S2—NJ; S3—WV, VA; limited colony sites bordering wetlands.

RANGE: Breeds over most of North America to central Mexico; winters central and southern South America; apparently the breeding range was restricted to western North America until bridges, barns, and other struc-tures pro-vided nest sites allowing eastward

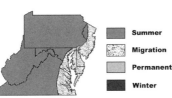

expansion during the past century.

KEY REFERENCES: Brown and Brown (1995).

Barn Swallow *Hirundo rustica*

DESCRIPTION: L = 18 cm (7 in); W = 33 cm (13 in); dark iridescent blue above; orange forehead and orange below with partial or complete blue breast band; long, deeply forked tail. **Immature** orange throat; white underparts; shorter tail.

HABITAT AND DISTRIBUTION: Common transient and summer resident* throughout: Open areas, usu-ally near water (requires mud for nest-building)

DIET: Forages by capturing flying insects in the air (usually < 10 above ground) (30 ft), sometimes following farm machinery or animals to capture flushed insects.

REPRODUCTION: Mostly monogamous but with some polygyny or extrapair paternity (22% in one pop-ulation); males generally arrive first at breeding sites followed within a few days by females; open platform nest, 13 cm x 8 cm x 5 cm deep (5 x 3 x 2 in), is located in a building, culvert, bridge abutment, or similar vertical wall or ledge with overhead protection (formerly caves or hollow trees); the character of nest sites often results in colonial nesting; the nest is built by both sexes of

mud balls, rolled at the mud source, carried in the mouth to the nest site, and stuck to the wall or other mud balls to create the platform; lining is grass and feathers; clutch—4–6; incubation (female)—14 days; fledging—21 days post-hatch-ing; post-fledging parental care—up to 2 weeks; two broods per season are common; adult males occasionally kill nestlings at neighboring nests; nonpair adults occasionally feed nestlings.

CONSERVATION STATUS: Barn Swallows have under-gone both short- (1980–1994) and long-term (1966–1994) declines in the Mid-Atlantic region for reasons that are not understood; successful breeding depends on the availability of mud, so breeding populations can be affected markedly by drought.

RANGE: Cosmopolitan—breeds over much of the Northern Hemisphere; winters in South America, Africa, north-

ern Australia, Micronesia; breeds occasionally on wintering grounds in South America, and has closely related resident populations in the African tropics.

KEY REFERENCES: Brown and Brown (1999).

FAMILY PARIDAE

Carolina Chickadee *Poecile carolinensis*

DESCRIPTION: L = 13 cm (5 in); W = 20 cm (8 in); black cap and throat contrasting with white cheek, gray back, and grayish-white breast and belly.

HABITAT AND DISTRIBUTION: Common resident* of the region except in northern PA and Appalachian highlands; overlaps with the Black-capped Chickadee in the foothills of the Appalachians: Deciduous and mixed woodlands, bottomland forest, southern lowland pine forest and savanna, residential areas.

DIET: Forages by gleaning arthropod eggs, larvae, and adults from twigs, branches, and bark of trees and shrubs, often hanging acrobatically; caches food for later use; takes seeds and fruit seasonally; associates in small foraging flocks with conspecifics and other species during the nonbreeding period.

REPRODUCTION: Monogamous; pairs form in winter flocks; breeding territory establishment occurs in March or April; nests in natural cavities or abandoned woodpecker holes, or excavates a cavity in a rotted snag (both sexes participate); female builds a nest within the cavity of moss, bark strips, hair, and plant fibers; clutch—6; incubation (female)—12–15 days; fledging—16–19 days post-hatching; post-fledging parental care—2–3 weeks; mostly single brood per season.

RANGE: Eastern United States from KS east to NJ south to FL and central TX.

Summer

Migration

Permanent

Winter

KEY REFERENCES: Mostrom et al. (2002).

Black-capped Chickadee *Poecile atricapilla*

DESCRIPTION: L = 13 cm (5 in); W = 20 cm (8 in); black cap, throat, and bib contrasting with white cheek; dark gray back, and grayish-white breast and belly; white edging to gray secondaries.

HABITAT AND DISTRIBUTION: Common resident* in PA (except extreme southwest and southeast corners), northern NJ, and the highlands of the Appalachian Mountains in VA, WV, and western MD; overlaps with the Carolina Chickadee in the foothills of the Appalachians; strays into neighboring lowlands in winter: Woodlands, residential areas.

DIET: Feeds mostly by gleaning and picking arthropods (often eggs or pupae) from bark cracks and surfaces while hanging upside down; also feeds on berries and seeds (mainly winter) and caterpillars (mainly summer).

REPRODUCTION: Monogamous; pair formation can occur any time of the year but peaks are in late summer or early fall and again in spring; pair bonds may persist for years; nests are built from late April to early June, placed in cavities, excavated in dead snags by both members of the pair, or in old woodpecker holes or nest boxes 1.5–7 m (5–23 ft) above the ground; the female lines the floor of the cavity with moss or other coarse plant material and then covers this with finer material, e.g., rabbit fur or plant down; clutch—6–8; incubation—12–13 days; fledging—16 days; post-fledging parental care, 3 to 4 weeks; normally one brood per season.

RANGE: Temperate and boreal North America south to CA,

Summer

Migration

Permanent

Winter

NM, OK, and NJ; south in Appalachians to NC.

KEY REFERENCES: Smith (1993).

Tufted Titmouse *Baeolophus bicolor*

DESCRIPTION: L = 15 cm (6 in); W = 23 cm (9 in); gray above; white below with buffy sides; gray crest; black forehead.

HABITAT AND DISTRIBUTION: Common resident* nearly throughout except at high elevations in the Appalachian Mountains; somewhat scarcer in northern PA: Deciduous forest.

DIET: Titmice use gleaning, pecking, and probing of bark, leaves, and leaf litter in trees, shrubs, and on the ground to search for arthropods, seeds, mast, and fruits; larger items, e.g., acorns, are held with the feet and hammered open with the bill.

REPRODUCTION: Monogamous; pairs form in winter flocks or persist through the year; nest construction begins in late March or April with first eggs laid in May; nest is placed in a cavity, often an abandoned woodpecker hole, and built of grass, leaves, bark strips, moss, and feathers lined with hair, plant down, and plant fibers; clutch—5–7; incubation—12–14 days; fledging—15–16 days post-hatching; post-fledging parental care—6 weeks, but some young remain with parents through the winter, sometimes serving as "helpers" the following year, assisting parents with raising of the brood for that season; single brood per season.

Summer
Migration
Permanent
Winter

RANGE: Eastern United States west to NE, IA, OK, and west TX, and in Mexico south to Hidalgo and northern Veracruz.

KEY REFERENCES: Grubb and Pravosudov (1994).

FAMILY SITTIDAE

Red-breasted Nuthatch *Sitta canadensis*

DESCRIPTION: L = 13 cm (5 in); W = 20 cm (8 in); gray above; orange buff below; dark gray cap with white line over eye; black line through eye. **Female and Immature** Paler buff below.

HABITAT AND DISTRIBUTION: Uncommon resident* in northern PA and highlands of the Appalachians in southern PA, western MD, VA, and WV; uncommon to rare and irregular winter visitor elsewhere: Boreal coniferous forest, particularly those with fir and spruce.

DIET: Inches along, often head down, probing bark for invertebrates, mostly on inner branches and trunks, or extracting seeds from conifer cones; regularly joins mixed-species foraging flocks during the nonbreeding period.

REPRODUCTION: Monogamous; breeding pairs seem to form during the winter months; nest cavity construction begins in late March or early April; the cavity, located 10 m (30 ft) or so above ground in a dead snag, is usually excavated by the female and lined with grasses, bark shreds, hair, feathers, pine needles, and leaves; both parents apply conifer resin to the cavity mouth, sometimes using a bark flake tool as an applicator, the presumed purpose of which is to deter possible predators;

clutch—6; incubation (female)—12–13 days; fledging—18–21 days post-hatching; post-fledging parental care—?; single brood per season is normal.

CONSERVATION STATUS: S1—MD; S2—VA; limited distribution in threatened highland forest.

RANGE: Breeds across the northern tier of Canada and the United States  south to NC in the Appalachians and to NM in the Rockies. Winters throughout breeding range and most of the United States to northern Mexico; apparent food shortages cause irregular migrations ("irruptions") into southern temperate areas in some winters.

KEY REFERENCES: Ghalambor and Martin (1999).

White-breasted Nuthatch *Sitta carolinensis*

DESCRIPTION: L = 15 cm (6 in); W = 28 cm (11 in); gray above; white below with black cap; rusty flanks. **Female** Cap is gray or dull black.

HABITAT AND DISTRIBUTION: Common (inland) to uncommon (along coast) resident* throughout: Deciduous and mixed forest, parklands, residential areas.

DIET: Forages on tree trunks and larger branches, generally working down the tree, probing bark for insect eggs, larvae, and adults; also feeds on plant seeds, wedging larger ones in crevices and hammering with the bill to open them.

REPRODUCTION: Monogamous with pair bonds often

maintained year-round; nests in natural tree cavities or old woodpecker holes, 5–20 m (16–70 ft) above ground; female builds nest, placing grass, rootlets, leaves, bark, hair, and feathers in the bottom of the cavity; egglaying takes place in late April and May; clutch—7; incubation—12–14 days; fledging—26 days post-hatching; post-fledging parental care—"several weeks."

CONSERVATION STATUS: S3—DE; conversion of bottomland forest to agriculture and residential.

RANGE: Southern tier of Canada and most of the United States; local  in Great Plains region; highlands of Mexico south to Oaxaca.

KEY REFERENCES: Pravosudov and Grubb (1993).

Brown-headed Nuthatch *Sitta pusilla*

DESCRIPTION: L = 13 cm (5 in); W = 20 cm (8 in); gray above; buffy white below; brown cap; white cheeks, throat, and nape.

HABITAT AND DISTRIBUTION: Common to uncommon resident* in the pine forests of southern DE, coastal MD, and southeastern VA: Pine forest.

DIET: Inches along, often upside down, probing bark, cones, and needle clusters on branches, limbs, and trunks for invertebrates (e.g., Orthoptera, spiders, and Coleoptera); known to use a tool (bark flake) to probe for arthropods under bark; forages mainly in the outer canopy; also feeds on pine seeds, especially in winter, when individuals often associate with mixed-species foraging flocks.

REPRODUCTION: Monogamous; breeds cooperatively, usually with one male (young of previous year?) assisting the pair with raising of the offspring; birds thought to remain paired throughout the year; nest cavity is usually excavated in a decayed pine snag 1–4 m (3–13 ft) above ground and lined with plant down, pine seed wings, arthropod silk,

loose, hanging bark provide nesting sites. Uncommon transient and winter resident (Oct–Mar) throughout: Breeds in a variety of mature forest types, seemingly wherever loose bark provides suitable nesting sites; most woodlands in winter.

DIET: Forages, usually in mixed-species flocks during the nonbreeding period, by creeping up tree trunks, beginning at the base of the trunk and working its way upward, probing bark crevices for eggs, larvae, and adults of insects, spiders, and pseudoscorpions.

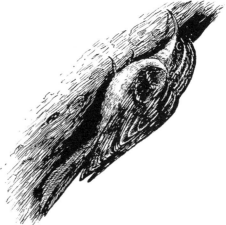

hair, and bark strips; clutch—4–6; incubation (female)—14 days; fledging—18–19 days post-hatching; post-fledging group care—5 weeks, but young may remain in small flocks with adults through the fall; mostly one brood per season.

CONSERVATION STATUS: S2—DE; S3—MD; the Brown-headed Nuthatch has undergone declines throughout its range during the past 40 years, apparently as a result of pine forest habitat loss.

RANGE: Southeastern United States from DE to east TX.

Summer
Migration
Permanent
Winter

KEY REFERENCES:
Norris (1958), Withgott and Smith (1998).

FAMILY CERTHIIDAE

Brown Creeper *Certhia americana*

DESCRIPTION: L = 13 cm (5 in); W = 20 cm (8 in); brown streaked and mottled with white above; whitish below; white eyeline; decurved bill.

HABITAT AND DISTRIBUTION: Uncommon to rare and local summer resident* (Apr–Sep) in PA, northern NJ, and the highlands of VA, WV, and MD. Also breeds locally in the Piedmont and coastal regions of MD and NJ wherever dead trees with

REPRODUCTION: Monogamous; breeding begins in May; hammock-like nest is placed between hanging bark slab and trunk of a large, dead tree, and built by the female; insect and spider silk are used to attach twigs and bark strips together and to the bark inner surface to form a base for the nest; a nest cup is built on this base of bark and wood fibers, spider egg cases, lichens, moss, leaf bits, hair, and feathers; clutch—5–6; incubation (female)—13–16 days; fledging—15–16 days post-hatching; post-fledging parental care—?, apparently forage and roost in family groups for some period after fledging; single brood per season.

CONSERVATION STATUS: SS1—DE; S3—WV, VA; mature forest breeding habitat declining.

RANGE: Breeds

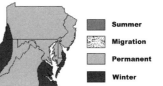

Summer
Migration
Permanent
Winter

in boreal, montane, and transitional zones of North America south through the highlands of Mexico and Central America to Nicaragua; winters nearly throughout in temperate regions of the continent.

KEY REFERENCES: Hejl et al. (2002).

FAMILY TROGLODYTIDAE

Carolina Wren *Thryothorus ludovicianus*

DESCRIPTION: L = 15 cm (6 in); W = 20 cm (8 in); a rich brown above; buff below with prominent white eyeline and whitish throat.

HABITAT AND DISTRIBUTION: Common resident* in lowlands and mid-elevations throughout except northern PA where uncommon to rare: Thickets, tangles, and undergrowth of moist woodlands, riparian forest, swamps, residential areas.

DIET: Forages low and on the ground by gleaning invertebrates (e.g., Lepidoptera, spiders, Hemiptera, Coleoptera) from leaf litter, bark, and leaves; some fruit, mast, and small vertebrates taken as well.

REPRODUCTION: Mostly monogamous, mating for life; breeding begins in April; partially or completely domed nest, 8–23 cm dia. (3–9 in), sometimes with a ramp leading to a side entrance, is usually placed in a low (1–2 m) (3–7 ft) natural or artificial cavity or cranny; built by both sexes of grass, leaves, twigs, bark shreds, and similar mate-

rials, and lined with fine grasses, moss, and hair; clutch—4–5; incubation (female)—15 days; fledging—12–14 days post-hatching; post-fledging parental care—27 days; 2 broods per season are common.

RANGE: Eastern North America from IA, MI, NY, and MA south

through the southeastern states, eastern Mexico, and Central America along the Caribbean slope to Nicaragua. The twenty-first century range of the bird represents a significant northward expansion.

KEY REFERENCES: Haggerty and Morton (1995).

Bewick's Wren *Thryomanes bewickii*

DESCRIPTION: L = 13 cm (5 in); W = 18 cm (7 in); brown above; whitish below with prominent white eyeline; long, active tail, barred black, brown, and white below.

HABITAT AND DISTRIBUTION: Rare and local summer resident* (May–Oct) in Appalachian highlands of VA and WV; formerly in MD: Riparian thickets, shrubby fields.

DIET: Forages by gleaning arthropods (e.g., Hemiptera, Coleoptera, Hymenoptera, Lepidoptera) from leaves, branches, and trunks or stems of low vegetation.

REPRODUCTION: Mainly monogamous with some polygyny; males arrive on breeding territory in late April or early May followed by the female 3–6 days later; nest is placed in a natural or artificial nook or cavity, 1–2 m (3–7 ft) above ground, and is built by both sexes of twigs, rootlets, leaves, and moss lined with feathers, hair, fine grasses, and plant down; clutch—5–6; incubation (female) 14–16 days; fledging—14–16 days post-hatching; post-fledging parental care—14 days; can fledge two broods per season.

CONSERVATION STATUS: S1—WV, VA, MD; SH—PA; breeding populations of the Bewick's Wren have

April followed within a few days by the female; male excavates a cavity nest in an old snag or appropriates an abandoned woodpecker hole or artificial nest box, 1–2 m (3–7 ft) above ground, and begins construction of cavity lining with twigs; female completes nest construction and lines the structure with plant down, moss, and fine grasses; clutch—4–8; incubation (female)—12–13 days; fledging—16–18 days post-hatching; post-fledging parental care—13 days; more than one brood per season is common.

declined sharply in the Mid-Atlantic region, perhaps as a result of reversion of old field habitat to forest or conversion to intensive agriculture.

RANGE: Breeds in most of temperate North America (except Atlantic Coastal Plain  and southern portions of Gulf states) south in highlands to southern Mexico; eastern United States populations have become scarce (extinct?) in recent years; winters in southern portion of breeding range.

KEY REFERENCES: Kennedy and White (1997).

House Wren *Troglodytes aedon*

DESCRIPTION: L = 13 cm (5 in); W = 18 cm (7 in); brown above, buff below; buff eyeline; barred flanks and tail.

HABITAT AND DISTRIBUTION: Common transient and summer resident* (Apr–Sep) throughout; rare winter resident (Oct–Apr) along the coast: Thickets, undergrowth, and tangles in riparian forest, woodlands, hedgerows, and residential areas.

DIET: Forages by gleaning in shrubs, undergrowth, and on the ground for arthropods (e.g., spiders, Coleoptera, Hemiptera, Collembola, and Diptera).

REPRODUCTION: Primarily monogamous with some polygyny; male arrives on breeding territory in

RANGE: Breeds in temperate North America from southern Canada south to south central United States; winters in southern United States south to southern Mexico. The Southern House Wren, now considered to be 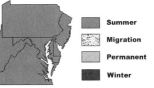 conspecific with the Northern, is resident over much of Mexico, Central and South America, and the Lesser Antilles.

KEY REFERENCES: Johnson (1998).

Winter Wren *Troglodytes troglodytes*

DESCRIPTION: L = 10 cm (4 in); W = 15 cm (6 in); brown above; brownish-buff spotted with dark brown and white below; buffy eyeline; barred flanks and belly; short tail, often cocked.

HABITAT AND DISTRIBUTION: Uncommon summer resident* (Apr–Sep) in northern PA, and highlands of southern PA, VA, MD, WV, and extreme northwestern NJ; uncommon winter resident (Oct–Mar) elsewhere: Thickets, tangles, undergrowth of mixed and coniferous forest, fens, bogs, and swamps; lowland riparian thickets in winter.

DIET: Forages by hopping and flitting along on fallen logs, branches, stumps, and understory vegetation, picking, probing, and gleaning arthropods (e.g., beetles, bugs, ants, millipedes, spiders); also takes fruits seasonally.

REPRODUCTION: Mostly monogamous; males arrive on breeding territory in April, followed by females; male builds several nests; female chooses one in which to raise offspring; domed nests (6–8cm dia. (2–3 in), with side entrance are placed in root masses of fallen trees, existing cavities, stumps, banks, or crevices, and built of mosses, bark strips, twigs, rootlets, wood shards, grass, feathers, and hair; female lines chosen nest with feathers and hair; clutch—5–7; incubation (female)14–17 days; fledging—15–17 days post-hatching; post-fledging parental care—9–18 days; double broods on occasion.

CONSERVATION STATUS: S2—VA, MD; S3—NJ; highland boreal bog breeding habitat is limited.

RANGE: Breeds from AK and British Columbia to Labrador,

▨	Summer
▤	Migration
▧	Permanent
■	Winter

south along the Pacific coast to central CA and in the Appalachians to north GA; winters from the central and southern United States to northern Mexico.

KEY REFERENCES: Hejl et al. (2002).

Sedge Wren *Cistothorus platensis*

DESCRIPTION: L = 10 cm (4 in); W = 15 cm (6 in); crown and back brown streaked with white; pale buff below; indistinct white eyeline.

HABITAT AND DISTRIBUTION: Uncommon to rare transient and winter resident (Sep–Apr) along the coast of DE, MD, and VA; rare and local summer resident* (Apr–Sep) at scattered marshy sites throughout the region: wet meadows, marshes, pond margins, bogs, estuaries.

DIET: Forages in sedges and grasses low or near the ground, probing soil and vegetation for insects and spiders.

REPRODUCTION: Monogamous, polygynous, or polyandrous in different populations; timing of Mid-Atlantic breeding initiation can be from May–July, perhaps because of the ephemeral nature of wet sedge breeding habitat; male builds several nests by weaving grass, sedge, and plant stems into a globular ball (8–13 cm dia. (3–5 in), with a side entrance attached to upright sedges or grasses; female selects one of the nests for raising of the offspring, and lines it with fine grasses, feathers, and hair; clutch—7; incubation (female)—13–14 days; fledging—12–14 days post-hatching; post-fledging parental care—?; single brood per season.

CONSERVATION STATUS: S1—PA, WV, VA, NJ, MD, DE; wetland breeding habitat is threatened by development.

RANGE: Breeds across the northeastern United States and south-eastern

Summer
Migration
Permanent
Winter

Canada from southeastern Saskatchewan to New Brunswick, south to VA and OK; winters along Atlantic and Gulf Coastal Plain from NJ to northern. Mexico. Scattered resident populations in Mexico, Central, and South America.

KEY REFERENCES: Herkert et al. (2001).

Marsh Wren *Cistothorus palustris*

DESCRIPTION: L = 13 cm (5 in); W = 18 cm (7 in); brown above; back variously streaked with white; white eyeline and throat; buff underparts; often cocks tail.

HABITAT AND DISTRIBUTION: Common to uncommon summer resident* (Apr–Sep) along the coast and in north-western PA; rare to casual winter resident (Oct–Mar) along the coast from DE southward; uncommon to rare and local summer resi-dent* (Apr–Sep) at scattered inland localities

in PA, MD, and VA: Cattail and bulrush marshes, wet grasslands; saltwater marshes.

DIET: Forages low in emergent aquatic vegetation by gleaning from plant, water, or soil surface for small invertebrates (e.g., Hymenoptera, Coleoptera, Homoptera, and spiders).

REPRODUCTION: Polygynous; males often have nests of 2 or more breeding females on their territories; males arrive on breeding territory in April or early May and build several ball-shaped nests (13 x 18 cm) (5 x 7 in) with a side entrance of grass or sedge blades (often cattails) and plant down woven into sturdy supporting stems of emergent sedges, bulrushes or cattails, usually 0.7–0.9 m (28–35 in) above water; female selects the nest she wants to lay her eggs in; clutch—3–6; incuba-tion 13–16 days; fledging—13–15 days post-hatching; post-fledging parental care—14–27 day; 2 broods per season apparently common; Marsh Wrens sometimes destroy the eggs and young of Marsh Wrens or other birds breeding in their vicinity.

CONSERVATION STATUS: S1—WV; S2—PA; wetland breeding habitat is threatened by development.

RANGE: Breeds across central and southern Canada and northern United States,

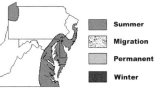

Summer
Migration
Permanent
Winter

southward along both coasts to northern Baja California and southeastern TX; rare and local in inland United States; winters along both coasts, and southern United States to southern Mexico; apparent resident population in south central Mexico.

KEY REFERENCES: Kroodsma and Verner (1997).

FAMILY REGULIDAE

Golden-crowned Kinglet *Regulus satrapa*

DESCRIPTION: L = 10 cm (4 in); W = 18 cm (7 in); grayish-green above, whitish below; dark wings with yellow edges of primaries and secondaries; white wingbars, white eyeline; male has orange crown bordered by yellow and black; female has yellow crown bordered in black; white line above eye. Continually flicks wings while foraging.

HABITAT AND DISTRIBUTION: Uncommon to rare and local summer resident* (Apr–Oct) in northern PA and highlands of southern PA, VA, WV, and MD; common to uncommon winter resident (Oct–Mar) throughout: Breeds in mature spruce-fir forest of the highland and northern portions of the region; found in a variety of woodland habitats in winter.

DIET: Forages by gleaning and hovering at branch tips, bark crevices, and conifer needle tufts for arthropod eggs, larvae, and small, soft-bodied adults; often associated with mixed-species foraging flocks during the nonbreeding period.

REPRODUCTION: Monogamous; adults arrive on breeding territory in May; pensile nest, 8 cm dia. (3 in), is placed high (8–20 m) (26–66 ft) in a horizontal branch fork of a spruce or fir and built by both sexes of moss, spider or insect silk, plant down, lichens, and bark strips lined with moss, lichens, feathers, and hair; clutch—9; incubation (female, fed by male)—15 days; fledging—18–19 days post-hatching; post-fledging parental care—16–17 days; two broods per season.

CONSERVATION STATUS: S2—VA, MD; mature highland forest breeding habitat is threatened.

RANGE: Breeds in boreal North America (except Great Plains) and in the Appalachians south to NC; south in Rockies to Guatemala; winters from southern Canada and the United States south through highland breeding range in Mexico and Guatemala.

KEY REFERENCES: Ingold and Galati (1997).

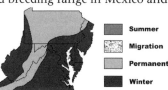

Ruby-crowned Kinglet *Regulus calendula*

DESCRIPTION: L = 10 cm (4 in); W = 18 cm (7 in); greenish above, whitish below; white wingbars, broken white eyering. Continually flicks wings while foraging; scarlet crown. **Immature and Female** Lack scarlet crown.

HABITAT AND DISTRIBUTION: Common to uncommon transient (Oct–Nov, Mar–May) and uncommon to rare (highlands) winter resident (Nov–Mar) throughout. Coniferous forests of the boreal zone in summer; a variety of woodlands during winter.

DIET: Forages by gleaning, hover-gleaning, and hawking arthropod eggs, larvae, and adults from leaf, branch, and bud surfaces and crannies, often toward the outer tips of canopy branches; pseudoscorpions, spiders, leaf hoppers, and aphids are among favored prey; some fruits and seeds eaten in winter.

RANGE: Breeds in boreal North America from AK to Labrador

south to northern NY, MI, Minnesota, and in the Rockies south to NM and AZ; winters over most of the United States, Mexico, and Guatemala; resident populations on Guadalupe Island (Baja California), and in Chiapas.

KEY REFERENCES: Ingold and Wallace (1994).

FAMILY SYLVIIDAE

Blue-gray Gnatcatcher *Polioptila caerulea*

DESCRIPTION: L = 13 cm (5 in); W = 18 cm (7 in); bluish-gray above, white below; tail black above with white outer feathers (tail appears white from below); white eyering. Adult male in breeding plumage (Apr–Aug) has black forehead and eye-line. Sexes nearly identical in nonbreeding plumage.

HABITAT AND DISTRIBUTION: Common transient and summer resident* (Apr–Sep) nearly throughout (scarce as a breeding bird in highlands and northern portions of region): Deciduous woodlands and riparian forest.

DIET: Gleans insects, spiders, and other small arthropods from leaves and branches in canopy or mid-story.

REPRODUCTION: Monogamous; pairs form shortly after arrival in late March or early April; open cup nest, 5–7 cm dia. (2–3 in) is saddled on the horizontal branch of a tree, 6–12 m up (20–40 ft); both pair members build the nest whose outer walls are lichen, spider or caterpillar silk, or bark flakes lined with fibrous plant material with the innermost portion constructed of plant down, hair, and feathers; clutch—4–5; incubation—13 days; fledging—13 days post-hatching; parental care post-fledging con-

	Summer
	Migration
	Permanent
	Winter

tinues 19–21 days; mostly single brooded, but some attempt second broods.

RANGE: Breeds in the temperate United States south through Mexico to Guatemala. Winters in the southern Atlantic (from VA south) and Gulf states; and in the west from CA, AZ, NM, and TX south throughout Mexico and Central America to Honduras. Resident in the Bahamas.

KEY REFERENCES: Ellison (1992).

FAMILY TURDIDAE

Eastern Bluebird *Sialia sialis*

DESCRIPTION: L = 18 cm (7 in); W = 31 cm (12 in); blue above, brick-red below with white belly. **Female** Similar but paler.

HABITAT AND DISTRIBUTION: Common summer resident* (Mar–Oct) throughout; common to uncommon winter resident (Nov–Feb), except in highlands and northern portions of the region where scarce or absent: Old fields, orchards, parkland, open woodlands, open country with scattered trees—appears limited by the availability of suitable nest sites (cavities in posts and nest boxes).

DIET: Forages by sallying from an exposed perch to capture arthropods from the ground or flying; also feeds on fruit as available.

REPRODUCTION: Mostly monogamous with some polygyny and polyandry; extra-pair paternity estimated at 20% in some populations; males may remain on breeding territory through the winter, arrive paired, or form pairs in March or April on the breeding territory; cavity nest is placed in an existing hole, e.g., an abandoned woodpecker nest or artificial nest box, 1–4 m (3–13 ft) above ground; female lines the cavity with grass, pine needles, hair, and feathers; clutch—6–7; incubation (female)—9–11 days; fledging—18–20 days post-hatching; post-fledging parental care—3 weeks; two broods per season are common.

RANGE: Breeds in eastern North America from southern Saskatchewan

to New Brunswick south to FL and TX; through highlands of Mexico and Central America to Nicaragua; Bermuda. Winters in southern portion of breeding range (central and eastern United States southward).

KEY REFERENCES: Gowaty and Plissner (1998).

Veery *Catharus fuscescens*

DESCRIPTION: L = 18 cm (7 in); W = 31 cm (12 in); russet above; throat buff with indistinct spotting; whitish belly.

HABITAT AND DISTRIBUTION: Uncommon transient (Sep, May) throughout; uncommon and local summer resident* (May–Sep) in the highlands of WV and VA, and wet woodlands of PA, MD, and NJ: mature deciduous and mixed forest.

DIET: Forages mainly by making short hops or flights on the ground or in vegetation to capture invertebrate prey (e.g., Coleoptera, Lepidoptera,

Hymenoptera); uses "foot quivering" (rapid foot shaking in litter) evidently to startle or reveal prey; also eats fruits when available.

REPRODUCTION: Monogamous; males arrive on breeding territories in May with females arriving a week or so later; female builds an open cup nest, 8–15 cm dia. (3–6 in),, on or near the ground at the base of a stump, tree, log, or hummock, of wet leaves, humus, bark shreds, and stems lined with rootlets and plant fibers; clutch—3–4; incubation (female)—10–14 days; fledging—12 days post-hatching; post-fledging parental care—two weeks?; single brood per season.

CONSERVATION STATUS: S2—DE; S3—NJ; Veeries have undegone significant short (1980–1994) and long-term (1966–1994) declines despite overall increases in breeding habitat, perhaps because of South American winter habitat loss.

RANGE: Breeds across southern Canada and northern United States, south in mountains to GA in the east and CO in the west;

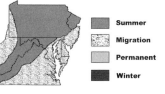

winters in northern South America.

KEY REFERENCES: Moskoff (1995).

Gray-cheeked Thrush *Catharus minimus*

DESCRIPTION: L = 20 cm (8 in); W = 31 cm (12 in); grayish-brown above, whitish below; dark spot-

ting at throat and neck diminishing to smudges at belly and flanks.

HABITAT AND DISTRIBUTION: Uncommon (fall) to rare (spring) transient throughout (Sep–Oct, May): Breeds in shrubby taiga and tundra; migration—riparian forest, deciduous and mixed woodlands and scrub.

DIET: Forages by hopping along on forest floor and picking invertebrates from litter or vegetation; uses "foot quivering" (rapid foot shaking in litter) evidently to startle or reveal prey; takes fruits seasonally.

RANGE: Breeds from northeastern Siberia across AK and the

northern tier of Canada; winters in northern South America.

KEY REFERENCES: Lowther et al. (2001).

Bicknell's Thrush *Catharus bicknelli*

DESCRIPTION: L = 20 cm (8 in); W = 31 cm (12 in); grayish-brown above, whitish below; heavy dark spotting at throat diminishing to smudges at belly and flanks.

HABITAT AND DISTRIBUTION: Uncommon (fall) to rare (spring) transient throughout (Sep–Oct, May): Breeds in spruce-fir forest; migration—riparian forest, deciduous and mixed woodlands and scrub.

DIET: Forages by hopping along on forest floor and picking invertebrates from litter or vegetation;

uses "foot quivering" (rapid foot shaking in litter) evidently to startle or reveal prey; takes fruits seasonally.

CONSERVATION STATUS: Marked declines recorded, for reasons not well understood.

RANGE: Breeds in the northeastern United States and southeastern

Canada; winters in Greater Antilles.

KEY REFERENCES: Rimmer et al. (2001).

Swainson's Thrush *Catharus ustulatus*

DESCRIPTION: L = 18 cm (7 in); W = 31 cm (12 in); brown or grayish-brown above, buffy below with indistinct dark spotting; lores and eyering buffy.

HABITAT AND DISTRIBUTION: Common to uncommon transient (Sep–Oct, May) throughout, more numerous in fall than spring; uncommon to rare and local summer resident* in highland spruce and hemlock forest of PA, WV, and Mount Rogers, VA: Breeding—thickets in coniferous forest, bogs, alder swamps; migration and winter—moist woodlands, riparian thickets.

DIET: Forages on or near the forest floor or from low in the understory, gleaning arthropod prey from litter and plant surfaces or sallying from low perches to capture prey on the ground; also feeds on fruits seasonally.

REPRODUCTION: Monogamous; males arrive on breeding territory in May followed by females;

open cup nest, 10–11 cm (4 in) is placed in a shrub or sapling in a forest thicket and built (female?) of grasses, bark shreds, plant stems, rootlets, lichens, moss, and twigs with a lining of fine grasses, skeletonized leaves, rootlets, lichens or moss; clutch—3–4; incubation (female)—10–14 days; fledging—12–14 days post-hatching; post-fledging parental care—?; single brood per season.

CONSERVATION STATUS: S1—WV, VA; SX—MD; S2—PA; short-term (1980–1994) and long-term (1966–1994) range-wide declines; montane breeding habitat is scarce and declining in MD and VA; reasons for range-wide declines are unknown, but may result from winter habitat loss.

RANGE: Breeds in boreal North America from AK across Canada to

Summer
Migration
Permanent
Winter

Labrador south to the northern United States; south in mountains and along coast to southern CA and in mountains to northern NM. Winters from southern Mexico to the highlands of South America.

KEY REFERENCES: Mack and Yong (2000).

Hermit Thrush *Catharus guttatus*

DESCRIPTION: L = 18 cm (7 in); W = 31 cm (12 in); brown or grayish-brown above; whitish or grayish below with brown spotting; whitish eyering; rusty tail, often flicked.

HABITAT AND DISTRIBUTION: Common to uncommon and local summer resident* in northern PA and NJ and highlands of southern PA, WV, western MD, and Mount Rogers, VA; common to uncommon transient (Sep–Oct, Apr–May) throughout; common to uncommon winter resident along the Coastal Plain, Piedmont and lower mountain valleys in the southern portion of the region: Breeding—mature spruce-fir and hemlock forest, especially small clearings and openings; Migration and winter—riparian thickets; broadleaf woodlands.

DIET: Forages on or near the ground for invertebrates and small vertebrates during the breeding season; consumes significantly more fruit and seeds (40–60% of diet) in nonbreeding periods.

REPRODUCTION: Monogamous; male arrives on breeding territory in late April to early May, followed shortly thereafter by the female; bulky, open cup nest, 10–15 cm (4–6 in) is placed in a depression on the ground (eastern populations), usually under dense, overhanging vegetation, and built by the female of grass and plant stems, mud, hair, lichens, moss, twigs, bark strips, and pine needles, and lined with moss, fine grass, hair, and plant down; clutch—3–6; incubation (female) 11–13 days; fledging—12 days post-hatching; post-fledging parental care—?; 2 broods likely in some parts of range.

CONSERVATION STATUS: S1—VA; S3—MD; loss of highland coniferous and mixed forest habitat.

RANGE: Breeds in boreal Canada and northern United States, south in

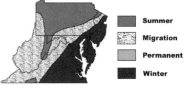

Summer
Migration
Permanent
Winter

mountains to southern CA, AZ, NM, and west TX. Winters from the southern United States north along coasts to southern British Columbia and NJ, and south through Mexico (excluding Yucatan Peninsula) to Central America. Resident population in Baja California.

KEY REFERENCES: Jones and Donovan (1996).

Wood Thrush *Hylocichla mustelina*

DESCRIPTION: L = 20 cm (8 in); W = 33 cm (13 in); russet above, white below with distinct black spots; reddish-brown crown and nape; white eyering.

HABITAT AND DISTRIBUTION: Common transient and summer resident* (May–Sep) throughout: Deciduous or mixed forest, riparian woodland.

DIET: Forages on forest floor by rummaging through leaf litter, humus, and detritus, probing, picking and gleaning for invertebrates (e.g., Coleoptera, Lepidoptera, Hymenoptera, isopods, millipedes) and small vertebrates (e.g., amphibians, reptiles); feeds on fruits when available.

REPRODUCTION: Mostly monogamous; males arrive on breeding territory in late April or early May, followed shortly thereafter by the female; open cup nest, 10–14 cm dia. (4–6 in), is placed normally < 6 m (20 ft) up in the crotch of a shrub or sapling or saddled on the horizontal branch of a tree, and is built mostly by the female of woven grasses, stems, and leaves with an interior mud cup lined with rootlets; clutch—3–4; incubation (female)—12–13 days; fledging—12–15 days post-hatching; post-fledging parental care—2–3 weeks; two broods per season are common.

CONSERVATION STATUS: The Wood Thrush has undergone both short (1980–1994) and long-term declines (1966–1994) for reasons that are not clear, but may be related to loss of winter habitat.

RANGE: Breeds in the eastern United States and southeastern Canada. Winters from southern Mexico to Panama and northwestern Colombia.

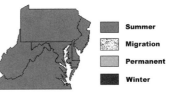

KEY REFERENCES: Rappole et al. (1989), Roth et al. (1996), Vega Rivera et al. (1998).

American Robin *Turdus migratorius*

DESCRIPTION: L = 26 cm (10 in); W = 43 cm (17 in); dark gray above, orange-brown below; white lower belly; dark head; white throat with dark streaks; partial white eyering; orange bill with dark tip.

HABITAT AND DISTRIBUTION: Common summer resident* (Mar–Oct) throughout; common to uncommon winter resident (Nov–Feb) along the Coastal Plain and Piedmont; uncommon to rare and irregular in winter elsewhere, generally in flocks: A wide variety of forest and parkland will serve as breeding habitat so long as there is rich, moist soil for foraging and mud available for nest construction; deciduous and mixed forest, scrub, parkland, riparian forest, oak woodlands, and residential areas in winter.

DIET: Forages on the ground for invertebrates, detected by sight and/or sound; also forages on branches in trees and shrubs for invertebrates and fruits; invertebrates constitute most of the diet in

spring and summer while fruits predominate in fall and winter.

REPRODUCTION: Mostly monogamous; males arrive on breeding territories in February or March, followed within a few days by the females; open cup nest, 14–15 cm dia. (6 in), is located on a horizontal branch or clump of a tree or shrub (lower in conifers early in the season, higher in deciduous trees later), and is built by the female in 3 layers: 1) outer layer of grass, twigs, feathers, rootlets, moss, and plant fibers; 2) mud layer brought in the bill from worm castings; 3) lining of fine grasses; clutch—3–4; incubation (female)—12–14 days; fledging—13 days post-hatching; post-fledging parental care—10–15 days; multiple broods per season are common.

RANGE: Breeds nearly throughout Canada and United States south

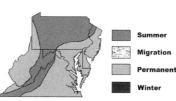

through central highlands of Mexico. Winters from southern half of breeding range into Guatemala; western Cuba; Bahamas. Resident population in Baja California and in the mountains of Mexico south to Guerrero.

KEY REFERENCES: Sallabanks and James (1999).

FAMILY MIMIDAE

Gray Catbird *Dumetella carolinensis*

DESCRIPTION: L = 23 cm (9 in); W = 31 cm (12 in); slate-gray throughout, black cap and tail, rusty undertail coverts.

HABITAT AND DISTRIBUTION: Common summer resident* (Apr–Oct) throughout; uncommon to rare winter resident (Nov–Mar) along the Coastal Plain from Cape May southward; rare elsewhere in the Piedmont and Coastal Plain in winter, often at feeders: Thickets in riparian areas; tangles, heavy undergrowth in coniferous and broadleaf woodlands; second growth, hedgerows.

DIET: Forages by gleaning or picking insects (e.g., Lepidoptera, Hymenoptera, Orthoptera, Coleoptera) and fruits from stems and branches in thickets or on the ground.

REPRODUCTION: Monogamous; male arrives on breeding territory in April, followed by the female; open cup nest, 14 cm dia. (6 in), is placed 1–3 m (3–10 ft) above ground in a shrub, and built mostly by the female of twigs, leaves, plant fibers, and bark shreds lined with a woven cup of fine grasses, hair, and leaf parts; clutch—3–4; incubation (female)—12–14 days; fledging—10–12 days post-hatching; post-fledging parental care—12 days; two broods per season are common.

RANGE: Breeds throughout eastern, central, and northwestern United States and southern Canada. Winters along central and southern Atlantic and Gulf coasts south through the Gulf and Caribbean lowlands of Mexico to Panama; Bahamas, Greater Antilles. Resident in Bermuda.

KEY REFERENCES: Camprich and Moore (1995).

Northern Mockingbird *Mimus polyglottos*

DESCRIPTION: L = 26 cm (10 in); W = 36 cm (14 in); gray body, paler below; black tail with white outer tail feathers; white wing bars and white patches on wings.

HABITAT AND DISTRIBUTION: Common permanent resident* nearly throughout except highlands and northwestern PA where a local summer resident*, withdrawing in winter: Old fields, hedgerows, agricultural areas, residential areas.

DIET: Captures invertebrate prey, mainly insects (e.g., beetles, grasshoppers, Hymenoptera) by making short runs along the ground or swoops from a perch; fruit composes a large portion of the diet, when available; other prey

includes earthworms and lizards. Often flies to the ground and spreads wings and tail in a very mechanical fashion while foraging.

REPRODUCTION: Mostly monogamous, at least for a given breeding season and occasionally for several years; males build open cup nests about 18 cm dia. (7 in) of twigs, often more than one, beginning in late March or April in the Mid-Atlantic; nests are placed in shrubs or saplings, 1–3 m (3–10 ft) above the ground; the female completes the selected nest, lining it with fine grasses, rootlets, and leaves, and lays 2–6 eggs; incubation 12–13 days; young are altricial and fledge at 12 days; parents care for young for up to 3 weeks post-fledging; the breeding season can last until September, and often more than one brood is raised.

CONSERVATION STATUS: The range of the Northern Mockingbird has been

	Summer
	Migration
	Permanent
	Winter

extending northward gradually in North America over the past century.

RANGE: Resident nearly throughout the United States and Mexico to Oaxaca; Bahamas, Greater Antilles.

KEY REFERENCES: Derrickson and Breitwisch (1992).

Brown Thrasher *Toxostoma rufum*

DESCRIPTION: L = 28 cm (11 in); W = 33 cm (13 in); rufous above, buff below with dark streaking; yellowish eye; long bill (= length of head); long tail.

HABITAT AND DISTRIBUTION: Common transient and summer resident* throughout (Apr–Sep); uncommon to rare winter resident (Oct–Mar) along the Coastal Plain: Tangles, undergrowth, and thickets of forests, old fields, hedgerows.

DIET: Forages on the ground by rummaging in litter with its bill, sweeping leaves and duff aside, and grasping exposed invertebrate prey; also feeds seasonally on fruits and seeds.

REPRODUCTION: Monogamous; males arrive on breeding territory in April; open cup nest, 19 cm (8 in) in dia., is placed low in thorny shrub tangles and built (both sexes) of twigs, leaves, roots, and plant fibers lined with rootlets; clutch—3–4; incubation (both sexes)—11–14 days; fledging—11–12 days post-hatching; post-fledging parental care—19 days; 2 broods per season in some populations.

RANGE: Breeds across eastern and central North America from

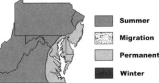

	Summer
	Migration
	Permanent
	Winter

southern Canada west to Alberta and south to east TX and southern FL. Winters in southern portion of breeding range.

KEY REFERENCES: Cavitt and Haas (2000).

FAMILY STURNIDAE

European Starling *Sturnus vulgaris*

DESCRIPTION: L = 23 cm (9 in); W = 38 cm (15 in); plump, short-tailed, and glossy black with purple and green highlights and long yellow bill in summer. **Winter** Dark billed and speckled with white.

HABITAT AND DISTRIBUTION: Common resident* of cities, towns, and farmlands throughout: Urban areas, farmland.

DIET: Forages on the ground for invertebrates, using an erect, waddling gait, in open pastures, lawns, and fields; also feeds on fruits, grains, and plant seeds when available; forms large roosting flocks during the nonbreeding season, flying from roosts to feeding areas each morning.

REPRODUCTION: Monogamous with polygyny common in some populations; pair formation begins in February and March; males choose nest sites, which are cavities in trees or buildings, or nest boxes; male initiates nest construction, lining the cavity with coarse vegetation; the female completes the nest, lining it with finer plant materials, rootlets, feathers, paper, string, plastic, or leaves; eggs are laid from Apr–Jun; clutch—4–5; incubation—12 days; fledging—21 days posthatching; parental care after fledging—10–12 days?

CONSERVATION STATUS: It is hypothesized that starlings have caused population declines for some species of cavity nesting birds by usurping limited nesting sites.

RANGE: New World—resident from southern Canada to central

Mexico; Bahamas and Greater Antilles; Old World—Breeds in temperate and boreal regions of Eurasia; winters in southern portions of breeding range into north Africa, the Middle East, and southern Asia. New World populations derive from birds released in Central Park, New York City in 1890 and 1891.

KEY REFERENCES: Cabe (1993).

FAMILY MOTACILLIDAE

American Pipit *Anthus rubescens*

DESCRIPTION: L = 18 cm (7 in); W = 28 cm (11 in); sparrow-like in size and coloration but sleek and erect in posture, walks rather than hops, and has thin bill; grayish-brown above; buffy below streaked with brown; whitish throat and eyeline; white outer tail feathers; dark legs; wags tail as it walks; undulating flight.

HABITAT AND DISTRIBUTION: Common to uncommon transient (Oct–Nov, Mar–Apr) throughout; common winter resident (Nov–Mar) along coast; uncommon to rare and irregular winter resident inland: Short grassland, plowed fields, swales, mudflats, roadsides; pond, stream, and river margins.

DIET: Forages on the ground or in shallow water by pecking prey from the surface or gleaning from low vegetation; main prey is arthropods (e.g., Diptera, Orthoptera, spiders), but occasionally feeds on seeds during migration.

CONSERVATION STATUS: Declines in winter populations have been reported, but possible cause is unknown.

RANGE: Breeds in Arctic regions of the Old and New World, and in mountainous areas and high plateaus of temperate regions; winters in temperate regions and high, arid portions of the tropics.

KEY REFERENCES: Verbeek and Hendricks (1994).

FAMILY BOMBYCILLIDAE

Cedar Waxwing *Bombycilla cedrorum*

DESCRIPTION: L = 18 cm (7 in); W = 28 cm (11 in); brown above and below with sharp crest; black face outlined in white; black throat; yellow wash on belly; tail tipped with yellow; red waxy tips to secondaries. **Immature** Faint streaking below, lacks waxy tips.

HABITAT AND DISTRIBUTION: Common summer resident* (Apr–Oct) nearly throughout; uncommon to rare along the Coastal Plain and Piedmont as a breeder; irregular winter visitor to fruiting trees throughout: Open coniferous and deciduous woodlands, bogs, swamps, and shrubby,

overgrown fields during the breeding season; cemeteries, arboreta, residential areas, or almost any site where fruiting trees are found during the nonbreeding period.

DIET: Feeds mostly on fruits (e.g., blackberry, dogwood, mountain ash, crabapple, hawthorn, serviceberry), but some insects are also taken by gleaning or sallying.

REPRODUCTION: Monogamous; birds apparently pair during migration and arrive on breeding territory in late May or June; Cedar Waxwings nest relatively late in the season (June) apparently to take advantage of fruit ripening; open cup nest, 12 cm dia. (5 in), is placed in the fork of an horizontal branch,2–4 m (7–13 ft) up in a shrub or tree and built of twigs, grass, plant down, moss, leaves, bark strips, and plant stems lined with rootlets, fine grasses, hair, pine needles, spider or insect silk, and pine needles; both sexes collect material for nest construction, but most building is done by the female; clutch—4; incubation (female)—12 days; fledging—15–16 days post-hatching; post-fledging parental care—3–4 days?; 2 broods per season may be common.

RANGE: Breeds across southern Canada and northern United States south to northern CA, KS, and NY; winters in temperate United States south through Mexico and Central America to Panama and the Greater Antilles.

KEY REFERENCES: Witmer et al. (1997).

FAMILY PARULIDAE

Blue-winged Warbler *Vermivora pinus*

DESCRIPTION: L = 13 cm (5 in); W = 20 cm (8 in); greenish-yellow above, yellow below; head yellow with black line through eye; bluish-gray wings and tail with white wing bars and tail spots;

female more greenish on head.

HABITAT AND DISTRIBUTION: Uncommon to rare transient (Aug–Sep, Apr–May) throughout; uncommon to rare and local summer resident* (May–Sep) nearly throughout except Coastal Plain, VA Piedmont, and Appalachian highlands, where scarce or absent: Deciduous scrub, old fields, shrubby swamps.

DIET: Forages in shrubs, brush, saplings, and trees, 3–8 m above ground, gleaning leaf and twig surfaces, probing and hanging upside down to search leaf clumps and vine tangles for insects and spiders.

REPRODUCTION: Mostly monogamous, but with some polygyny and helpers at the nest; males arrive on breeding territory in late April or early May followed within a week by females; bulky, open cup nest, 8–14 cm dia. (3–6 in), is placed on or near the ground in dense vegetation and built mostly by the female of bark strips, leaves, grasses, and plant stems, and lined with fine grasses, plant fibers, and hair, partially canopied with leaves in some nests; clutch—4–5; incubation—11–12 days; fledging—9–10 days post-hatching; post-fledging parental care—?; single brood per season.

CONSERVATION STATUS: S1—DE; S2—VA; old field breeding habitat is disappearing in the region.

RANGE: Breeds in the eastern United States. Winters from southern Mexico to Panama. Interbreeds with Golden-winged Warbler to produce hybrid Brewster's (patterned

Summer
Migration
Permanent
Winter

like Blue-winged Warbler but with whitish underparts) and Lawrence's (patterned like Golden-wing but yellow below) warblers.

KEY REFERENCES: Gill et al. (2001).

Golden-winged Warbler *Vermivora chrysoptera*

DESCRIPTION: L = 13 cm (5 in); W = 20 cm (8 in); gray above, white below; golden crown and epaulets; black throat and ear patch; white tail spots. **Female and immature** are patterned like male but with gray throat and ear patch.

HABITAT AND DISTRIBUTION: Uncommon to rare transient (Aug–Sep, May) throughout; uncommon to rare and local summer resident* (May–Sep) in northwestern NJ, PA (except Coastal Plain), WV, western MD, and the VA mountains: Deciduous scrub, old fields.

DIET: Pecks and probes for moth and other insect larvae; occasionally sallies or hawks after flying insects.

REPRODUCTION: Monogamous; males arrive and establish breeding territories in May; females arrive 2–7 days later and build an open cup nest, 9–15 cm dia. (4–6 in) usually on the ground at the base of a clump of vegetation; nest is constructed of coarse plant material (bark, dried leaves) on the outside with a cup of finer plant material, often reddish, inside; clutch—5; incubation—11 days; fledging—10 days post-hatching; duration of post-fledging parental care up to 31 days.

CONSERVATION STATUS: S2—WV; S3—NJ, MD, VA; breeding populations have declined or disappeared in many parts of the range, and there is evidence of genetic swamping resulting from hybridization with the closely related Blue-winged Warbler.

RANGE: Breeds in the northeastern and north central United States. Winters from southern Mexico to Colombia and Venezuela.

KEY REFERENCES: Confer (1992), Bent (1953), Gill (1997).

Tennessee Warbler *Vermivora peregrina*

DESCRIPTION: L = 13 cm (5 in); W = 20 cm (8 in); olive above, dingy white below with gray cap, white eye stripe and dark line through eye. Immatures are tinged with yellow.

HABITAT AND DISTRIBUTION: Common to uncommon fall transient (Sep) and uncommon to rare spring transient (May) nearly throughout, scarcer along the Coastal Plain and Piedmont: Breeds in coniferous forest; deciduous and mixed forest on migration.

DIET: Forages mainly in the outer canopy of trees and shrubs by gleaning and probing leaves for arthropods; often joins mixed-species foraging flock during nonbreeding periods, and feeds on small fruits and nectar opportunistically.

RANGE: Breeds across boreal North America. Winters from southern Mexico to northern South America.

KEY REFERENCES: Morton (1980), Rimmer and McFarland (1998).

Orange-crowned Warbler *Vermivora celata*

DESCRIPTION: L = 13 cm (5 in); W = 20 cm (8 in); greenish-gray above, dingy yellow faintly streaked with gray below; grayish head with faint whitish eye stripe; orange crown visible on some birds at close range.

HABITAT AND DISTRIBUTION: Rare transient and winter visitor (Oct–Apr) along the immediate MD and VA coast: Old fields, coastal scrub.

DIET: Forages by flitting along twig and branch tips and gleaning or probing leaves, leaf clusters, moss, and bark; prey is mostly arboreal arthropods (e.g., caterpillars, Hemiptera, Coleoptera), but some fruits are used during migration or on the wintering grounds.

RANGE: Breeds in western and northern North America. Winters from

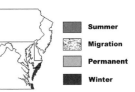

the southern United States southward into Mexico, Belize and Guatemala.

KEY REFERENCES: Sogge et al. (1994).

Nashville Warbler *Vermivora ruficapilla*

DESCRIPTION: L = 10 cm (4 in); W = 18 cm (7 in); olive above, yellow below; gray head; white eyering; rufous cap. **Female** Dingier; lacks reddish cap.

HABITAT AND DISTRIBUTION: Uncommon to rare and local summer resident* (May–Sep) in northwestern NJ, PA (except southeastern and extreme western portions), western MD, VA highlands (very rare), and highland bogs of Tucker and Grant counties in WV; uncommon (inland) to rare (coast) transient (Sep–Oct, May) throughout: Coniferous bogs; scrub; old fields, overgrown strip mines, clearcuts, and Pine Barrens.

DIET: Forages by deliberately gleaning, mostly in leaves and tassels at tips of branches, for insects (e.g., Lepidoptera, Hemiptera, Coleoptera.

REPRODUCTION: Monogamous; birds arrive on breeding territory in May; open cup nest, 7–11 cm dia. (3–4 in), is placed on the ground in dense undergrowth, and built by the female of leaves, bark strips, pine needles, and grass lined with fine grass, pine needles, rootlets, and hair, often with moss around the rim; clutch—4–5; incubation (female, perhaps with some help from male—male occasionally feeds female at nest)—11–12 days; fledging—9–11 days post-hatching; post-fledging parental care—?; single brood per season.

CONSERVATION STATUS: S1—WV, VA, MD, NJ; S3—

PA; mature highland forest breeding habitat is threatened.

RANGE: Breeds across extreme north central and northeastern

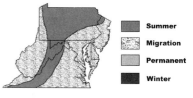

■	Summer
▨	Migration
▨	Permanent
■	Winter

United States, south central and southeastern Canada, and northwestern United States. Winters from south TX to Honduras.

KEY REFERENCES: Williams (1996).

Northern Parula *Parula americana*

DESCRIPTION: L = 13 cm (5 in); W = 20 cm (8 in); bluish above; bluish head with black lores and white partial eyering; yellow throat; dark grayish band across breast rimmed below with rust; white belly; 2 partial or complete white wingbars; greenish-yellow on back. **Female and Immature** Lack collar.

HABITAT AND DISTRIBUTION: Common to uncommon transient and summer resident* (Apr–Oct) nearly throughout, although scarce or absent as a breeder in NJ and much of extreme northern PA: Swampy or moist deciduous, mixed, or coniferous forest.

DIET: Forages mostly in forest canopy by gleaning tips of leaves and twigs by stretching, hovering, hanging, and flycatching; main prey are arthropods (e.g., Lepidoptera, Coleoptera, spiders).

REPRODUCTION: Mostly monogamous; birds arrive on breeding territory in May; pendulous nest, 7 cm dia. (3 in), is placed at the end of a tree

branch, 2–30 m (7–100 ft) up, in a clump of epi-phyte, e.g., Spanish moss or beard moss; a side entrance is made, and the nest is constructed inside the clump by interweaving strands of the epiphyte with fine grasses, and lining the cup with pine needles and plant down; clutch—3–5; incubation (mostly by female)—12–14 days; fledging—10–11 days post-hatching; post-fledging parental care—?; two broods per season have been reported.

CONSERVATION STATUS: S1—DE; S3—NJ; floodplain forest conversion to agriculture and residential.

RANGE: Breeds in the eastern United States and south-eastern Canada.

■	Summer
▦	Migration
▨	Permanent
■	Winter

Winters from southern Mexico to Panama; West Indies.

KEY REFERENCES: Moldenhauer and Regelski (1996).

Yellow Warbler *Dendroica petechia*

DESCRIPTION: L = 13 cm (5 in); W = 20 cm (8 in); yellow throughout, somewhat dingier on the back; yellow tail spots; male variably streaked rusty below.

HABITAT AND DISTRIBUTION: Common transient and summer resident* (May–Sep) throughout: Old fields, riparian thickets; relatively open areas with scattered trees or shrubs during migration.

DIET: Forages by gleaning arthropods from bark and leaf surfaces at nearly all levels in occupied habitats; takes some fruit during the nonbreeding period.

REPRODUCTION: Mostly monogamous with some polygyny; males arrive on breeding territory in May, followed shortly thereafter by females; open cup nest, 8 cm dia. (3 in), is located in the fork of a shrub or sapling, 1–3 m (3–10 ft) above ground, and is built by the female of grass, plant down and fibers, and arthropod silk lined with hair, feathers, and plant down; females will often build a second nest on top of the first if Brown-headed Cowbirds parasitize them by laying an egg in their nest; clutch—4–5; incubation (female)—11–12 days; fledging—8–10 days post-hatching; post-fledging parental care—17–21 days; normally one brood per season.

CONSERVATION STATUS: Breeding populations of Yellow Warblers appear to have increased in size in the Mid-Atlantic region despite decreases in old field, wetland, and scrub breeding habitats.

RANGE: Breeds across most of North America; winters from extreme

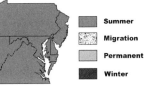

■	Summer
▦	Migration
▨	Permanent
■	Winter

southern United States through Mexico and Central America to northern and central South America. Resident races in mangroves of West Indies, Central and South America.

KEY REFERENCES: Lowther et al. (1999).

Chestnut-sided Warbler *Dendroica pensylvanica*

DESCRIPTION: L = 13 cm (5 in); W = 20 cm (8 in); greenish streaked with black and white above; white below with chestnut sides (more extensive in male); yellow cap; white neck, cheek and wing bars; black eye stripe and malar stripe. **Winter and Immature** Spring green above; white below; whitish-yellow eyering, wingbars and tail spots; **Winter Male** Some chestnut along flanks.

HABITAT AND DISTRIBUTION: Common to uncommon summer resident* (May–Sep) in northern NJ, PA (except southeast and southwest where scarce and local), western MD, and the highlands of VA and eastern WV: Scrub, thickets, second growth, hedgerows.

DIET: Forages mostly by using rapid short hops and flights to search the underside of leaves for arthropods (e.g., Lepidoptera, Diptera, spiders); some fruits on occasion.

REPRODUCTION: Mostly monogamous; males arrive on breeding territory in May, females arrive a week or so later; open cup nest, 7–9 cm dia. (3–4 in), is placed in the crotch or fork of a low, 1–2 m (3–7 ft) shrub, and built by the female of grass, stems, plant down, and bark shreds lined with fine grasses, hair, and rootlets; clutch—4; incubation (female)—11–12 days; fledging—10–11 days post-hatching; post-fledging parental care—28 days; single brood per season is normal.

CONSERVATION STATUS: S1—DE; S3—NJ; reversion of abandoned farm land to forest or conversion to housing or intensive farming has resulted in disappearance of scrub second growth habitat favored by this bird.

RANGE: Breeds in southeastern Canada and the northeastern United States

▨	Summer
▧	Migration
▒	Permanent
■	Winter

south in the Appalachians to GA and AL. Winters from southern Mexico to Panama.

KEY REFERENCES: Greenberg (1984), Richardson and Brauning (1995).

Magnolia Warbler *Dendroica magnolia*

DESCRIPTION: L = 13 cm (5 in); W = 20 cm (8 in); black back, gray cap; yellow below broadly streaked with black; yellow rump; white wing and tail patches. **Female and Winter Male** More brownish above than breeding male; breast streaking is faint and grayish.

HABITAT AND DISTRIBUTION: Common to uncommon and local summer resident* (May–Sep) in northern NJ, highlands and northern portions of PA, western MD (Garrett County), Highland County, VA (Mount Rogers), and extreme highlands of the Allegheny Mountains in eastern WV; Common (fall) to uncommon (spring) transient (Sep, May) elsewhere: Thickets in coniferous forest, especially young spruce and hemlock, on the breeding ground; various woodlands and scrub at other times of the year.

DIET: Forages at midlevel in canopy, gleaning and making short sallies or jumps for arthropods (e.g., Coleoptera, spiders, Lepidoptera), taken mainly from the underside of overhanging leaves.

REPRODUCTION: Monogamous; males arrive on breeding territories in May followed 7–10 days later by females; open cup nest, 8–12 cm dia. (3–5 in), is placed low, < 3 m (10 ft) in dense stands of young spruce, fir, or hemlock at a site where horizontal branches from the trunk form a small platform; both sexes build the nest, laying a base of

twigs covered with coarse grass and plant stems, and lined with rootlets; clutch—4; incubation (female)—11–13 days; fledging—9–10 days post-hatching; post-fledging parental care—up to 25 days; single brood per season.

CONSERVATION STATUS: S1—NJ; S2—VA; S3—MD; small breeding populations in parts of the Mid-Atlantic region are dependent on a threatened habitat (young hemlock and spruce/fir stands).

RANGE: Breeds across much of boreal Canada and the north-eastern United States south in the Appalachians to VA and WV. Winters from central Mexico to Panama; West Indies.

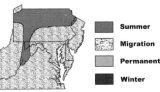

Summer
Migration
Permanent
Winter

KEY REFERENCES: Hall (1994).

Cape May Warbler *Dendroica tigrina*

DESCRIPTION: L = 13 cm (5 in); W = 20 cm (8 in); greenish with black streakings above; dark streaking on crown; yellow breast with black streaks; yellowish rump; chestnut face with black line through eye; white undertail coverts, wing patch, and tail spots. **Female and Immature** Dingier and has yellow face.

HABITAT AND DISTRIBUTION: Common to uncommon fall transient (Sep), uncommon to rare spring transient (May) throughout: Coniferous, mixed, and deciduous forest, riparian and oak woodland, and thickets.

DIET: Forages by gleaning and hawking for arthropods, mostly in canopy; also feeds on fruit and nectar during the nonbreeding seasons.

RANGE: Breeds in boreal regions of central and eastern Canada and northeastern United States Winters in the West Indies.

Summer
Migration
Permanent
Winter

KEY REFERENCES: Baltz and Latta (1998).

Black-throated Blue Warbler *Dendroica caerulescens*

DESCRIPTION: L = 13 cm (5 in); W = 20 cm (8 in); dark grayish-blue above; black face, throat, and sides; white breast, belly, and spot on primaries. **Female** Brownish above, dingy white below with whitish eye stripe, dark ear patch and (usually) with a white patch on the primaries.

HABITAT AND DISTRIBUTION: Common transient and summer resident* (May–Oct) in the highlands of VA, WV, PA, western MD, and northern NJ (scarce); common transient (Sep–Oct, May) elsewhere: Mixed woodlands, often with laurel or rhododendron understory, during the breeding season; various woodlands on migration and in winter.

DIET: Gleans, picks, and snatches arthropod prey from leaf surfaces, often by hovering or fluttering from a perch; in summer, caterpillars and beetles are major food items; in winter small fruits are occasionally taken.

REPRODUCTION: Monogamous with some serial polygyny; males arrive on breeding territory in late April or early May followed within a few days by the female; open cup nest, 8–10 cm dia. (3–4 in), is placed in the crotch of a shrub or sapling, usually within 2 m (7 ft) of the ground; built by the female of bark strips and shards held together with spider silk and, perhaps, saliva and lined with moss, hair, and rootlets; eggs laid in May; clutch—4; incubation—12–13 days; fledging—9–10 days post-hatching; post-fledging parental care—2–4 weeks; may produce two or three broods per season.

CONSERVATION STATUS: S3—MD; highland mixed forest is declining.

RANGE: Breeds from south-eastern Canada and the north-eastern

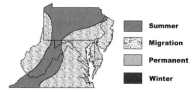

United States south along the Appalachian chain to northern GA. Winters in the Caribbean basin.

KEY REFERENCES: Holmes (1994), Graves et al. (1997), Graves (1997).

Yellow-rumped Warbler *Dendroica coronata*

DESCRIPTION: L = 13 cm (5 in); W = 20 cm (8 in); blue-gray streaked with black above; black breast and sides; white belly; yellow cap, rump and shoulder patch; white wing bars. Breeding males of the eastern and northern race have white throat; western race has yellow throat. **Female** Brownish above, dingy below with faint streaking; yellow rump.

HABITAT AND DISTRIBUTION: Common to uncommon transient and uncommon to rare and local summer resident* (Apr–Sep) in the highlands of northern PA, western MD (Garrett County), and mountains of eastern WV; common transient

(Sep–Oct, Apr–May) throughout; common winter resident (Sep–May) along the Coastal Plain and Piedmont: Breeds in coniferous and mixed forest; various woodlands and scrub on migration and in winter.

DIET: Forages on arthropods gleaned from trunks and lower branches of forest understory during the breeding period; gleans, hawks, and sallies for insects in shrubs and trees in single- or mixed-species flocks during the nonbreeding period when the diet can also included significant amounts of fruit (e.g., bayberry, wax myrtle, juniper).

REPRODUCTION: Monogamous; male arrives on breeding territory in late April or early May followed 5–7 days later by the female; open cup nest, 8–9 cm dia. (in), is placed on the horizontal branch of a conifer (e.g., hemlock, spruce, or pine, 1–15 m above ground and built (female) of woven twigs, pine needles, grass, hair, rootlets, moss, and lichens and lined with hair and feathers; clutch—4–5; incubation (female)—12–13 days; fledging—10–14 days post-hatching; post-fledging parental care—14 days; mostly one brood per season.

CONSERVATION STATUS: S3—PA, WV; highland coniferous forest is threatened by several factors.

RANGE: Breeds across north-

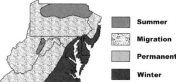

ern boreal North America and in mountains of the west south to southern Mexico. Winters from central and southern United States to Panama and the West Indies.

KEY REFERENCES: Hunt and Flaspohler (1998).

Black-throated Green Warbler
Dendroica virens

DESCRIPTION: L = 13 cm (5 in); W = 20 cm (8 in); green above; black bib; green crown; golden face; white belly. **Female** Usually has some gray or black across breast.

HABITAT AND DISTRIBUTION: Common to uncommon and local summer resident* (May–Oct) in the highlands of VA, WV, PA, western MD, northern NJ (scarce); also breeds in VA coastal lowlands of extreme southeastern corner; uncommon to rare transient (Sep–Oct, May) elsewhere: Breeds in coniferous and mixed forest; found in broadleaf forest and scrub on migration.

DIET: Gleans leaf surfaces for adult and larval insects, often hovering to inspect the undersides of leaves; main prey is caterpillars in summer.

REPRODUCTION: Monogamous; males arrive on breeding territories in mid-April to early May; females arrive several days later; both pair members build the open cup nest, 8–10 cm dia. (3–4 in), often in the crotch formed by 2 or more small branches joining the trunk of a conifer, usually 1–3 m (3–10 ft) above ground; outside

cup of nest is made of stems, bark, twigs, and spider silk, while the lining is moss, feathers, and hair; clutch—4; incubation—12 days; fledging—11 days post-hatching; post-fledging parental care—"one month;" one brood per season.

CONSERVATION STATUS: S3—NJ; coastal breeding populations—*Dendroica virens waynei* are threatened by conversion of floodplain forest to agriculture and residential.

RANGE: Breeds across central and southern Canada and north central and northern

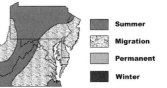

Summer
Migration
Permanent
Winter

United States, also south in the Appalachians to north GA and along the Coastal Plain from MD to SC. Winters from south TX and south FL through Mexico and Central America to Panama; West Indies.

KEY REFERENCES: Morse (1993).

Blackburnian Warbler *Dendroica fusca*

DESCRIPTION: L = 13 cm (5 in); W = 20 cm (8 in); black above, white below; orange bib; black and orange facial pattern. **Female** Patterned like male but black areas of male are grayish in female, orange areas are yellowish.

HABITAT AND DISTRIBUTION: Common to uncommon transient and uncommon to rare and local

summer resident* (Apr–Sep) in the highlands of northern PA, western MD, and mountains of VA and eastern WV; uncommon (inland) to rare (coast) transient (Sep, May) elsewhere, more numerous in fall: Breeds in spruce-fir, mixed, and deciduous forest (in the Appalachians); found in various forest types on migration and in winter.

DIET: Forages for arboreal arthropods (e.g., caterpillars, spiders, Coleoptera), often by moving along branches from trunk to tip, and gleaning from leaf surfaces; some fruits taken during the non-breeding period.

REPRODUCTION: Monogamous; male arrives on territory in late April or early May followed by female in a few days; open cup nest, 8–9 cm dia. (3–4 in), is placed high, usually > 10 m (33 ft) on a horizontal branch, often a hemlock or spruce; female builds nest of bark strips, plant fibers, rootlets, and twigs lined with lichen, moss, fine grasses, plant down, and hair; clutch—4; incubation—12–13 days; fledging—?; post-fledging parental care—?; one brood per season.

CONSERVATION STATUS: S1—MD; S2—VA, NJ; threatened by loss of mature hemlock forest caused by infestation with the wooly adelgid, and by destruction of spruce/fir forest by acid rain.

RANGE: Breeds in southeastern Canada and the northeastern United States

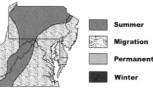

▨	Summer
▧	Migration
▢	Permanent
■	Winter

south in the Appalachians to GA. Winters in Costa Rica, Panama, and northern South America.

KEY REFERENCES: Morse (1994).

Yellow-throated Warbler *Dendroica dominica*

DESCRIPTION: L = 13 cm (5 in); W = 20 cm (8 in); dark gray above, white below with yellow bib; black mask; white eye stripe (yellow in some individuals); black streaks on sides.

HABITAT AND DISTRIBUTION: Common to uncom-

mon transient and summer resident* (Apr–Sep) in lowlands of southeastern (scarce and local) and southwestern PA, western WV, and the Piedmont and Coastal Plain of MD, DE, and VA: Conifers, cypress, sycamores; riparian woodland; pine-oak forest.

DIET: Normally forages in the canopy by creeping along branches, picking, probing, and gleaning deliberately from bark surfaces and cracks, pine cones, and leaf clumps for arthropods (e.g., Lepidoptera, Coleoptera, Diptera, Orthoptera, spiders).

REPRODUCTION: Monogamous; pairs arrive on breeding territory in April; open cup nest, 7 cm dia. (3 in), is placed in canopy, 10–15 m up (33–50 ft) in a hanging clump of Spanish moss (*Tillandsia*) where available or on a horizontal limb; nest is built mostly by the female of grasses, bark strips, and plant stems, and lined with plant down and feathers; clutch—4; incubation (female)—12–13 days; fledging—10 days? post-hatching; post-fledging parental care—2–3 weeks?; 2 broods per season are common.

CONSERVATION STATUS: S2—DE; conversion of floodplain forest to agriculture and residential.

RANGE: Breeds in the eastern United States. Winters from the Gulf coast of the United States south through Central America to Costa Rica; Bahamas and Greater

▨	Summer
▧	Migration
▢	Permanent
■	Winter

Antilles. The breeding range seems to be expanding northward.

KEY REFERENCES: Hall (1996).

Pine Warbler *Dendroica pinus*

DESCRIPTION: L = 15 cm (6 in); W = 23 cm (9 in); olive-yellow above, yellow below with faint grayish streaking; whitish belly; yellow eye stripe and eyering; white wingbars; white tail spots.

HABITAT AND DISTRIBUTION: Common to uncommon summer resident* (Mar–Sep) in pines nearly throughout, although scarce or absent in northern and western PA and northern NJ; uncommon to rare winter resident (Oct–Mar) along the Coastal Plain: Pine forest.

DIET: Forages during the breeding season high in pines, gleaning arthropod prey from pine needles and bark; during the nonbreeding period, birds forage at nearly all levels in pines, as well as on the ground, and, in addition to arthropods, also feed on pine, and perhaps other seeds.

REPRODUCTION: Monogamous; pair formation information lacking, but breeding activity apparently begins in April in the Mid-Atlantic region; open cup nest, 7 cm dia. (3 in), is placed high on the horizontal branch, fork, or terminal needle tuft of a pine and built mostly by the female of plant stems, grass, bark strips, pine needles, twigs, rootlets, and arthropod silk lined with feathers, hair, and plant down; clutch—4; incubation (female)—12–13 days; fledging—10 days post-hatching; post-fledging parental care—?; two or three broods per season are possible, but data are few.

CONSERVATION STATUS: S3—PA; timbering of native pine forest.

RANGE: Breeds in the eastern half of the United States and south central and

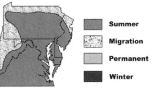

southeastern Canada; Bahamas; Hispaniola. Winters in southern portion of breeding range.

KEY REFERENCES: Rodewald et al. (1999).

Prairie Warbler *Dendroica discolor*

DESCRIPTION: L = 13 cm (5 in); W = 18 cm (7 in); greenish-yellow above streaked with chestnut; yellow below with black markings on sides; yellow face with black eyeline and chin stripe; yellowish wingbars. **Female and Immature** Similarly patterned but much dingier. Wags tail while foraging.

HABITAT AND DISTRIBUTION: Common to uncommon summer resident* (May–Sep) in lowlands nearly throughout, although scarce and local or absent in extreme northern PA: Scrubby coniferous and deciduous second growth, clearcuts, old fields, pine plantations.

DIET: Forages very actively low (1–3 m) (3–10 ft) in shrubby habitats by gleaning arthropods from leaves and branches; takes some fruits during the nonbreeding period.

REPRODUCTION: Mostly monogamous with some polygyny; males arrive on breeding territory in April followed within 5 days or so by the female; open cup nest, 6 cm dia. (2 in), is located in the

fork of a shrub or sapling, 2–3 m (7–10 ft) above ground on average, and is built (female) of arthropod silk, plant fibers, leaves, wasp nest bits, snake skin, and plant down lined with hair, feathers, sporophytes, and fine grasses; clutch—3–5; incubation (female)—12 days; fledging—8–11 days post-hatching; post-fledging parental care—3–6 weeks; about 20% attempt second broods in some populations.

CONSERVATION STATUS: Prairie Warblers have shown long-term (1966–1994) declines in the Mid-Atlantic region, perhaps due to reversion of old field habitats to forest or conversion to intensive agriculture.

RANGE: Breeds in the eastern half of the United States; winters in southern FL and the Caribbean basin.

KEY REFERENCES: Nolan (1979), Nolan et al. (1999).

Palm Warbler *Dendroica palmarum*

DESCRIPTION: L = 13 cm (5 in); W = 18 cm (7 in); olive with dark streaks above; yellowish or creamy below with brownish streaks; yellow undertail coverts; chestnut cap in breeding plumage; yellow eyestripe and eyering; wags tail frequently.

HABITAT AND DISTRIBUTION: Common to uncommon transient (Sep–Oct, Apr) throughout, more numerous in fall; rare winter resident, mainly along the coast: woodland thickets, grassland, scrub, old fields, brushy second growth, savanna, marshes.

DIET: Forages by walking on the ground and gleaning arthropods (e.g., Orthoptera, Odonata, Coleoptera, Diptera, spiders) from grass or other low plants or by flitting from branch to branch to glean or hawk arthropods; occasionally feeds on nectar when available.

RANGE: Breeds across central and eastern Canada and extreme north central

and northeastern United States. Winters along the Atlantic and Gulf Coastal Plain, northern Caribbean Basin, and the U.S. Pacific coast.

KEY REFERENCES: Wilson (1996).

Bay-breasted Warbler *Dendroica castanea*

DESCRIPTION: L = 16 cm (6 in); W = 23 cm (9 in); gray above with dark streaking; chestnut cap, throat and sides; black mask; beige neck patch; two white wingbars; dark tail with white patches (from below). **Breeding Female** Similar to male but dingier; head is whitish with brown, dark-streaked cap; whitish below, often with some chestnut on sides and flanks. **Winter Male** Greenish above with dark streaking on crown and back; dark wings with two white wingbars; whitish below with some chestnut on flank; dark legs, buffy undertail coverts. **Winter Female and Immature** Like winter male but duller and lacks chestnut on flanks.

HABITAT AND DISTRIBUTION: Common to uncommon transient (Sep, May) throughout, more numerous in fall: Breeds in coniferous forest; various woodlands on migration and in winter.

DIET: Forages by gleaning from needles, leaflets, leaves, and twigs for arthropods (e.g., Lepidoptera, Coleoptera, Diptera, Hymenoptera); also takes fruits when available.

CONSERVATION STATUS: Populations undergo sharp

increases and decreases, presumably in response to spruce budworm outbreaks, a major food source during the breeding season.

RANGE: Breeds across central and eastern Canada and extreme north central and northeastern United States. Winters in Panama, Colombia, and northwestern Venezuela.

KEY REFERENCES: Williams (1996).

Blackpoll Warbler *Dendroica striata*

DESCRIPTION: L = 15 cm (6 in); W = 23 cm (9 in); grayish-green streaked with black above, white below; black cap; white cheek; black chin stripe and streakings on side; flesh-colored legs. **Female, Winter Male, and Immature** Olive cap streaked with black; faint white eye stripe;

greenish cheek; whitish below with variable amounts of gray streaking.

HABITAT AND DISTRIBUTION: Common transient (Sep–Oct, May) throughout; singing stragglers continue to pass through on migration well into June; more common in fall than in spring except in VA, where it is scarcer in fall: Breeds in boreal spruce forest; uses various forest and scrub sites on migration and in winter.

DIET: Forages at various heights in trees and shrubs by gleaning arthropods from foliage and twigs; takes some fruit on occasion during migration.

RANGE: Breeds in northern boreal regions of North America. Winters in South America to northern Argentina.

KEY REFERENCES: Hunt and Eliason (1999).

Cerulean Warbler *Dendroica cerulea*

DESCRIPTION: L = 13 cm (5 in); W = 20 cm (8 in); a delicate bluish-gray above, white below; dark bar across chest and streakings down side; white wingbars. **Female** Olive tinged with blue above; whitish below with grayish streakings; blue-gray crown; creamy eye stripe.

HABITAT AND DISTRIBUTION: Uncommon summer resident* (May–Sep) in WV, the northern Piedmont and mountains of VA, PA (except extreme north, where scarce); rare and local summer resident* in the Coastal Plain: Mature deciduous forest and mixed forest, often tall oaks or sycamores, formerly in old-growth bottomland forests.

DIET: Forages along branches and twigs in canopy

trees, gleaning arthropods from upper and lower leaf surfaces and bases.

REPRODUCTION: Monogamous; male arrives on breeding territory in May followed within a week by the female; open cup nest, 7 cm dia. (3 in), is placed on the horizontal limb of a deciduous tree in the midstory or upper canopy and built (female) of bark strips, spider web, and grasses, and lined with fine grass and hair; clutch—4; incubation (females)—11–12 days; fledging—10–11 days post-hatching; post-fledging parental care—no information.

CONSERVATION STATUS: S1—DE; S3—VA, MD, NJ; long-term declines (1966–1994) in Mid-Atlantic region; reasons are not known but may be related to loss of winter habitat in the South American Andes.

RANGE: Breeds in southeastern Canada and the eastern United States except southeastern Coastal Plain. Winters along the east slope of the Andes in northwestern South America.

- Summer
- Migration
- Permanent
- Winter

KEY REFERENCES: Hamel (2000).

Black-and-white Warbler *Mniotilta varia*

DESCRIPTION: L = 13 cm (5 in); W = 23 cm (9 in); boldly striped with black and white above and below. **Female** Similar to male but with faint grayish streaking below.

HABITAT AND DISTRIBUTION: Common to uncommon transient and summer resident* (Apr–Sep) nearly throughout; scarcer as a summer resident on the Coastal Plain: Deciduous and mixed forest.

DIET: Forages by clambering along tree trunks and branches, nuthatch fashion, picking, probing, and gleaning arthropods (e.g., caterpillars, spiders, ants, beetles) from bark or, less commonly, leaves; occasionally eats fruit when available.

REPRODUCTION: Monogamous; male arrives on breeding territory in early to mid-April, followed by the female; nest is placed on the ground, often in a depression at the base of a stump, tree, or rock, and built of leaves, shreds of bark and plant stems, pine needles, and rootlets lined with fine grasses, hair, and moss; clutch—4–5; incubation (female)—10–12 days; fledging—young leave the nest before they can fly at 8–12 days post-hatching; post-fledging parental care—?; single brooded.

CONSERVATION STATUS: S3—DE; conversion of floodplain forest to agriculture and residential.

RANGE: Breeds across Canada east of the mountains and in the eastern

- Summer
- Migration
- Permanent
- Winter

half of the United States. Winters from extreme southern United States, eastern Mexico, and Central America to northern South America; West Indies.

KEY REFERENCES: Kricher (1995).

American Redstart *Setophaga ruticilla*

DESCRIPTION: L = 13 cm (5 in); W = 23 cm (9 in); black above and below with brilliant orange patches on tail, wings, and sides of breast; white belly and undertail coverts; often fans tail and droops wings while foraging. **Female** Grayish-brown above, whitish below with yellow patches on tail, wings, and sides of breast. **Immature Male** Like female but with salmon-colored patches on sides of breast.

HABITAT AND DISTRIBUTION: Common transient and summer resident* (May–Sep) throughout: Deciduous and mixed forest.

DIET: Forages mostly by gleaning arthropod prey from leaves, branches, and trunks of trees during the breeding period; hawking and sallying for flying insects are the predominant foraging methods during the nonbreeding period, when redstarts often join mixed-species foraging flocks.

REPRODUCTION: Mostly monogamous but with some polygyny; male arrives on breeding territory in early May followed in a week or so by the female; open cup nest, 7 cm dia. (3 in), is placed in a fork against the trunk, or on horizontal branches, of a sapling or tree and built by the female of woven bark strips, grass, spider and insect silk, plant down, lichens, and rootlets lined with hair, feathers, rootlets, and fine grasses; clutch—4; incubation (female)—11 days; fledging—9 days post-hatching; post-fledging parental care—3 weeks?; single brood per season.

CONSERVATION STATUS: S1—DE; conversion of floodplain forest to agriculture and residential.

RANGE: Breeds across Canada south of the Arctic region and in

 Summer

Migration

Permanent

Winter

the eastern half of the United States except the southeastern Coastal Plain. Winters from central Mexico south to northern South America; West Indies.

KEY REFERENCES: Ficken and Ficken (1967), Rappole and Warner (1980), Sherry and Holmes (1997).

Prothonotary Warbler *Protonotaria citrea*

DESCRIPTION: L = 15 cm (6 in); W = 23 cm (9 in); golden head and breast; yellow-green back; blue-gray wings; yellow underparts; white undertail coverts. **Female** Similar but yellow rather than orange on head.

HABITAT AND DISTRIBUTION: Common summer resident* (May–Sep) along the Coastal Plain, uncommon to rare and local elsewhere; in WV found mainly along the Ohio River and tributaries; in NJ mainly along the Delaware River, Delaware Bay, and neighboring bottomlands; in inland MD, mainly along the Potomac; in PA, mainly at Pymatuning State Park (nest boxes) and the Delaware River: Wooded swamps; mangroves in winter.

DIET: Forages by picking invertebrate prey (e.g., Lepidoptera, Diptera, molluscs, and spiders) from stumps, logs, trunks, and foliage in swamp mid- and understory, < 7 m (23 ft); takes some fruits and nectar during the nonbreeding period.

REPRODUCTION: Mostly monogamous (1% polygyny); male arrives on breeding territory in April followed within a few days by the female; cavity nest is placed in a natural or abandoned

woodpecker hole, usually about 2 m (7 ft) above water, and lined, mostly by the female, with moss, rootlets, plant down, leaves, grasses, and bark strips; clutch—4–5; incubation (female)—12–14 days; fledging—10 days post-hatching; post-fledging parental care—35 days; two broods per season are common in southern populations.

CONSERVATION STATUS: S2—PA, WV; drainage and pollution of wetlands.

RANGE: Breeds in the eastern half of the United States; winters in mangroves

along coast from southeastern Mexico, Caribbean basin of Central America, Costa Rica, Panama, and northern South America.

KEY REFERENCES: Petit (1999).

Worm-eating Warbler *Helmitheros vermivorum*

DESCRIPTION: L = 15 cm (6 in); W = 23 cm (9 in); olive above, buffy below; crown striped with black and buff.

HABITAT AND DISTRIBUTION: Uncommon transient and local summer resident* (May–Sep) nearly throughout; rare as a breeder on the VA Coastal Plain and in northern and western PA: Deciduous and mixed forest, often on west-facing slopes.

DIET: Forages by gleaning and probing leaves, buds, leaf clusters, and dead leaf clumps in trees or on the ground for arthropods.

REPRODUCTION: Mostly monogamous, some polygyny; male arrives on breeding territory in May followed some days later by the female; open cup nest is placed on the ground in low dense vegetation, usually on a slope or bank against a trunk or ledge, and built (female) of leaves and lined often with dark orange or reddish moss sporophytes, hair, fine grass, and pine needles; clutch—4–6; incubation (female)—12–14 days; fledging—10 days post-hatching; post-fledging parental care—3 weeks; single brood per season.

CONSERVATION STATUS: S3—DE, NJ; conversion of floodplain forest to agriculture and residential.

RANGE: Breeds in the eastern United States. Winters from southern Mexico to Panama; West Indies.

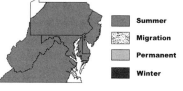

KEY REFERENCES: Rappole and Warner (1980), Hanners and Patton (1998).

Swainson's Warbler *Limnothlypis swainsonii*

DESCRIPTION: L = 15 cm (6 in); W = 23 cm (9 in); brown above, buff below with chestnut cap; whitish eye stripe.

HABITAT AND DISTRIBUTION: Uncommon to rare and local summer resident* (May–Aug) in southern VA and southwestern WV; scattered records elsewhere: Canebrakes, bottomland forests, swampy thickets; also in rhododendron thickets in the southern Appalachians.

DIET: Feeds mainly on the ground by rummaging through leaf litter for arthropod prey, e.g., Coleoptera, spiders, ants, Orthoptera, and millipedes.

REPRODUCTION: Monogamous; male arrives on breeding territory in mid to late April followed several days later by the female; open cup nest, 9–23 cm dia. (4–9 in), is placed low, < 1 m (3 ft), in dense understory vegetation normally supported by shrub branch or vine tangles, and built

by the female of dead leaves, twigs, and plant stems lined with grass, rootlets, pine needles, Spanish moss, and hair; clutch—3–4; incubation (female)—13–15 days; fledging—10–12 days post-hatching; post-fledging parental care—2–3 weeks; single brood per season.

CONSERVATION STATUS: S1—MD; SH—DE; S2—VA, WV; this species may have undergone a significant decline in the late 1800s, perhaps from loss of winter habitat in Cuba.

RANGE: Breeds in the south-eastern United States. Winters in the Bahamas,  Greater Antilles, eastern Mexico, and Yucatan.

KEY REFERENCES: Meanley (1971), Brown and Dickson (1994).

Ovenbird *Seiurus aurocapilla*

DESCRIPTION: L = 15 cm (6 in); W = 26 cm (10 in); olive above, white below with heavy dark streaks; orange crown stripe bordered in black; white eye-ring.

HABITAT AND DISTRIBUTION: Common transient and summer resident* (May–Sep) throughout: Deciduous and mixed forest.

DIET: Walks on forest floor, flicking leaves and duff while foraging for invertebrates; caged birds cast pellets.

REPRODUCTION: Monogamous with some serial polygyny; male arrives on breeding territory in late April or early May; female arrives a few days later; domed nest, 16–23 cm dia. (6–9 in), is placed on the forest floor, and is built by the female of grasses, stems, bark shreds, and leaves lined with hair; clutch—4–5; incubation—11–14 days; fledging—7–10 days post-hatching; post-fledging parental care—20–30 days; single brood per season normally; one record of two broods.

CONSERVATION STATUS: Declines have been reported from some parts of the range, for reasons unknown but perhaps related to winter habitat destruction.

RANGE: Central and eastern Canada and central and eastern United States  south to east KS and north GA. Winters from southern FL and southern Mexico south through Central America to northern Venezuela; West Indies.

KEY REFERENCES: Van Horn and Donovan (1994).

Northern Waterthrush *Seiurus noveboracensis*

DESCRIPTION: L = 15 cm (6 in); W = 26 cm (10 in); brown above, white or yellowish below with dark streaking on throat and breast; prominent creamy or yellowish eyestripe.

HABITAT AND DISTRIBUTION: Uncommon to rare and local summer resident* (May–Aug) in northern PA and in southern PA in the mountains, also the highlands of western MD, northern NJ (rare), and eastern WV; uncommon transient (Aug–Sep, May) elsewhere: Swamps, bogs, swales, ponds, rivers, lakes, usually near stagnant water.

DIET: Forages by wading in shallow water or walking on mud, logs, and vegetation along shore, gleaning invertebrates (e.g., insects, snails, crustacea, spiders) from water, vegetation, or soil surface.

REPRODUCTION: Monogamous with some serial polygyny; males arrive on breeding territory in late April or early May, followed by females; nest, 11 cm dia. (4 in), is placed in the roots and soil of fallen trees or in a streambank hole or cavity, generally open from the side and with a roof of roots, ferns, or soil; female builds nest of mosses, liverwort gametophytes, and leaves lined with grasses, rootlets, and hair; clutch—4–5; incubation (female)—12 days; fledging—9 days post-hatching; post-fledging parental care—4 weeks; single brood per season.

CONSERVATION STATUS: S1—VA; S2—WV, MD; S3—PA; threatened by loss of mature highland forest habitat.

RANGE: Breeds in boreal North America south of the Arctic Circle. Winters from central Mexico south through Central America to northern South America; Caribbean basin.

	Summer
	Migration
	Permanent
	Winter

KEY REFERENCES: Eaton (1995).

Louisiana Waterthrush *Seiurus motacilla*

DESCRIPTION: L = 15 cm (6 in); W = 26 cm (10 in); brown above, creamy white below with dark streaking on breast; prominent white eyestripe.

HABITAT AND DISTRIBUTION: Common to uncommon summer resident* (Apr–Aug) in lowlands and mid-elevations along streams nearly throughout; rare and local in NJ, mainly in the north: Streams, rivers, swales, ponds.

DIET: Forages on the ground or in shallow, running water, < 2 cm (1 in), bobbing as it walks, by picking prey from substrate or by pecking and pulling substrate material (e.g leaves) to expose prey; aquatic and soil insects are the main prey, but other invertebrates (e.g., spiders, isopods, earthworms) or vertebrates (e.g., tadpoles) are also taken.

REPRODUCTION: Monogamous; males arrive on breeding territory in late March or early April followed shortly thereafter (1–7 days) by the female; the open cup nest, 13–18 cm dia. (5–7 in), is placed usually in a shallow cavity of a stream bank, and is built by both sexes with leaves, pine needles, twigs, and stems in a mud matrix, lined with rootlets, fine plant stems, hair, and moss; clutch—5; incubation (female)—12–14 days; fledging—10–11 days post-hatching; post-fledging parental care—3–4 weeks; single brood per season.

CONSERVATION STATUS: S3—DE; conversion of floodplain forest to agriculture and residential.

RANGE: Breeds in the eastern United States; winters

from Mexico
to northern
South
America;
West Indies.

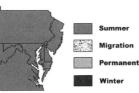

KEY REFERENCES:
Eaton (1958), Robinson (1995).

Kentucky Warbler *Oporornis formosus*

DESCRIPTION: L = 13 cm (5 in); W = 20 cm (8 in); greenish-brown above, yellow below; yellow spectacles (forehead, eyestripe, eyering); black crown, lores, earpatch. Black more or less replaced by greenish-brown in female.

HABITAT AND DISTRIBUTION: Common to uncommon summer resident* (May–Aug) in lowlands and mid-elevations nearly throughout; rare and local in NJ and scarce or absent from northern PA: Moist deciduous forest.

DIET: Forages by hopping (not walking) on the ground, picking arthropods from overhanging vegetation.

REPRODUCTION: Mainly monogamous with some polygyny as well as DNA evidence indicating extrapair fertilizations; male arrives on breeding territory in late April or early May followed by the female within 1–2 days; open cup nest, 7–9 cm dia. (3–4 in), is placed on or near the ground at the base of a shrub and is built by the female of leaves, twigs, and woven grass blades lined with fine grasses and rootlets; clutch—3–5; incubation (female)—11–13 days; fledging—8–9 days post-

hatching; post-fledging parental care—up to one month; some pairs attempt two broods per season.

CONSERVATION STATUS: S3—DE, NJ; conversion of floodplain forest to agriculture and residential.

RANGE: Breeds in the eastern United States. Winters from southern Mexico to northern Colombia and northwestern Venezuela.

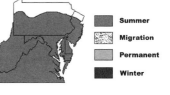

KEY REFERENCES: Rappole and Warner (1980), McDonald (1998).

Connecticut Warbler *Oporornis agilis*

DESCRIPTION: L = 15 cm (6 in); W = 23 cm (9 in); olive above, yellow below; gray hood; complete, white eyering. **Female** has a brownish-yellow head.

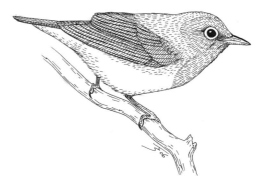

HABITAT AND DISTRIBUTION: Rare fall transient (Sep–Oct) throughout; scattered records in spring (May): Bogs, thickets.

DIET: Forages low and on the ground, gleaning arthropods from detritus or plant surfaces; takes some fruits on occasion.

RANGE: Breeds in central Canada and extreme

north central United States. Winters in northern South America.

KEY REFERENCES: Pitocchelli et al. (1997).

Mourning Warbler *Oporornis philadelphia*

DESCRIPTION: L = 15 cm (6 in); W = 23 cm (9 in); olive above, yellow below; gray hood with black on breast. **Female** Brownish-yellow head and partial eyering.

HABITAT AND DISTRIBUTION: Uncommon to rare and local summer resident* (May–Sep) in northern PA (mainly northwest), the Appalachians of eastern WV, Garrett County, MD, and highlands of Highland, Augusta, and Bath counties in VA; rare transient (Sep, May) elsewhere: Dense thickets and tangles of second growth woodlands.

DIET: Forages low by hopping between branches and stems, also on the ground; gleans arthropods (e.g., caterpillars, beetles, spiders) from branches and leaves of understory plants and leaf litter.

REPRODUCTION: Monogamous, so far as known; male arrives and establishes breeding territory in May with pair formation occurring shortly there-after; female selects nest site and builds nest, 10–22 cm dia. (4–9 in) on or near the ground in low, dense vegetation (e.g., blackberry, blueberry, elderberry, and cherry thickets); nest is built of bark shreds, leaves, sedges, grasses, and stems lined with fine grasses, roots, and hair; eggs are laid in late May or June; clutch—3–5; incuba-tion—12 days; fledging—9 days post-hatching;

post-fledging parental care—3 weeks; single brood per season.

CONSERVATION STATUS: S1—VA, MD; S3—WV, PA; threatened by loss of mature highland forest habitat.

RANGE: Breeds in central and eastern Canada and extreme

north central and northeastern United States. Winters from Nicaragua to northern South America.

KEY REFERENCES: Pitocchelli (1993).

Common Yellowthroat *Geothlypis trichas*

DESCRIPTION: L = 13 cm (5 in); W = 18 cm (7 in); olive-brown above; yellow below; black mask. **Female** Brownish above, bright yellow throat fading to whitish on belly; brownish on sides.

HABITAT AND DISTRIBUTION: Common transient and summer resident* (Apr–Sep) throughout; uncom-mon to rare winter resident (Oct–Apr), mainly along the southern Coastal Plain: Dense low veg-etation in marshes, stream borders, estuaries, meadows, riparian areas, open woodlands; reed beds bordering rivers, ponds, and streams.

DIET: Forages by skulking in low vegetation, gleaning arthropods from foliage.

REPRODUCTION: Mostly monogamous, with some polygyny; males arrive on breeding territory in April, followed within a week or so by the female;

bulky open cup nest, 9 cm dia. (4 in), is located on or near the ground (later nests often higher) in dense herbaceous vegetation, and built (female) of grass, leaves, and plant stems lined with fine grasses, bark fibers, and hair; clutch—4; incubation (female)—12 days; fledging—11–12 days post-hatching; post-fledging parental care—3 weeks; more than one brood per season occurs in some populations.

CONSERVATION STATUS: Both short- (1980–1994) and long-term (1966–1994) declines have been recorded for Common Yellowthroats in the Mid-Atlantic region, perhaps due to reductions in breeding or wintering wetland habitats.

RANGE: Breeds from Canada south throughout the continent to the south-

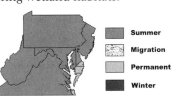

▣	Summer
▨	Migration
▤	Permanent
■	Winter

ern United States, and in the highlands to southern Mexico; winters from the southern United States through Mexico and Central America to Costa Rica; Bahamas, Greater Antilles.

KEY REFERENCES: Klicka (1994), Guzy and Ritchison (1999).

Hooded Warbler *Wilsonia citrina*

DESCRIPTION: L = 13 cm (5 in); W = 20 cm (8 in); olive above, yellow below with black hood, yellow forehead and face; white tail spots. **Female** Similar to male, but usually lacks hood and has a greenish cap, yellow forehead and eyestripe.

HABITAT AND DISTRIBUTION: Common to uncommon summer resident* (May–Sep) nearly throughout, although local in eastern PA, NJ, and along the Coastal Plain: Dense thickets, tree falls in lowland deciduous, swamp, and riparian forest.

DIET: Forages by sallying, hawking, hover-gleaning, or gleaning from perches in shrubs, saplings, or trees; often fans tail exposing white spots; main prey items are flying insects during the nonbreeding season, but caterpillars, spiders, beetles, and grasshoppers can be important during the breeding period.

REPRODUCTION: Monogamous, but extrapair copulations account for significant numbers of offspring in some populations; males arrive on breeding territories in early to mid-May with females arriving a few days later; open cup nest is placed 0.3–1 m (1–3 ft) above the ground in the crotch of a shrub, usually in a patch of shrubs; female builds nest of bark shreds, grasses, plant down, and spider web lined with fine grasses and hair; clutch—4; incubation—12 days; fledging—young crawl off from nest at 8–9 days post-hatching, and can fly 2–3 days later; post-fledging parental care—4–5 weeks; multiple broods are common in some populations.

CONSERVATION STATUS: S1—DE; S2—NJ; conversion of floodplain forest to agriculture and residential.

RANGE: Breeds in the eastern United States; winters from southern Mexico to Panama.

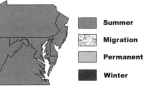

▣	Summer
▨	Migration
▤	Permanent
■	Winter

KEY REFERENCES: Ogden and Stutchbury (1994).

Wilson's Warbler *Wilsonia pusilla*

DESCRIPTION: L = 13 cm (5 in); W = 18 cm (7 in); olive-yellow above, yellow below with a black cap; **Female** Often has only a partially black or completely greenish-yellow crown.

HABITAT AND DISTRIBUTION: Uncommon to rare

transient (Sep–Oct, May) nearly throughout, although scarcer along the southern Coastal Plain: Wet woodlands, brushy thickets (willow, alder, aspen, dogwood).

DIET: Forages actively by gleaning, hovering, and sallying in the mid- to upper canopy level of shrubs or trees for arthropods.

CONSERVATION STATUS: Wilson's Warblers have undergone significant short- (1980–1994) and long-term (1966–1994) population declines for reasons that are not understood, but may relate to destruction of winter habitat (tropical forest).

RANGE: Breeds in boreal regions of northern and western North America; winters from southern CA and TX south to Panama.

KEY REFERENCES: Buskirk et al. (1972), Powell (1980), Rappole and Warner (1980), Ammon and Gilbert (1999).

Canada Warbler *Wilsonia canadensis*

DESCRIPTION: L = 13 cm (5 in); W = 20 cm (8 in); slate-gray above, yellow below with black "necklace" across breast; yellow lores and eyering.
Female Similar but necklace is usually fainter.

HABITAT AND DISTRIBUTION: Common to uncommon and local summer resident* (May–Aug) in northern PA and in southern PA in the mountains, also the highlands of VA, western MD, northern NJ, and the Appalachians of eastern WV; uncommon to rare transient (Aug–Sep, May) elsewhere: A variety of coniferous and mixed forests and thickets.

DIET: Forages at middle and lower levels of forest vegetation mainly by hawking and sallying for flying insects and by gleaning arthropods from foliage while hopping along branches; participates in mixed-species foraging flocks during nonbreeding periods.

REPRODUCTION: Monogamous; no information on pair formation—probably pair members arrive on breeding territory in mid-May in the Mid-Atlantic region; open cup nest, 9–14 cm dia. (4–6 in), is placed on the ground in low dense vegetation, often in a nook or hole in a bank, rock, stump, or upturned tree roots, and built (female) of grass, bark strips, leaves, plant stems, moss, pine needles, and plant down lined with hair, rootlets, leaves, and fine grasses; clutch—4–5; incubation (female)—12 days; fledging—10 days? post-hatching; post-fledging parental care—at least 1 week; single brood per season.

CONSERVATION STATUS: S3—NJ, MD; the Canada Warbler has declined sharply over the past 40 years for reasons that are not understood, but may be related to

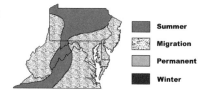

destruction of winter habitat (cloud forest) in the foothills of the Andes.

RANGE: Breeds in eastern and central Canada from Labrador to northeastern British Columbia and in boreal United States from Minnesota to New England; south in the Appalachians to northern GA. Winters in northern South America.

KEY REFERENCES: Conway (1999).

Yellow-breasted Chat *Icteria virens*

DESCRIPTION: L = 18 cm (7 in); W = 26 cm (10 in); a nearly thrush-sized warbler; brown above; yellow throat and breast; white belly and undertail coverts; white eyering and supraloral stripe; lores black or grayish; white malar stripe.

HABITAT AND DISTRIBUTION: Common but local summer resident* (May–Sep) nearly throughout; scarce in northern PA and in highlands; rare winter visitor along the Coastal Plain: Dense thickets, brushy pastures, old fields, hedgerows, forest undergrowth. Very shy; has flight song.

DIET: Forages in low, dense vegetation by gleaning arthropods from leaves and branches; takes fruit seasonally.

REPRODUCTION: Mostly monogamous, some polygyny; males arrive on breeding territory in May followed within a few days by females; bulky, open cup nest, 14 cm (6 in), is placed in the crotch of a low shrub or sapling in dense vegetation and built (female) of grasses, leaves, bark strips, and plant stems, lined with fine grasses and plant stems, pine needles, rootlets, and hair; clutch—3–5; incubation (female)—11–12 days; fledging—8–10 days post-hatching; post-fledging parental care—?; some double-brooding.

CONSERVATION STATUS: S3—DE,NJ; old field breeding habitats are being replaced by forest, urban development, or intensive agriculture.

RANGE: Breeds in scattered regions nearly throughout the United

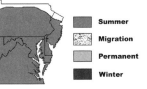

States and southern Canada south to central Mexico. Winters from central Mexico to Panama.

KEY REFERENCES: Eckerle and Thompson (2001).

FAMILY THRAUPIDAE

Summer Tanager *Piranga rubra*

DESCRIPTION: L = 20 cm (8 in); W = 33 cm (13 in); red. **Female and Immature Male** Tawny brown above, more yellowish below; second-year males and some females are blotched with red.

HABITAT AND DISTRIBUTION: Common to uncommon and local summer resident* (May–Sep) in western WV, extreme southwestern PA (Green County), and along the Coastal Plain of DE, MD, and VA; uncommon in the Piedmont and valleys and mid-elevations of the mountains of MD and VA; rare or absent elsewhere: Open deciduous and mixed woodlands.

DIET: Forages by hawking for large flying insects (e.g., Hymenoptera, Orthoptera), hover-gleaning, or gleaning arthropods from leaf or branch surfaces, or tearing apart bee and wasp nests to capture larvae; also feeds on fruit when available.

REPRODUCTION: Monogamous; males arrive on breeding territory in late April or early May followed shortly thereafter by the female; open cup nest, 9–11 cm (4 in) is usually placed in a clump of leaves on the horizontal branch of a deciduous tree some distance, 4 m (13 ft), from the trunk; the female builds the nest of leaves, stems, and grass lined with fine grasses; clutch—3–4; incubation (female)—11–12 days; fledging—9–11 days post-hatching; post-fledging parental care—2–4 weeks; single brood per season.

CONSERVATION STATUS: S2—DE, PA; conversion of floodplain forest to agriculture and residential.

RANGE: Breeds across eastern and southern portions of the United States and

northern Mexico; winters from southern Mexico through Central America to northern South America.

KEY REFERENCES: Robinson (1996).

Scarlet Tanager *Piranga olivacea*

DESCRIPTION: L = 20 cm (8 in); W = 33 cm (13 in); red with black wings. **Winter male** Greenish above with black wings and tail—splotched with red during molt. **Female and Immature Male** Olive above, yellowish below.

HABITAT AND DISTRIBUTION: Common transient and summer resident* (May–Sep) throughout, more numerous as a breeder inland than along the Coastal Plain: Deciduous and mixed forest; oak and riparian woodland.

DIET: Forages relatively high, 10 m (33 ft) on average in forests by hawking and gleaning invertebrates (e.g., Lepidoptera, Coleoptera, Diptera) from foliage, branches, trunks, and, occasionally from

the forest floor; also feeds on fruits, especially during the nonbreeding period when it often joins mixed-species foraging flocks.

REPRODUCTION: Monogamous; male arrives on breeding territory in late April or early May followed within a week or so by the female; open cup nest, 9–13 cm dia. (4–5 in), is placed away from the trunk on a horizontal branch, often in a tuft of leaves, and built (female) of twigs, plant stems, grass, bark shreds, rootlets, and pine needles lined with fine grasses, rootlets, plant strips, and pine needles; clutch—4; incubation (female)—13–14 days; fledging—9–12 days post-hatching; post-fledging parental care—at least two weeks; one brood per season is normal.

CONSERVATION STATUS: Scarlet Tanager populations likely are threatened by winter habitat loss (evergreen forest) along the east slope of the northern Andes in South America.

RANGE: Breeds in the northeastern United States and southeastern Canada; winters in northern South America.

KEY REFERENCES: Mowbray (1999), Vega Rivera et al. (2003).

FAMILY EMBERIZIDAE

Eastern Towhee *Pipilo erythrophthalmus*

DESCRIPTION: L = 20 cm (8 in); W = 28 cm (11 in); red eye; black head, breast, and back; black wings with white patches; rufous sides; white belly; tail black and rounded with white corners. **Female** Patterned similarly but brown instead of black.

HABITAT AND DISTRIBUTION: Common transient and summer resident* (Mar–Oct) throughout; common to uncommon winter resident (Nov–Feb) in southwestern WV and the Coastal Plain and Piedmont of DE, MD, and VA, rare (mainly at feeders) or absent elsewhere in the region in winter: Undergrowth and thickets of deciduous and mixed woodlands, old fields, and scrub.

DIET: This bird spends most of its time on the ground, using backward kick-hops to scatter duff and expose seeds and invertebrate prey, although it will glean arthropods and pick fruits in trees.

REPRODUCTION: Monogamous; male arrives on breeding territory in mid-late April followed by the female; open cup nest, 12 cm dia. (5 in), is normally placed on the ground (first nest) or low in the understory (later nests) and is built by the female of leaves, bark strips, plant stems, and twigs lined with fine grasses, rootlets, pine needles, and hair; clutch—3–5; incubation—12–13 days; fledging—10–11 days post-hatching; post-fledging parental care—3–4 weeks; two broods per season apparently common in Mid-Atlantic region.

CONSERVATION STATUS: Short (1980–1994) and long (1966–1994) term declines have been recorded, presumably resulting from reversion of old field and scrub habitats to forest or conversion to intensive agriculture.

RANGE: Breeds from extreme southern Canada across the United States

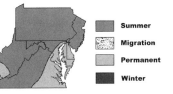

Summer

Migration

Permanent

Winter

(except most of TX) and in the highlands of Mexico and Guatemala; winters from central and southern United States to Mexico and Guatemala.

KEY REFERENCES: Greenlaw (1996).

Bachman's Sparrow *Aimophila aestivalis*

DESCRIPTION: L = 15 cm (6 in); W = 20 cm (8 in); grayish streaked with brown above; pale gray below; dark malar stripe; gray cheek; central crown stripe bordered by brown.

HABITAT AND DISTRIBUTION: Rare and local resident* of southeastern VA: Open pine woods and savanna with dense ground cover; brushy, overgrown fields and pastures. Secretive.

DIET: Forages by walking, running, or hopping on ground and gleaning seeds, especially those of grasses, e.g., *Panicum*, and insects from litter and low vegetation.

REPRODUCTION: Monogamous; males begin territorial singing in February or March; construction of open cup or domed nest, 13 cm dia. (5 in), begins in April or May; nest is placed in a small scrape or depression in the ground, often at the base of an overhanging grass clump, shrub, or seedling, and built (female) of grass, plant stems and strips, and rootlets lined with fine grass and hair; clutch—3–4; incubation (female)—12–14 days; fledging—9–10 days post-hatching; post-fledging parental care—21–28 days; 2 broods per season are common.

CONSERVATION STATUS: SX—PA; SH—MD, WV; S1—VA; populations of this species expanded into the Mid-Atlantic region in the early 1800s, but disappeared during the mid-1900s, presumably due to disappearance of shrubby second growth associated with small family farms.

RANGE:
Southeastern
United States.
KEY REFERENCES:
Dunning
(1993).

American Tree Sparrow *Spizella arborea*

DESCRIPTION: L = 15 cm (6 in); W = 23 cm (9 in); streaked brownish above; dingy white below with dark breast spot; rufous crown; 2 white wingbars.

HABITAT AND DISTRIBUTION: Common to uncommon transient and winter resident (Nov–Mar)

nearly throughout, scarcer along the Coastal Plain: Open, shrubby areas; weedy fields, grassland, overgrown pasture, gardens, feeders.

DIET: Forages on the ground by picking and gleaning seeds and invertebrates from litter or vegetation; also perches on plant stems to pick fruits and seeds.

RANGE: Breeds
in bog, tundra, and willow thickets
of northern
North America; winters in southern Canada, northern and central United States.

KEY REFERENCES: Naugler (1993).

Chipping Sparrow *Spizella passerina*

DESCRIPTION: L = 13 cm (5 in); W = 20 cm (8 in); a small sparrow, streaked rusty brown above, dingy white below with 5 whitish wingbars; rufous cap (somewhat streaked in winter); whitish eyebrow; black eyeline. **Immature** Streaked crown; gray or buffy eyebrow; brown cheek patch.

HABITAT AND DISTRIBUTION: Common summer resident* (Apr–Oct) throughout; rare winter visitor, mainly along the Coastal Plain: Open pine forests, woodlands, orchards, parks, suburbs, and cemeteries with scattered coniferous trees.

DIET: Forages mostly on the ground by scratching in detritus or gleaning low in vegetation for plant seeds or arthropods (breeding season); often in single-species flocks during the nonbreeding period.

REPRODUCTION: Mostly monogamous with some polygyny; male arrives on breeding territory in April, 1–2 weeks before the female; open cup nest, 11 cm dia. (4 in), is usually placed in small, dense, conifers, 1–3 m (3–10 ft) above ground, and built by the female of woven grasses and rootlets lined with hair and fine grasses; clutch—3–4; incubation (female)—10–12 days; fledging—9–12 days post-hatching; post-fledging parental care—3 weeks; two broods per season are normal.

RANGE: Breeds over most of North America south of the tundra, south through Mexico and Central America to Nicaragua; winters along coast and in southern portions of the breeding range.

KEY REFERENCES: Middleton (1998).

Clay-colored Sparrow *Spizella pallida*

DESCRIPTION: L = 13cm (5 in); W = 20 cm (8 in); streaked brown above, buffy below; streaked crown with central gray stripe; grayish eyebrow; buffy cheek patch outlined in dark brown; gray nape; 2 white wingbars.

HABITAT AND DISTRIBUTION: Rare fall transient (Sep–Oct) in coastal MD and VA; scattered breed-

ing records in western PA: Grasslands with scattered low shrubs, brushy pastures.

DIET: Forages mostly on the ground by picking seeds and invertebrates from the ground and low vegetation.

RANGE: Breeds from central Canada to the north central United States; winters from TX south to Guatemala.

KEY REFERENCES: Knapton (1994).

Field Sparrow *Spizella pusilla*

DESCRIPTION: L = 15 cm (6 in); W = 20 cm (8 in); pink bill; streaked brown above, buff below; crown with gray central stripe bordered by rusty stripes; 2 whitish wingbars.

HABITAT AND DISTRIBUTION: Common summer resident* (Mar–Nov) throughout; common to uncommon winter resident (Dec–Feb) in southwestern WV, Piedmont, and Coastal Plain; scarce or absent in winter in PA, northern NJ, and highland areas: Old fields, brushy pastures.

DIET: Forages on the ground or in low vegetation, <1 m (3 ft) mainly for grass and other small plant seeds; insects, e.g., caterpillars, can be a large part of the diet during the breeding season.

REPRODUCTION: Monogamous; males may remain on the breeding territory through the winter in southern parts of the region; otherwise, they arrive by April; females arrive later, in mid-April to early May; open cup nest, 12 cm dia. (5 in), is placed on or near the ground early in the season, but higher up, < 1 m (3 ft), in shrubs or saplings later in the season; female builds nest of interwoven grasses lined with fine grasses, hair, and rootlets; clutch—4; incubation—11–12 days; fledging—7–8 days post-hatching; post-fledging parental care—26–34 days; multiple broods per season are common.

CONSERVATION STATUS: Field Sparrows have suffered both short (1980–1994) and long (1966–1994)-term declines, presumably caused by land-use changes resulting in significant reductions in old field habitat.

RANGE: Breeds across the eastern half of the U. S. and southeastern Canada; winters in the southern half of the breeding range south to FL and northeastern Mexico.

Summer
Migration
Permanent
Winter

KEY REFERENCES: Carey et al. (1994).

Vesper Sparrow *Pooecetes gramineus*

DESCRIPTION: L = 15 cm (6 in); W = 26 cm (10 in); grayish streaked with brown above; white with brown streaks below; rusty shoulder patch; white outer tail feathers; white eyering.

HABITAT AND DISTRIBUTION: Common to uncommon transient and summer resident* (Mar–Oct) across most of the region; rare and local along the Coastal Plain and Piedmont as a breeder; uncommon to rare in winter (Nov–Feb), scarcer inland: Grasslands, old fields, pastures, scrub, agricultural fields.

DIET: Forages by hopping along on the ground and scratching with both feet in litter to expose seeds and insects, which are gleaned with a rapid pecking motion.

REPRODUCTION: Mostly monogamous; males arrive on breeding territory in April with females following within a week; open cup nest, 8–10 cm dia. (3–4 in), is placed on the ground, usually in a

small depression, hidden by low vegetation, and built (female) of grasses, plant fibers, stems, sedges, rootlets, mosses, and bark strips, and lined with hair, rootlets, down feathers, fine grasses and pine needles; clutch—3–5; incubation (mostly female)—12–13 days; fledging—9–14 days post-hatching; post-fledging parental care—20–29 days; double broods in some parts of range, e.g., WV.

CONSERVATION STATUS: S1—NJ; S3—MD, DE, WV; grassland breeding habitat is limited, perhaps because of land-use changes associated with forest regrowth and development.

RANGE: Breeds across much of temperate North America; winters in the southern United States and Mexico.

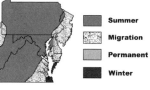

Summer
Migration
Permanent
Winter

KEY REFERENCES: Jones and Cornely (2002).

Lark Sparrow *Chondestes grammacus*

DESCRIPTION: L = 18 cm (7 in); W = 28 cm (11 in); streaked light and dark brown above; dingy white or grayish below; distinctive chestnut, white, and black face pattern; white throat; black breast spot; white corners on black, rounded tail.

HABITAT AND DISTRIBUTION: Rare fall transient (Sep), mainly along the VA coast; recent breeding records from Wayne County, WV; scattered records from other places and seasons: Grasslands, pastures, coastal prairie and dunes, agricultural fields.

DIET: Forages on the ground in open areas for plant seeds (75%) and insects (25%).

RANGE: Breeds in the Canadian prairie states and across most of the  United States except eastern, forested regions, south into northern Mexico; winters from the southern United States to southern Mexico.

KEY REFERENCES: Martin and Parrish (2000).

Savannah Sparrow *Passerculus sandwichensis*

DESCRIPTION: L = 15 cm (6 in); W = 23 cm (9 in); buff striped with brown above; whitish variously streaked with brown below; yellow or yellowish lores; whitish or yellowish eyebrow; often with dark, central breast spot. Plumage is highly variable in amount of yellow on face and streaking

on breast according to subspecies, several of which winter in the region.

HABITAT AND DISTRIBUTION: Common to uncommon transient and summer resident* (Mar–Nov) in PA, the northern panhandle of WV, and highlands of eastern WV, VA, and western MD; rare and local as a breeder in NJ saltmarshes; common to uncommon transient and winter resident (Sep–Apr), mainly along the coast: Grasslands, pastures, agricultural fields, coastal marshes. Often in flocks.

DIET: Forages mainly on the ground for seeds, fruits, arthropods, and other small invertebrates.

REPRODUCTION: Chiefly monogamous, but polygyny is common in some populations; males arrive and establish breeding territories in April and May; pairs form with female arrival somewhat later; female builds nest, 7–8 cm dia.(3 in) in May or early June; the open cup nest is constructed on the ground, usually in a clump of grass or at the base of a shrub, and often concealed by a canopy of herbaceous growth; the nest is built of coarse grass blades lined with finer, interwoven grasses often with a few feathers; clutch—4; incubation—12–13 days; fledging—9–11 days after hatching; post-fledging parental care—3 weeks; multi-brooded populations common in some parts of the range.

CONSERVATION STATUS: S2—NJ; S3—VA, MD; grassland breeding habitat is limited, perhaps because of land-use changes associated with forest regrowth and development.

RANGE: Breeds throughout the northern half of North America south to the central 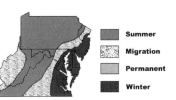 United States; also breeds in the central highlands of Mexico and Guatemala; winters in the coastal and southern United States south to Honduras.

KEY REFERENCES: Wheelwright and Rising (1993).

Grasshopper Sparrow *Ammodramus savannarum*

DESCRIPTION: L = 13 cm (5 in); W = 20 cm (8 in); a small, relatively short-tailed sparrow; streaked brown above; creamy-buff below; buffy crown stripe; yellow at bend of wing; yellowish or buffy lores and eyebrow.

HABITAT AND DISTRIBUTION: Common to uncommon and local transient and summer resident* (Apr–Sep) throughout; rare to casual in winter (Oct–Mar): Prairie, pasture, grassland, savanna.

DIET: Forages by walking or running (in a hunched posture) along the ground in low, sparse vegetation or bare areas to capture invertebrates (e.g., Orthoptera) or pick up grass and forb seeds.

REPRODUCTION: Mostly monogamous; males arrive on breeding territory in April followed 3–5 days later by females; domed nest with side entrance (11–14 cm dia. (4–5 in), is placed in a scrape or depression on the ground in low, dense vegetation; nest is built by the female of grasses woven into surrounding vegetation and lined with fine grasses, sedges, and hair; clutch—3–5; incubation (female)—11–13 days; fledging—9 days posthatching; post-fledging parental care—1–2 weeks?; 2 broods per season are common.

CONSERVATION STATUS: S2—NJ; S3—DE, WV; Grasshopper Sparrows have shown long-term declines (1966–1994) in the Mid-Atlantic region and range-wide; suggested reasons included reversion of grasslands to forests and conversion to intensive farming.

RANGE: Breeds across the northern and central United States; southern

Mexico to northwestern South America; Greater Antilles; winters in the southern United States, Mexico, and elsewhere within its tropical breeding range.

KEY REFERENCES: Vickery (1996).

Henslow's Sparrow *Ammodramus henslowii*

DESCRIPTION: L = 13 cm (5 in); W = 18 cm (7 in); streaked rusty and dark brown above with rusty wings; whitish breast with dark streaks; whitish belly; buffy head with dark brown crown and malar stripes. **Immature** Breast streaking is faint or absent.

HABITAT AND DISTRIBUTION: Uncommon to rare and local transient and summer resident* (Apr–Sep) in scattered localities throughout the region. The principal known concentration is in western PA, with lesser numbers in western MD and western WV. Formerly more numerous in the region. However, the bird is difficult to locate, and numbers may be higher than suspected: Tall grass prairie, wet meadows, sedge marshes, weedy fields, hayfields.

DIET: Forages on or near the ground in dense foliage, gleaning insects, weed seeds, and grass seeds, the proportion varying according to site and season.

REPRODUCTION: Monogamous; breeding begins in May; open cup nest, 9–13 cm dia. (4–5 in), is placed on or near the ground in a dense clump of grasses or forbs, and built (female) of woven grass blades, with finer grasses in the lining; clutch—3–5; incubation (female)—10–11 days; fledging—9 days post-hatching; post-fledging parental care—?; no data on number of broods per season.

CONSERVATION STATUS: S1—WV, VA, NJ, MD; SH—DE; tall grassland breeding habitat is considered threatened by changes in farming practices.

RANGE: Breeds in the north-eastern and north central United States and south-eastern Canada; winters in the southeastern United States.

KEY REFERENCES: Herkert et al. (2002).

LeConte's Sparrow *Ammodramus leconteii*

DESCRIPTION: L = 13 cm (5 in); W = 18 cm (7 in); streaked brown above; whitish below with dark streaking on sides; whitish central crown stripe bordered by dark brown stripes; buffy-yellow eyebrow stripe; gray cheek patch outlined by buffy-yellow.

HABITAT AND DISTRIBUTION: Rare winter visitor (Oct–Apr) to extreme southeastern VA: Tall grass-

lands, wet meadows, prairie, saltmarsh, rank fields.

DIET: Apparently forages low or on the ground in dense grass stands for seeds of grasses and forbs, although few foraging observations have been reported; also takes insects.

RANGE: Breeds in central Canada (British Columbia to Quebec) and extreme north central United States (Montana to MI); winters in the southeastern United States.

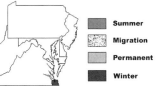

KEY REFERENCES: Lowther (1996).

Nelson's Sharp-tailed Sparrow *Ammodramus nelsoni*

DESCRIPTION: L = 15 cm (6 in); W = 20 cm (8 in); streaked dark brown above; buffy throat and buffy-orange breast; belly whitish; gray crown stripe bordered by dark brown crown stripes; buffy-orange eyebrow; gray nape.

HABITAT AND DISTRIBUTION: Uncommon and local winter resident (Oct–Apr) along the Coastal Plain: Wet grasslands, coastal and inland marshes.

DIET: Forages on the ground or clinging to low vegetation by pecking, probing, and gleaning from mud or vegetation; feeds mainly on arthropods (e.g., insects, spiders, amphipods), but also occasionally on molluscs or other invertebrates and plant seeds.

RANGE: Breeds in central Canada and northeastern and north central United States (ND, MN, ME); winters along the coast of the southeastern United States from VA to TX and northeastern Mexico.

KEY REFERENCES: Greenlaw and Rising (1994).

Saltmarsh Sharp-tailed Sparrow *Ammodramus caudacutus*

DESCRIPTION: L = 15 cm (6 in); W = 20 cm (8 in); streaked dark brown above; buffy breast and flanks with faint brown streaks; belly whitish; gray crown stripe bordered by dark brown crown stripes; buffy-orange eyebrow; gray cheek outlined by buffy-orange stripes above and below; white throat; gray nape.

HABITAT AND DISTRIBUTION: Common to uncommon and local resident* of saltmarshes along the immediate coast: Coastal saltmarsh.

DIET: Forages on the ground or clinging to low vegetation by pecking, probing, and gleaning from mud or vegetation; feeds mainly on arthropods (e.g., insects, spiders, amphipods), but also occasionally on molluscs or other invertebrates and plant seeds.

REPRODUCTION: Polygynous with no pair bond; nest is placed in dense tufts of grass in open marsh, a few cm to a meter or more above the ground or water; the female builds the open cup or partially domed nest, 10 cm dia. (4 in), of woven, coarse grass blades lined with finer grasses; clutch—3–5; incubation—11–12 days;

fledging—10–11 days post-hatching; post-fledging parental care—?; may produce more than one brood per season.

CONSERVATION STATUS: S2—VA; S3—DE, MD; saltmarsh habitats are disappearing in many areas due to development.

RANGE: Breeds along the Atlantic Coast of the United States from ME to NC; winters along the Atlantic Coast from NY to the central east coast of FL.

KEY REFERENCES: Greenlaw and Rising (1994).

Seaside Sparrow *Ammodramus maritimus*

DESCRIPTION: L = 15 cm (6 in); W = 20 cm (8 in); yellow lores; white throat; grayish streaked with dark brown above; buffy breast and whitish belly streaked with brown; a stocky bird with longish bill and short tail.

HABITAT AND DISTRIBUTION: Common summer resident* along the immediate coast, scarcer in winter: Saltmarshes; different micro-habitat requirements for nesting (dense stands of sturdy marsh grass) and foraging (open mud and smooth cordgrass) mean that adults may commute some distance between nesting and foraging sites.

DIET: Forages mainly by picking and probing while walking in mud or clambering through stems of open, smooth cordgrass stands or other saltmarsh types; main prey items are arthropods (e.g., Orthoptera, Lepidoptera, and spiders).

REPRODUCTION: Monogamous; males may remain on or near breeding territories throughout the year in the Mid-Atlantic region; females arrive by April or early May; partially canopied or completely domed nest, 10 cm dia. (4 in), is built of woven grasses by the female, usually in saltmarsh grass stems, 0.1–0.3 m (4–12 in) above ground; clutch—3–4; incubation (female)—12 days; fledging—young clamber off the nest 9–11 days post-hatching, but are unable to fly for another 10–12 days; post-fledging parental care—20 days; some pairs raise more than one brood per season.

RANGE: Resident along the coast of the eastern United States from ME to TX; northern populations (ME to VA) are partly or totally migratory.

KEY REFERENCES: Post and Greenlaw (1994).

Fox Sparrow *Passerella iliaca*

DESCRIPTION: L = 18 cm (7 in); W = 28 cm (11 in); hefty, for a sparrow—nearly thrush-sized; streaked dark or rusty brown above; whitish below with heavy dark or rusty streakings that often coalesce as a blotch on the breast; grayish eyebrow and neck; rusty cheek; rusty rump and tail.

HABITAT AND DISTRIBUTION: Common to uncommon transient (Nov, Mar) and uncommon to rare winter resident (Dec–Feb) along the Coastal Plain,

Piedmont, southwestern WV, or at feeders: Thickets and undergrowth of deciduous, mixed, and coniferous woodlands; hedgerows; scrub, brush piles, weedy stream borders, bushy tangles.

DIET: Forages in towhee fashion, jump-kicking its way through understory duff and leaf litter for invertebrates (e.g., Coleoptera, Lepidoptera, Homoptera, Arachnida), seeds, and fruits; also picks seeds and fruits from shrubs, vines, and trees.

RANGE: Breeds across the northern tier of North America and in the mountains of the west; winters in the coastal and southern United States.

KEY REFERENCES: Weckstein et al. (2002).

Song Sparrow *Melospiza melodia*

DESCRIPTION: L = 15 cm (6 in); W = 23 cm (9 in); streaked brown above; whitish below with heavy brown streaks and central breast spot; gray eyebrow; dark whisker and postorbital stripe.

HABITAT AND DISTRIBUTION: Common resident* nearly throughout; scarce in highlands; withdraws to dense stream-border thickets and marshes in winter: Shrubby thickets, swamp borders, inland and coastal marshes, riparian thickets (reeds, sedges), wet meadows, brushy fields.

DIET: Forages on or near the ground in dense undergrowth for invertebrates, seeds, and fruits by scratching, picking, and gleaning; mostly animal matter in summer and vegetable matter in winter.

REPRODUCTION: Mostly monogamous with some polygamy; adult males may remain on territory throughout the year while females often migrate; pairs form in February and March with initial nest construction in late March or early April; open cup nest is placed on or near the ground in dense vegetation; later nests average higher in vegetation; bulky nest, 13–16 cm dia. (5–6 in), is built (female) of grass, plant stems, bark strips, and leaves lined with fine grasses, rootlets, and hair; clutch—3–5; incubation (female)—12–13 days; fledging—10 days post-hatching; post-fledging parental care—12–20 days; multiple broods per season are common.

RANGE: Breeds across temperate and boreal North America; winters in temperate breeding range, southern United States, and northern Mexico; resident population in central Mexico.

KEY REFERENCES: Nice (1937, 1943), Arcese et al.(2002).

Lincoln's Sparrow *Melospiza lincolnii*

DESCRIPTION: L = 15 cm (6 in); W = 20 cm (8 in); streaked brown above; patterned gray and brown face; white throat and belly; distinctive finely streaked, buffy breast band.

HABITAT AND DISTRIBUTION: Rare transient and winter resident (Oct–May) in southeastern VA: Brushy fields, hedgerows, riparian thickets.

DIET: Forages by kicking and scratching through ground litter for plant seeds and arthropods; also gleans or picks prey from leaves and twigs.

RANGE: Breeds across the northern tier of North America and in the mountains of the west; winters in the coastal and southern United States south to Honduras.

KEY REFERENCES: Ammon (1995).

Swamp Sparrow *Melospiza georgiana*

DESCRIPTION: L = 15 cm (6 in); W = 20 cm (8 in); rusty crown; gray face; streaked brown above with rusty wings; whitish throat but otherwise grayish below with tawny flanks—faintly streaked. **Winter** Crown is brownish with central stripe gray rather than rusty.

HABITAT AND DISTRIBUTION: Uncommon and local summer resident* (May–Sep) in much of PA and NJ; local breeding populations in highlands and coastal marshes of MD and DE, Highland County, VA, and the Appalachian highlands of WV; common transient and winter resident (Sep–Apr) along the Piedmont and Coastal Plain; transient (Sep–Oct, Mar–Apr) elsewhere: Bogs, coastal and inland marshes, wet grasslands, brushy pastures.

DIET: Forages by walking on moist ground or in shallow water, picking invertebrate prey (e.g., Odonata, Orthoptera, Hymenoptera, Coleoptera) or seeds from substrate; also takes fruit when available.

REPRODUCTION: Mainly monogamous with some polygyny; male arrives on breeding territory in April followed 2–3 weeks later by the female; bulky open cup nest, 11 cm dia. (4 in), is placed low in dense marsh vegetation with sturdy shrub, cattail, or grass stems for support; female builds nest of grass and sedge blades, twigs, leaves, and rootlets lined with woven fine grasses, hair, rootlets, and plant down; clutch—4; incubation (female)—12–14 days; fledging—9–11 days post-hatching; post-fledging parental care—15 days; two broods per season common.

CONSERVATION STATUS:—S1—VA; S3—WV, DE; wetland drainage and pollution.

RANGE: Breeds in central and eastern Canada and north central and north-

eastern United States; winters along the Pacific coast and in eastern and south central United States south to southern Mexico.

KEY REFERENCES: Mowbray (1997).

White-throated Sparrow *Zonotrichia albicollis*

DESCRIPTION: L = 18 cm (7 in); W = 23 cm (9 in); white throat; alternating black and white (or black and buff) crown stipes; yellow lores; streaked brown above; grayish below.

HABITAT AND DISTRIBUTION: Uncommon and local summer resident* (Apr–Sep) in northern PA; also breeding records from the highlands of eastern WV and northwestern NJ; common transient and winter resident (Oct–Apr) nearly throughout, scarce or absent from highlands in winter: Coniferous bogs, coniferous and mixed forest and second growth, beaver meadows (breeding); woodland thickets, brushy fields, feeders (migration, winter).

DIET: Forages on the ground by picking arthropods

or seeds exposed by kicking or tossing away leaf litter, and by gleaning along branches in trees and shrubs; main prey items are arthropods (e.g., caterpillars, spiders, and Coleoptera) during the breeding period, but seeds and fruits predominate during fall migration and in winter, when birds forage mostly in small, loose flocks; plant buds can be important food items in spring.

REPRODUCTION: Monogamous; males arrive on breeding territories in late April or early May with females arriving 1–2 weeks later; open cup nest, 7–14 cm dia. (3–6 in), is placed on or near the ground in dense understory vegetation, usually along the edge of a forest clearing or opening; female builds nest of coarse grass, twigs, wood chips, pine needles, and roots lined with moss, hair, fine grasses and rootlets; clutch—4–5; incubation (female)—11–14 days; fledging—young leave nest 7–12 days post-hatching, and are able to fly 4–5 days later; post-fledging parental care—at least 2 weeks; multiple broods are common in some populations.

CONSERVATION STATUS: S3—PA; long-term (1966–1989) breeding population declines have been recorded for reasons that are not understood.

RANGE: Breeds across most of boreal Canada and northeastern and north central United States; winters in the eastern,

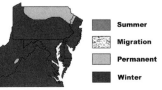

southern, and Pacific coastal regions of the United States and northern Mexico.

KEY REFERENCES: Falls and Kopachena (1994).

White-crowned Sparrow *Zonotrichia leucophrys*

DESCRIPTION: L = 18 cm (7 in); W = 26 cm (10 in); black and white striped crown; gray neck, breast, and belly; streaked gray and brown back; pinkish bill. **Immature** Crown stripes are brown and gray.

HABITAT AND DISTRIBUTION: Common to uncommon transient (Oct–Nov, Mar–Apr) throughout; uncommon to rare and local winter resident (Nov–Mar) in southwestern PA, western WV, the Piedmont, and Coastal Plain, often at feeders and multiflora rose hedges: Thickets in coniferous and deciduous woodlands, brushy fields.

DIET: Forages, often in small flocks, by picking up plant matter (mostly seeds) from the ground or while perched in low forbs, grasses, or shrubs; arthropods form an increasing portion of the diet during northward migration and on the breeding ground.

RANGE: Breeds in northern and western North America; winters across

Summer
Migration
Permanent
Winter

most of the U. S. south to central Mexico.

KEY REFERENCES: Chilton et al. (1995).

Dark-eyed Junco *Junco hyemalis*

DESCRIPTION: L = 15 cm (6 in); W = 26 cm (10 in); entirely dark gray except for white belly and outer tail feathers, and pinkish bill. **Female** Similar but brownish rather than gray.

HABITAT AND DISTRIBUTION: Common to uncommon resident* in northern PA and in southern PA in the mountains, also the highlands (> 1,000 m) of western MD, northern NJ (rare), eastern WV, and VA; common transient and winter resident (Oct–Apr) elsewhere: Coniferous and mixed forests (breeding); open mixed woodlands, grasslands, roadsides, hedgerows, agricultural fields, feeders (migration, winter).

DIET: Forages mostly near or on the ground by picking and gleaning small seeds, < 5 mm (0.2 in) and invertebrates from the substrate; vegetable matter composes the bulk of the diet for most of the year but animal matter can be up to 50% of diet in summer.

REPRODUCTION: Monogamous; male arrives on breeding territory in late March or April with females arriving 10–14 days later; open cup nest, 11–12 cm dia. (4–5 in), is usually placed on the ground in a crevice, cranny, stump root ball, or natural depression, often with overhanging vegetation, and built (female) of roots, leaves, bark strips, moss, grass, and fern rhizomes lined with fine grasses, hair, and moss setae; clutch—3–5; incubation (female)—12–13 days; fledging—9–12 days post-hatching; post-fledging parental care—14 days; two broods per season are common in some populations.

CONSERVATION STATUS: S1—NJ; S2—MD; loss of mature highland forest habitat.

RANGE: Breeds across northern North America and in the eastern and western

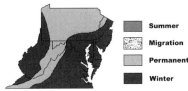

U. S., south in the mountains; winters from southern Canada south through the United States to northern Mexico.

KEY REFERENCES: Nolan et al. (2002).

Lapland Longspur *Calcarius lapponicus*

DESCRIPTION: L = 15 cm (6 in); W = 28 cm (11 in); black head and breast with broad, white or buff postorbital stripe; yellow bill; rusty nape; streaked brown above; white belly; tail is all dark except for outermost tail feathers; very long hind claw. **Winter Male and Female** Brownish crown; buffy eyebrow, nape, and throat with darker brown mottlings on breast and flanks; buffy cheek outlined by darker brown.

HABITAT AND DISTRIBUTION: Rare to casual transient and winter visitor (Nov–Feb) throughout: Plowed, recently manured, and stubble fields, overgrazed pasture, open areas where grass and weed seeds are readily found.

DIET: Forages by walking on bare or sparsely vegetated ground and picking up weed and grass seeds and grains; also takes insects as available.

RANGE: Breeds in tundra of extreme northern North America and

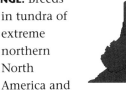

Eurasia; winters in temperate regions of the Old and New World.

KEY REFERENCES: Hussell and Montgomerie (2002).

Snow Bunting *Plectrophenax nivalis*

DESCRIPTION: L = 18 cm (7 in); W = 28 cm (11 in); white head, rump, and underparts; black back; wings white with black primaries. **Winter Male and Female** White is tinged with buff and dark back is mottled with white.

HABITAT AND DISTRIBUTION: Uncommon to rare and irregular winter visitor (Nov–Feb) from PA and NJ to northern VA, and along the coast; casual elsewhere: Lake shores, shingle beaches, sand dunes, saltmarsh, snowy fields, pastures, agricultural areas (especially recently manured fields), roadsides, farmyards.

DIET: Forages by walking along on the ground and pecking and gleaning seeds and invertebrates from the litter or plant stems, often in the melt zone bordering snow patches; also forages on beaches, running after retreating waves to capture exposed invertebrates.

RANGE: Breeds circumpolar in tundra; winters in

southern boreal and northern temperate regions of Old and New World.

KEY REFERENCES: Lyon and Montgomerie (1995).

Northern Cardinal *Cardinalis cardinalis*

DESCRIPTION: L = 23 cm (9 in); W = 31 cm (12 in); red with crest; black face; red bill. **Female and Immature Male** Crested like male but greenish-brown, paler below, bill brownish or reddish.

HABITAT AND DISTRIBUTION: Common resident* throughout: Thickets and tangles of open deciduous forest and second growth, hedgerows, old fields, residential parklands.

DIET: Forages by picking fruits, seeds, and arthropods from trees, shrubs, and ground litter; annual diet is 71% plant and 29% animal, with a much heavier emphasis on the animal portion during the breeding period.

REPRODUCTION: Monogamous, with some birds remaining paired throughout the year; extrapair paternity has been measured at 9–35% in some populations; breeding activities begin late March or early April; open cup nest, 11 cm dia. 4(in), is placed low, avg. 1.5 m (60 in) in dense shrubbery and built by the female of twigs, bark strips, leaves, rootlets, and pine needles and lined with fine grasses; clutch—2–3; incubation (female)—11–13 days; fledging—9–10 days post-hatching; post-fledging parental care—25–56 days; two or more broods per season are a common occurrence.

RANGE: Eastern United States and southeastern Canada; southwestern United States, Mexico, Guatemala, and Belize.

KEY REFERENCES: Halkin and Linville (1999).

Rose-breasted Grosbeak *Pheucticus ludovicianus*

DESCRIPTION: L = 20 cm (8 in); W = 33 cm (13 in); black head, back, wings, and tail; brownish in winter; red breast; white belly, rump, wing patches, and tail spots. **Female** Mottled brown above, buffy below heavily streaked with dark brown; white eyebrow and wingbars. **First Year Male** Like female but shows rose tints on breast and underwing coverts. **Second Year Male** Patterned much like adult male but splotched brown and black on head and back.

HABITAT AND DISTRIBUTION: Uncommon and local summer resident* (May–Sep) in PA, northern NJ, and the highlands, > 1000 m (3300 ft) of western MD, eastern WV, and VA; common to uncommon transient (Sep, May) elsewhere: Deciduous and mixed forest and parklands.

DIET: Mostly insectivorous during spring migration and early summer, gleaning insects from leaves and branches (e.g., Coleoptera, Hymenoptera, Hemiptera, Homoptera, Lepidoptera); fruits, seeds, flowers, and buds increasingly important in late summer and fall.

REPRODUCTION: Monogamous; males arrive on breeding territory in May a few days before females; both males and females sing, sometimes even while on the nest; open, loose, cup nest, 15 cm dia. (6 in), is usually placed in the fork of a deciduous tree about 6 m (20 ft) above the ground, and built (both sexes) of sticks, twigs, grasses, plant stems, and leaves, and lined with rootlets, fine grasses, and hair; clutch—3–5; incubation (both sexes)—12–13 days; fledging—10–12 days post-hatching; post-fledging parental care—3 weeks; mostly one brood per season.

CONSERVATION STATUS: S3—MD; short-term declines (1980–1994) have occurred in our region for reasons not understood.

RANGE: Breeds in northeastern United States, central and southeastern

Canada; winters from southern Mexico south through Central America to northern South America and western Cuba.

KEY REFERENCES: Wyatt and Francis (2002).

Blue Grosbeak *Passerina caerulea*

DESCRIPTION: L = 18 cm (7 in); W = 28 cm (11 in); dark blue with 2 rusty wingbars. **Female and Immature Male** Brownish above, paler below with tawny wingbars, often has a blush of blue on shoulder or rump.

HABITAT AND DISTRIBUTION: Common to uncommon summer resident* (May–Sep) in the Coastal Plain and Piedmont from southern NJ south, valleys of the mountain region of VA, and western WV: Thickets, scrub, brushy pastures, hedgerows.

DIET: Flicks and fans tail while gleaning prey from vegetation; also forages on the

ground; feeds on insects (e.g., Orthoptera, Coleoptera), plant seeds, and snails.

REPRODUCTION: Presumed monogamous; pair formation occurs in April; female probably builds open cup nest, 8 cm dia. (3 in) of bark shreds, twigs, leaves, and snake skins lined with fine grass, rootlets, and hair, and placed in dense understory vegetation, especially brambles and vine tangles; clutch—3–5; incubation—11–12 days; fledging—10 days post-hatching; post-fledging parental care—?; multi-brooded, at least in southern part of range.

CONSERVATION STATUS: S3—WV; land-use changes resulting in significant reductions in old field habitat.

RANGE: Breeds from central and southern U. S. through Mexico and Central America to Costa Rica; winters from

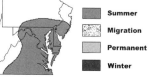

northern Mexico to Panama; rarely Cuba.

KEY REFERENCES: Ingold (1993).

Indigo Bunting *Passerina cyanea*

DESCRIPTION: L = 15 cm (6 in); W = 23 cm (9 in); indigo blue. **Female and Immature Male** Brown above, paler below with faint streaking on breast. **Winter Adult and Second Year Male** Bluish with variable amounts of brown on back and wings.

HABITAT AND DISTRIBUTION: Common transient and summer resident* (May–Sep) nearly throughout: Thickets, hedgerows, brushy fields; occasionally open forests where dense herbaceous and low shrub layers occur.

DIET: Small invertebrates (caterpillars, spiders, beetles, etc.), seeds, and berries during the breeding season; principally grass and weed seeds during the nonbreeding season.

REPRODUCTION: Mostly monogamous but with variable amounts of serial polygyny (15%) and extra-pair fertilizations (20%—40%); males arrive and establish breeding territories in May, females arrive a few days later; female selects a site 0.3–1.0 m (12–39 in) up in the crotch of a low shrub (e.g., blackberry, goldenrod, or nettle) and builds an open cup nest, 8 cm dia. (3 in), of

leaves, grasses, and bark lined with fine grasses; clutch—3–4; incubation—12–13 days; fledging— 9–12 days post-hatching; post-fledging parental care—3 weeks after fledging; re-nesting or attempts to raise more than one brood can extend the nesting period into September, but most breeding is complete by mid-August.

CONSERVATION STATUS: Indigo Buntings have shown both short- and long-term population declines in the eastern United States, based on National Breeding Bird Survey data, perhaps resulting from once extensive old field habitat being replaced by closedcanopy forest, high-intensity agriculture, and residential areas.

RANGE: Breeds from extreme southeastern and south central Canada south

Summer
Migration
Permanent
Winter

through eastern and southwestern United States; winters from central Mexico to Panama and the West Indies.

KEY REFERENCES: Payne (1992).

Dickcissel *Spiza americana*

DESCRIPTION: L = 15 cm (6 in); W = 23 cm (9 in); patterned like a miniature meadowlark—black bib (gray in winter) and yellow breast; streaked brown above; grayish head with creamy eyebrow; rusty red wing patch. **Female and Immature Male** Patterned like male but paler yellow below and without black bib.

HABITAT AND DISTRIBUTION: Rare, irregular, and local summer resident* (May–Sep) at scattered localities that change from year to year: Grasslands, agricultural fields.

DIET: Feeds mostly on animal matter (70%) during the breeding period and vegetable matter (up to 99%), mostly seeds picked from plant seed heads or gleaned from the ground, during the non-breeding period. Sorghum and rice are favorite

foods in many parts of both its breeding and wintering ranges.

REPRODUCTION: Polygynous; when birds do breed in the region, males arrive on territories in May; females arrive shortly thereafter; several females may settle on the territory of a single male, depending on the number of good nest sites available; bulky, open cup nest is placed low, < 1m (3 ft) in dense shrubs, grass clumps, or herbaceous vegetation, and built (female) of plant stems, grass blades, and leaves lined with fine grasses, rootlets, and hair; clutch—3–6; incubation (female)—12–13 days; fledging—8–10 days post-hatching; post-fledging care (female)—14 days; usually one brood per season, although it has been suggested that birds may breed at one locality, and then migrate to breed at another locality.

CONSERVATION STATUS: S1—NJ; S2—PA, WV, MD; Dickcissels were common in the Mid-Atlantic states during the early and mid-1800s, but disappeared as regular breeders in the region in the late

1800s for reasons that are not well understood.

RANGE: Breeds across the eastern and central United States and south central Canada; winters mainly in northern South America.

KEY REFERENCES: Temple (2002).

FAMILY ICTERIDAE

Bobolink *Dolichonyx oryzivorus*

DESCRIPTION: L = 18 cm (7 in); W = 31 cm (12 in); black with creamy nape, white rump and shoulder patch. **Winter Male and Female** Streaked brown and yellow buff above; yellow-buff below; crown with central buff stripe bordered by dark brown stripes; buff eyebrow.

HABITAT AND DISTRIBUTION: Uncommon to rare and local summer resident*, nesting in hayfields of PA, northwestern NJ, western MD, the eastern panhandle of WV, and Highland County, VA; common fall (Aug–Sep) and uncommon spring (May) transient throughout: Prairie, grasslands, grain and hay fields, brushy pastures.

DIET: Forages by walking on the ground or clambering on stems, picking and gleaning seeds and invertebrates.

REPRODUCTION: Serially polygynous; males arrive on breeding territory in May, followed by females some days later; walled and partially or completely canopied nest with side entrance is placed in a shallow depression on the ground in dense herbaceous growth, and built by the female of woven grass, leaves, and stems, lined with fine grasses; clutch—5; incubation (both sexes, at least for primary female on male's territory)—11–13 days; fledging—10–11 days post-hatching, but cannot fly until about 13 days; post-fledging parental care—28 days; mostly single brood per season but some second broods are attempted.

CONSERVATION STATUS: S1—VA; S2—WV, NJ; S3—MD; Bobolink breeding populations have declined sharply in recent years; two different explanations have been offered: 1) loss of breeding habitat due to reforestation and/or changes in farming practices, and 2) loss of winter habitat (Di Giacomo et al. 2003).

RANGE: Breeds in the northern United States and southern Canada; winters in southern South America.

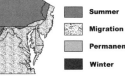

KEY REFERENCES: Bollinger and Gavin (1992), Martin and Gavin (1995).

Red-winged Blackbird *Agelaius phoeniceus*

DESCRIPTION: L = 20 cm (8 in); W = 36 cm (14 in); black with red epaulets bordered in orange. **Female and Immature Male** Dark brown above, whitish below heavily streaked with dark brown; whitish eyebrow and malar stripes. **Second Year Male** Intermediate between female and male—black blotching, some orange on epaulet.

HABITAT AND DISTRIBUTION: Common transient and summer resident* (Mar–Oct) throughout; common winter resident (Nov–Feb) in coastal marshes; also winters elsewhere in lowlands of the region, generally in large flocks: Inland and coastal marshes (breeding); brushy fields; tall grasslands; grain and hay fields; grain storage areas (winter).

DIET: Forages by picking and gleaning plant matter and invertebrates from the ground or while

perched in vegetation; also uses "gaping" to expose arthropods and other prey for capture; feeds primarily on plant matter, mainly grain and other seeds, during the nonbreeding season and animal matter, mostly insects, during the breeding season, although the proportion can vary according to availability.

REPRODUCTION: Polygynous; males arrive on breeding territory in March or early April, followed by females; open cup nest, 10–12 cm dia. (4–5 in), is placed often in sturdy emergent vegetation (e.g., bulrush, cattail, or reeds), and is built by the female of coarse, wet grass interwoven with living support stems, wet leaves, and decayed wood, lined with fine grass; clutch—3–4; incubation (female) 11–13 days; fledging—12 days; post-fledging parental care—up to 5 weeks?; single brood per season normally.

RANGE: Breeds nearly throughout North America from the Arctic

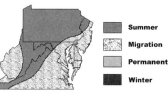

▓	**Summer**
░	**Migration**
▒	**Permanent**
█	**Winter**

Circle south to Costa Rica; winters from temperate portions of breeding range south; resident populations in Bahamas and Cuba.

KEY REFERENCES: Orians (1985), Yasukawa and Searcy (1995).

Eastern Meadowlark *Sturnella magna*

DESCRIPTION: L = 26 cm (10 in); W = 38 cm (15 in); streaked brown and white above; yellow below with a black or brownish "V" on the breast; crown striped with buff and dark brown; tail is dark in center, white on outer edges.

HABITAT AND DISTRIBUTION: Common transient and summer resident* (Mar–Oct) throughout; common to uncommon or rare winter resident (Nov–Feb), seen mainly in agricultural fields and marshes in lowlands: Grasslands, savanna, grain and hay fields, overgrown pastures.

DIET: Forages by walking on the ground, picking and probing for insects (e.g., Orthoptera, Lepidoptera, Coleoptera), grain, and seeds.

REPRODUCTION: Polygynous; males usually have 2 mates; males arrive on breeding territory in late February or March, females arrive 2–4 weeks later; open, partially domed, or completely roofed nest with side entrance tunnels is placed on the ground in open grassland, and built by the female of woven grass, plant stems, and bark strips (14–21 cm dia. (6–8 in),; clutch—4–5; incubation (female)—13–14 days; fledging—young leave the nest at 10–12 days, but cannot fly until 21 days post-hatching; post-fledging parental care—2 weeks; successful raising of 2 broods per season is uncommon.

CONSERVATION STATUS: S3—DE, NJ; Eastern Meadowlark breeding populations have shown both short (1980–1994) and long-term (1966–1994) declines in Mid-Atlantic, presumably due to replacement of grasslands and old fields by forest.

RANGE: Breeds from southeastern Canada and the eastern

▓	**Summer**
░	**Migration**
▒	**Permanent**
█	**Winter**

and southern U. S. west to AZ; Mexico through Central and northern South America; Cuba; winters through most of breeding range except northern portions.

KEY REFERENCES: Lanyon (1995).

Yellow-headed Blackbird
Xanthocephalus xanthocephalus

DESCRIPTION: L = 26 cm (10 in); W = 41 cm (16 in); black body with yellow head and breast; white wing patch. **Female and Immature Male** Brown body; yellowish breast and throat; yellowish eyebrow.

HABITAT AND DISTRIBUTION: Rare transient, mainly in fall (Aug–Oct), and winter visitor along the immediate coast; scarcer inland: Marshes, brushy pastures, agricultural fields. Favorite sites for these and several other blackbird species are cattle feedlots and grain elevators in winter.

DIET: During the nonbreeding period, Yellow-headed Blackbirds migrate (diurnally), roost, and forage in large, mixed flocks with other icterid species (e.g., Red-winged Blackbird, Common Grackle, Brewer's  Blackbird, and Brown-headed Cowbird); they feed primarily by walking on the ground in agricultural fields, feed lots, and grain elevators, and picking up grains, seeds, and arthropods; they will also perch on plant stems to feed on unharvested grain or other plants with seed heads.

RANGE: Breeds in south central and southwestern Canada, north central and western United States; winters from southern CA, AZ, NM, and TX south to southern Mexico.

KEY REFERENCES: Orians (1985), Twedt and Crawford (1995).

Rusty Blackbird *Euphagus carolinus*

DESCRIPTION: L = 23 cm (9 in); W = 36 cm (14 in); entirely black with creamy yellow eye. Breeding **Female** Grayer. **Winter Male** Black is tinged with rusty; often shows a buffy eyebrow. **Winter Female** Rusty above, buffy below with prominent buffy eyebrow.

HABITAT AND DISTRIBUTION: Common to uncommon transient (Oct–Nov, Mar–Apr) throughout; uncommon to rare and local winter visitor (Dec–Feb) mainly along the coast from Cape May south, and in bottomland swamps and marshes: Deciduous and coniferous forests, swamps, wooded edges of marshes.

DIET: During the nonbreeding period, this bird often forages in small flocks in forested or open wetlands, and in pastures, fields, and feedlots; feeds by walking on the ground or in shallow water and picking, probing, or "gaping" ; main food items in fall and winter are grains, plant seeds, mast, fruit, aquatic invertebrates, and small aquatic vertebrates; less likely to

join huge winter roosts or feeding flocks than other "blackbird" species.

RANGE: Breeds in boreal coniferous forest and bogs across the northern tier of North America; winters in the eastern United States.

KEY REFERENCES: Orians (1985), Avery (1995).

Brewer's Blackbird *Euphagus cyanocephalus*

DESCRIPTION: L = 23 cm (9 in); W = 38 cm (15 in); entirely black with purplish gloss on head (in proper light); yellow eye; some fall males are tinged rusty. **Female** Dark brown above with dark brown eye, slightly paler below.

HABITAT AND DISTRIBUTION: Rare fall and winter visitor (Nov–Mar) mainly along the immediate coast of southern VA; scarcer elsewhere in the region: Grasslands, pastures, agricultural fields, feedlots, grain elevators.

DIET: Forages in large flocks during the nonbreeding period, mostly by walking on bare or sparsely vegetated ground and picking up seeds and grains; also takes insects and fruits seasonally.

RANGE: Breeds in the western and central United States and Canada; winters in the breeding range from southwestern Canada and the western United States southward through the southern United States to central Mexico.

Summer
Migration
Permanent
Winter

KEY REFERENCES: Martin (2002).

Common Grackle *Quiscalus quiscula*

DESCRIPTION: L = 31 cm (12 in); W = 43 cm (17 in); entirely black with purple gloss on head in proper light; tail long and rounded; cream-colored eye. **Female** Dull black; whitish eye; tail not as long as male's.

HABITAT AND DISTRIBUTION: Common summer resident* (Mar–Oct) throughout; common although local winter resident (Nov–Feb) in western WV, Piedmont, Coastal Plain, and at scattered lowland localities elsewhere in the region, often concentrated in large flocks at roosts and feeding sites: Open woodlands, swamps, marshes, urban areas, agricultural fields, pastures, feeders, grain storage areas.

DIET: Forages mostly by walking on the ground and gleaning or probing for plant and animal food items from the soil or detritus; more animal than plant material is eaten during the breeding period, while the reverse is true during the nonbreeding period, when birds often move in large flocks between roosting and foraging sites.

REPRODUCTION: Some males are monogamous while others are polygynous; pairs form with arrival of females on the breeding territory in April; bulky, open cup nest, 17–22 cm dia. (7–9 in), is placed 3–6 m (10–20 ft) up in the fork of a conifer, deciduous tree or shrub and built mostly by the female of twigs, plant stems, leaves, grass, bark strips, and moss, lined first with mud and then fine grasses and hair; sometimes nests are built in loose colonies with other grackles; clutch—5; incubation—13–14 days; fledging—12–15 days post-hatching; post-fledging parental care—"several weeks"; mostly one brood per season.

RANGE: Breeds east of the Rockies in Canada and United States; winters in the southern half of the breeding range.

KEY REFERENCES: Peer and Bollinger (1997).

Boat-tailed Grackle *Quiscalus major*

DESCRIPTION: Male L = 43 cm (17 in); W =59 cm (23 in); Female L = 33 cm (13 in); W = 46 cm (18 in); entirely black with purplish gloss on head (in proper light); tail wedge-shaped and longer than body; creamy eye. **Female** Brown above, paler below; buffy eyebrow; brown eye; wedge-shaped tail not as long as male's.

HABITAT AND DISTRIBUTION: Common but local summer resident* (Apr–Oct) along the immediate coast; winter resident (Nov–Mar) from Cape May southward, although in lower numbers: Coastal marshes, beaches, coastal residential and urban areas, fields, feedlots.

DIET: Forages by walking on the ground or in shallow water, pecking, probing, "gaping," and rummaging in benthos, soil, litter, and debris for a wide variety of food items including crustacea, molluscs, aquatic insects, small aquatic vertebrates, tubers, seeds, fruits, and organic garbage.

REPRODUCTION: harem polygyny—one male paired with several nesting females; males and females begin visits to colony sites in March or April with first nests initiated 3–4 weeks later; nests are often

located in marshland, with several females building nests within a few meters of each other in sturdy emergent vegetation, shrubs, or trees, usually near or over water; open cup nest, 17–22 cm dia. (7–9 in), is built of grass, Spanish moss, and leaves woven around the supporting branches or stems, floored with grass, mud, and wet, decayed vegetation, and lined with fine grass and pine needles; clutch—3; incubation (female)—13 days; fledging—13 days post-hatching; post-fledging female care—up to three weeks; two broods per season are common.

CONSERVATION STATUS: S3—MD; saltmarsh habitats are disappearing in many areas due to development.

RANGE: Coastal eastern United States from NY to TX; northern populations at least partly migratory.

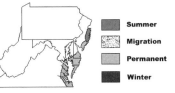

KEY REFERENCES: Post et al. (1996).

Brown-headed Cowbird *Molothrus ater*

DESCRIPTION: L = 18 cm (7 in); W = 33 cm (13 in); black body; brown head. **Female** Brown above, paler below with grayish streakings.

HABITAT AND DISTRIBUTION: Common transient and summer resident* (Apr–Oct) throughout; common although local winter resident (Nov–Mar) in western WV, Piedmont, Coastal Plain, and at scattered lowland localities elsewhere in the region, often in large flocks: Pastures, agricultural fields, feedlots, grain elevators, scrub, open woodlands.

DIET: Forages on the ground for native plant seeds and grains; also insects (25% of diet).

REPRODUCTION: A number of mating systems have been observed in different populations, including monogamy, polygyny, polyandry, and promiscuity; egglaying occurs mainly from mid-May to mid-June; cowbirds are social parasites, laying their eggs in other birds' nests; at least 220 host species have been reported, 144 of which are

known to have successfully raised cowbird young; incubation—11–12 days; fledging—8–13 days post-hatching; care by adoptive parents after fledging—25–39 days.

CONSERVATION STATUS: This species is hypothesized to have caused population declines for several species (e.g., Kirtland's Warbler) by reducing nesting success, and control measures have been put in place in some instances.

RANGE: Breeds across most of North America from south of the Arctic to central Mexico; winters in the southern half of its breeding range.

KEY REFERENCES: Lowther (1993).

Orchard Oriole *Icterus spurius*

DESCRIPTION: L = 18 cm (7 in); W = 26 cm (10 in); black hood, wings and tail; chestnut belly, wing patch, lower back, and rump. **Female** Greenish-yellow above, yellow below with 2 white wing-bars; blue-gray legs. **Second Year Male** Like female but with black throat and breast.

HABITAT AND DISTRIBUTION: Common to uncommon summer resident* (May–Aug) at lowlands and mid-elevations from southern PA and NJ south throughout the region; uncommon to rare and local in northern areas and highlands: Riparian woodlands, orchards, brushy pastures, scrub.

DIET: During the breeding season, adults glean insects (e.g., Orthoptera and Coleoptera) and spiders from leaves and twigs of shrubs and trees; during the nonbreeding seasons, the species often forages in large flocks for fruits and seeds.

REPRODUCTION: Monogamous; males arrive on breeding territory in late April or early May followed shortly thereafter by females; pendulous nest is suspended from the outer branches of a deciduous shrub or tree, and built mainly by the female of woven grasses lined with fine grasses, hair, and plant down; clutch—5; incubation (female) 12–14 days; fledging—14 days post-hatching; post-fledging parental care—1 week+?; adult males apparently begin migration in July, shortly after young achieve independence; single brood per season.

CONSERVATION STATUS: Long term decline (1966–1994) recorded for the region, presumably resulting from loss of old field habitat.

RANGE: Breeds across the eastern and central United States south into

central Mexico; winters from central Mexico south to northern South America.

KEY REFERENCES: Scharf and Kren (1996).

Baltimore Oriole *Icterus galbula*

DESCRIPTION: L = 20 cm (8 in); W = 31 cm (12 in); black hood, back, and wings; tail black at base and center but outer terminal portions orange; orange belly, rump, and shoulder patch; white wing bar and wing patch. **Female and**

Immature Orange-brown above, yellow-orange below with varying amounts of black on face and throat; white wing bars.

HABITAT AND DISTRIBUTION: Common transient (Aug–Sep, May) throughout; common to uncommon summer resident* (May–Aug) nearly throughout except along Coastal Plain where scarce and local; rare along coast in winter (feeders): Riparian woodland (sycamore, cottonwood, willow), orchards, open deciduous woodlands, hedgerows, second growth.

DIET: Forages by picking, probing, and gleaning arthropods from leaves and bark, generally high in the canopy; often feeds in small single-species flocks during the nonbreeding period on fruits and nectar in addition to arthropods.

REPRODUCTION: Monogamous with some extrapair paternity; male arrives on breeding territory in May followed 2–4 days later by the female; pendant nest, 7–8 cm (3 in) wide, 8–11 cm (3–4 in) deep, is suspended high in the outer branches of a tree (e.g., sycamore or elm) and built by the female of woven hair, grass, plant stems, and bark strips lined with plant down and feathers; clutch—4–5; incubation—11–14 days; fledging—12–13 days post-hatching; post-fledging parental care—2 weeks?

RANGE: Breeds in the eastern U. S. and southeastern Canada; winters from central Mexico to northern South America; also in Greater Antilles.

KEY REFERENCES: Rising and Flood (1998).

FAMILY FRINGILLIDAE

Pine Grosbeak *Pinicola enucleator*

DESCRIPTION: L = 23 cm (9 in); W = 36 cm (14 in); rosy head, breast, and rump; gray flanks and belly; black back, wings, and tail; white wingbars; heavy, black grosbeak bill. **Female** Gray body; head tinged with yellow; whitish shading below eye; white wingbars.

HABITAT AND DISTRIBUTION: Rare to casual and irregular winter visitor (Nov–Mar) throughout, more numerous in northern portions and highlands of the region. Breeds in sub-Arctic and boreal coniferous forest, often bordering streams or tarns; found in hemlocks, pines, sumac, mountain ash, and similar fruiting trees in winter, often in small flocks.

DIET: Forages on buds, seeds, and fruits of spruce, pine, juniper, elm, mountain ash, and a wide variety of other tree, shrub, and herbaceous species; feeds insects to its nestlings.

CONSERVATION STATUS: Numbers of winter invasions into temperate regions may have declined over the past half century, perhaps from clear cutting of boreal forests in Canada causing population reductions.

RANGE: Breeds in boreal forest of the Old and New Worlds, south to NM in the

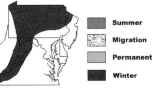

Rockies; winters in southern portions of the breeding range south into north temperate regions of the Old and New World.

KEY REFERENCES: Adkisson (1999).

Purple Finch *Carpodacus purpureus*

DESCRIPTION: L = 15 cm (6 in); W = 26 cm (10 in); rosy head and breast with darker broad, postorbital stripe; whitish belly; brown above suffused

with rose; rose rump; tail is notched; undertail coverts white. **Female and Immature Male** Brown above; white below heavily streaked with brown; brown head with white eyebrow and malar stripes, and broad, brown postorbital stripe; undertail coverts white.

HABITAT AND DISTRIBUTION: Common to uncommon summer resident* (May–Aug) in northern PA and in southern PA in the mountains, also the highlands, above 1000 m (3300 ft), of western MD, northwestern NJ (rare), eastern WV, and VA (Highland Co., Mt. Rogers area); common to uncommon and local transient and winter resident (Oct–Apr) elsewhere: Coniferous and mixed forest and parklands; also fields, hedgerows, shrubby areas, and feeders in winter.

DIET: Forages mostly by picking buds, seeds, and fruits from trees.

REPRODUCTION: Monogamous; nest construction begins in April; open cup nest is placed on a coniferous or deciduous tree branch, 1–18 m (3–60 ft) above ground, and is built, mostly by the female, of twigs and roots lined with fine grasses; clutch—4; incubation (mostly by female)—12–13 days; fledging—13–16 days posthatching; post-fledging parental care—?; second broods per season have been recorded, but frequency is unknown.

CONSERVATION STATUS: S1—VA; S3—WV, MD, NJ; threats to limited highland forest breeding habitat.

RANGE: Breeds across Canada, northern United States,

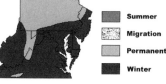

western coastal mountains south into Baja California, and the Appalachians south to VA; winters throughout much of the United States except in the Great Plains, Rockies, and western deserts.

KEY REFERENCES: Wootton (1996).

House Finch *Carpodacus mexicanus*

DESCRIPTION: L = 15 cm (6 in); W = 26 cm (10 in); brown above with red brow stripe; brown cheeks; rosy breast; whitish streaked with brown below. **Female and Immature** Brown above; buffy head and underparts finely streaked with brown; buffy preorbital stripe in some.

HABITAT AND DISTRIBUTION: Common to uncommon resident* nearly throughout; some southward migration in winter: Scrub, old fields; agricultural, residential, and urban areas; usually nests in scrubby conifers, including ornamentals; common at feeders.

DIET: 97% vegetable matter, picking seeds, fruits, and buds from plant stems or off the ground; even young in the nest are fed plant products.

REPRODUCTION: Monogamous; pairs form in winter

flocks in January and February; nests, 12 cm dia. (5 in) are bulky, open cups built by the female of coarse plant material lined with rootlets, grasses, and feathers, and placed in dense conifers, ornamental evergreens, or, occasionally, on ledges, nooks, or crannies of buildings.

RANGE: Resident from southwestern Canada throughout much of the United States except Great Plains, south to southern Mexico.

Summer

Migration

Permanent

Winter

KEY REFERENCES: Hill (1993).

Red Crossbill *Loxia curvirostra*

DESCRIPTION: L = 18 cm (7 in); W = 28 cm (11 in); red with dark wings and tail; dark lores and eyeline; crossed bill. **Female and Immature** Yellowish with dark wings and tail.

HABITAT AND DISTRIBUTION: Rare and irregular winter visitor (Oct–Apr); travels in small flocks widely during the nonbreeding period, wherever appropriate conifer nut crops are available; much more numerous in some years than others; has bred* in the region: Mature conifer forests; a variety of coniferous species during nonbreeding periods.

DIET: Conifer seeds, including red and black spruce, white, pitch, loblolly and scrub pine, and eastern hemlock in the Mid-Atlantic; Benkman (1993) determined that there are at least 8 different "types" of the Red Crossbill in its North American range, each apparently best adapted to use of conifer seeds of specific size and hardness; types

1–3 are the main ones found in our region .

CONSERVATION STATUS: S1— VA; spruce-fir breeding habitat for the Red Crossbill is rare and threatened by acid rain.

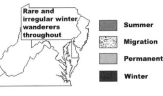

Rare and irregular winter wanderers throughout

Summer

Migration

Permanent

Winter

RANGE: Resident in boreal regions of the Old and New World, south in the mountains of the west through Mexico and the highlands of Central America to Nicaragua; winters in breeding range and irregularly south in north temperate regions of the world.

KEY REFERENCES: Adkisson (1996).

White-winged Crossbill *Loxia leucoptera*

DESCRIPTION: L = 18 cm (7 in); W = 28 cm (11 in); rosy-red body with dark band across back; black tail and wings with broad white wingbars; crossed bill. **Female and Immature** Brown tinged with yellow above; yellowish below; dark wings with white wingbars.

HABITAT AND DISTRIBUTION: Rare and irregular winter visitor (Oct–Apr); travels in small flocks widely during the nonbreeding period, wherever conifer nut crops are available; more numerous in some years than others: Coniferous and mixed forest.

DIET: Mainly spruce and tamarack seeds, but also other conifer seeds; occasionally buds and seeds of other types of plants; insects in summer.

RANGE: Resident in boreal northern tier of North

America rang-
ing south in
winter into
northern
temperate
zone.

Rare and
irregular winter
wanderers
throughout

Summer

Migration

Permanent

Winter

KEY REFERENCES: Benkman (1992).

Common Redpoll *Carduelis flammea*

DESCRIPTION: L = 13 cm (5 in); W = 23 cm (9 in);
front half of crown red; back of head and back
streaked brown and white; black face and chin;
yellowish bill; white below tinged with rose on
throat and breast; brown streaks on flanks and
belly. **Female** Like male but with little or no rose
on breast.

HABITAT AND DISTRIBUTION: Rare and irregular winter visitor (Nov–Mar) in northern portions and
highland meadows of the region; casual elsewhere. Travels widely during the nonbreeding
period in small, active, noisy flocks; more numerous in some years than others: Open woodlands,
agricultural fields, brushy pastures, feeders.

DIET: Forages by hanging, chickadee-like, and picking small seeds from terminal twigs, catkins, and
shoots of shrubs, weeds, grasses, and forbs.

RANGE: Breeds
in the high
Arctic of both
Old and New
World; win-
ters in boreal
and northern temperate regions.

Summer

Migration

Permanent

Winter

KEY REFERENCES: Knox and Lowther (2000).

Pine Siskin *Carduelis pinus*

DESCRIPTION: L = 13 cm (5 in); W = 23 cm (9 in);
streaked brown above; whitish below with brown
streaks; yellowish wingbar and yellow wing patch
and base of tail.

HABITAT AND DISTRIBUTION: Uncommon to rare and
local summer resident* (Apr–Sep) in spruce forests
of northern and highland PA; has bred* in other
parts of the region (NJ, WV, VA), but only sporad-
ically; common to uncommon or rare and irregular winter visitor (Nov–Mar), scarcer and less
regular in southern parts of the region. Travels in
small flocks widely during the nonbreeding
period; more numerous in some years than others: Coniferous and mixed forests; feeders in winter.

DIET: Forages on a wide variety of small annual and
perennial plant seeds.

REPRODUCTION: Monogamous; breeding pairs form
in winter flocks, and nesting can begin nearly any
time in spring from Feb–May; open cup nest, 9
cm dia. (4 in), is placed 1–15 m (3–50 ft) above
ground on the horizontal branch of a conifer,
often as part of a loose colony with nests of other
breeding pairs; female builds nest of twigs, grass,
leaves, rootlets, and bark strips lined with plant
down, hair, fine grasses, feathers, and moss;
clutch—3–5; incubation (female)—13 days; fledging—15 days post-hatching; post-fledging
parental care—?; multiple broods per season seem
probable, but are not documented.

CONSERVATION STATUS: S1—WV; threats to limited
highland forest breeding habitat.

RANGE: Breeds in boreal regions of northern North America, and

in western mountains south through the United States and highlands of Mexico to Veracruz; winters in all except the extreme northern portions of the breeding range and in most of the temperate United States.

KEY REFERENCES: Dawson (1997).

American Goldfinch *Carduelis tristis*

DESCRIPTION: L = 13 cm (5 in); W = 23 cm (9 in); yellow body; black cap, wings, and tail; white at base of tail and wing bar; yellow shoulder patch.
Female and Winter Male Brownish above; yellowish or buff breast; whitish belly; dark wings with white wingbars.

HABITAT AND DISTRIBUTION: Common summer resident* (Apr–Oct) throughout; common to uncommon winter resident (Nov–Mar) in lowlands and mid-elevations, often in flocks at feeders and in brushy fields; numbers vary widely in winter, depending on food availability and weather: Grasslands, brushy pastures, old fields; feeders in winter and spring. Usually in flocks; dipping-soaring flight, like a roller coaster; almost always giving characteristic flight call—"ker-chik ker-chik-chik-chik."

DIET: Forages mostly in flocks on the ground or hanging from plant stems, feeding mainly on herbaceous plant (e.g., Compositae, Gramineae) and tree (e.g., alder, birch, cedar, and elm) seeds—thistles (*Cirsium*) are a favorite; some insects on occasion.

REPRODUCTION: Mostly monogamous with some serial polyandry; pairs form in winter flocks but may change by beginning of breeding; nesting commences relatively late in the summer, with the female constructing the nest, 8 cm dia. (3 in), in late June or early July; the open cup nest is placed in the crown of a low shrub, usually with an overhanging canopy, and built of twigs, spider silk, plant strips, and rootlets lined with a thick layer of plant down; clutch—5; incubation—12–14 days; fledging—12–17 days post-hatching; post-fledging parental care—about 3 weeks.

CONSERVATION STATUS: Long-term (1966–1994) declines in the Mid-Atlantic region for unknown reasons, perhaps related to conversion of field habitats to forest and intensive farming.

RANGE: Breeds across southern Canada, northern and central U. S. to southern

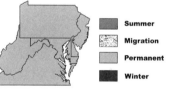

CA and northern Baja California in the west; winters in the central and southern U. S. and northern Mexico.

KEY REFERENCES: Middleton (1993).

Evening Grosbeak *Coccothraustes vespertinus*

DESCRIPTION: L = 20 cm (8 in); W = 33 cm (13 in); a large finch with heavy, yellowish or whitish bill; yellow body; black crown and brownish head with yellow forehead and eyebrow; black tail and wings with white wing patch. **Female** Grayish above, buffy below; dark malar stripe; white wing patch.

HABITAT AND DISTRIBUTION: Common to uncommon or rare and highly irregular winter visitor (Nov–Mar) throughout, often in flocks in conifers

and at feeders; numbers vary widely in winter, depending on food availability and weather: Coniferous, mixed, and deciduous forest, parkland, residential areas; often at feeders in winter.

DIET: Forages in flocks, feeding in the tops and outer branches of mast-producing trees during the fall and winter, and on buds in the spring (e.g., maple, box elder, ash, apple, tulip poplar, elm).

RANGE: Breeds in boreal portions of Canada and the northern and western

United States, south in western mountains to western and central Mexico; winters in breeding range and in temperate and southern United States.

KEY REFERENCES: Gillihan and Byers (2001).

FAMILY PASSERIDAE

House Sparrow *Passer domesticus*

DESCRIPTION: L = 15 cm (6 in); W = 26 cm (10 in); a chunky, heavy billed bird; brown above with heavy dark brown streaks; dingy gray below; gray cap; chestnut nape; black lores, chin, and bib. **Female** Streaked buff and brown above; dingy gray below; pale buff postorbital stripe.

HABITAT AND DISTRIBUTION: Common resident* throughout: Urban areas, pastures, agricultural fields, feed lots, farms, grain elevators.

DIET: Almost entirely plant seeds, including grasses, weeds, cereal grains (e.g., corn, wheat, oats, sorghum); some insects (up to 10% of diet in June).

REPRODUCTION: Monogamous with pair formation occurring in the fall (Sep–Oct); male defends nest site, usually a cavity, nook, or cranny, but occasionally in trees; if the nest is in a cavity, the cavity is lined with dried plant material with an inner cup lined with finer material (roots, grasses, feathers); nests in trees are balls of coarse plant material 30–40 cm dia. (12–16 in) with a side entrance to an inner chamber; clutch size—4–6; incubation—11 days; fledging—14 days post-hatching; young appear to be independent of adults 7–10 days post-fledging; multiple broods per season are common (up to 4).

RANGE: Resident in boreal, temperate, and subtropical regions of the Old and

New World; currently expanding into tropical regions. Introduced into the Western Hemisphere in 1850.

KEY REFERENCES: Lowther and Cink (1992).

Class Mammalia—Mammals

Mammals first appear in the fossil record over 200 million years ago in the late Triassic or early Jurassic Period of the Mesozoic Era, probably derived from therapsid reptiles. They are characterized by a 4-chambered heart, 4 limbs tipped with claws, giving birth to live young (except the monotremes, which lay eggs), feeding young with milk produced by the mother, ability to thermoregulate (i.e., "warm blooded), and hair. There are 24 living orders of mammals, and about 4600 species. Eighty-one species of mammals are found regularly during one or more seasons in the Mid-Atlantic region.

ORDER DIDELPHIMORPHIA; FAMILY DIDELPHIDAE

Virginia Opossum *Didelphis virginiana*

DESCRIPTION: L = 0.6–0.8 m (24–32 in); house-cat-sized with pale whitish, gray, or black pelage; cone-shaped head with small, naked ears; mostly naked, prehensile tail; hand-like feet with 5 digits on all 4 paws.

STATUS AND HABITAT: Common resident throughout in both wooded and open habitats, although preferred sites are wooded with open water.

HABITS: Mainly nocturnal, spending daylight hours in dens, e.g., abandoned woodchuck or skunk holes, hollow logs, or crawl spaces under houses; less active in winter when individuals may spend days or weeks in dens. When attacked by predators, individuals may attempt to fight or flee, but if captured often become catatonic and emit foul odors from anal glands ("playing possum").

DIET: Omnivorous, feeding on invertebrates, fruits, carrion, birds' eggs, snakes (shows some immunity to pit viper venom), small mammals, and garbage.

REPRODUCTION: Breeds Jan–Sep; males occupy large home ranges, 100 ha (247 acres) often overlapping home ranges of 2 or more females, 50 ha (124 acres); litter size—8 with 1–2 litters/year; gestation—13 days; young are born small and naked, and crawl at birth from the female's vulva to the abdominal pouch where they attach to one of 13 nipples, remaining inside the pouch for 2 months; weaning—96–106 days.

RANGE: Eastern North America from southern ME, southern Canada, and MI west to southern SD south through Mexico and Central America to Costa Rica; introduced along the West Coast from British Columbia to Baja California.

KEY REFERENCES: McManus 1974.

ORDER INSECTIVORA; FAMILY SORICIDAE

Cinereus Shrew *Sorex cinereus*

DESCRIPTION: L = 9–10 cm (3.5–3.9 in); a small shrew with relatively thick, soft fur; brown above, paler

below; long, narrow snout; tiny ears (hidden beneath pelage); small, beady eyes; *Sorex cinereus fontinalis*, the Maryland Shrew, is sometimes recognized as a separate species.

HABITAT AND DISTRIBUTION: Frequents litter, rotting logs, rock piles, and stumps of woodlands, old fields, and wet meadows; found nearly throughout the region except for the VA Coastal Plain and Piedmont.

HABITS: Mainly nocturnal, but active day or night throughout the year; consumes up to 3 times its body weight in food daily; some mammalian predators find this shrew distasteful.

DIET: Forages in loose litter for soil invertebrates, locating prey by touch and smell.

REPRODUCTION: Breeds Mar–Sep; 2–3 litters/year; nest is built of leaves and grasses, 8 cm dia. (3 in), and placed under a rock or log; gestation—18 days; litter size—7; weaning—3 weeks; male remains with the family until weaning; sexual maturity at 5–6 months.

RANGE: Northern North America, south in mountains to GA, NM, AZ.

KEY REFERENCES: Linzey (1998), Cannings and Hammerson (2004).

Long-tailed Shrew *Sorex dispar*

DESCRIPTION: L = 11–14 cm (4.3–5.5 in); slate-gray pelage; long, pointed snout; long, thick tail, dark above and lighter below; whitish feet.

HABITAT AND DISTRIBUTION: Talus slopes, mossy boulder fields, and rocky stream borders in highland coniferous and mixed forest; mountains of VA, WV, PA, and MD.

HABITS: Circadian; frequents crevices and tunnels under and between rocks.

DIET: Beetles, spiders, centipedes, and other invertebrates.

REPRODUCTION: Details poorly known for this rarely encountered species; breeds—Apr–Aug; nest/den location—subterranean gaps among rocks and boulders; litter size—2–5.

CONSERVATION STATUS: S1—NJ; S2—MD, WV; S3—VA, PA; restricted distribution and lack of information raise concerns regarding conservation.

RANGE: Appalachian Mountains from Nova Scotia to NC and TN.

KEY REFERENCES: Kirkland (1981), DeGraaf and Yamasaki (2001).

Smoky Shrew *Sorex fumeus*

DESCRIPTION: L = 11–13cm (4–5 in); a medium-sized, long-tailed shrew with long, narrow snout; light brown in summer with yellowish underbelly; dark gray in winter, paler below; feet are yellowish; tail is brownish above, yellowish below; similar in appearance to the Long-tailed Shrew (*S. dispar*).

HABITAT AND DISTRIBUTION: Forages under rotting logs, rocks, and litter, often using mouse or mole tunnel systems, in cool, wet coniferous and mixed forests in highlands of WV, PA, western VA, MD, and northeastern NJ.

HABITS: Circadian; active throughout the year.

DIET: Soil and litter invertebrates.

REPRODUCTION: Breeds Mar–Sep; 2–3 litters/year; nest is ball- shaped, constructed of leaf bits and other vegetation, and located under rocks, logs, litter, or in small burrowing mammal tunnels; gestation—20 days; litter size—5–6; weaning—30 days.

CONSERVATION STATUS: S2—MD; limited range and secretive habits obscure status.

RANGE: Northeastern North America from southern Quebec south in the Appalachians to GA and west to WI.

KEY REFERENCES: Owens (1984), Merritt (1987).

Pygmy Shrew *Sorex hoyi*

DESCRIPTION: L = 8–9 cm (3–3.5 in); smallest land mammal in the world; the short, smooth pelage is grayish to reddish-brown, darker in winter; long, pointed snout; short tail; formerly placed in the genus *Microsorex*.

HABITAT AND DISTRIBUTION: Forest and field habitats with dense litter cover, especially under or in rotted logs; probably found throughout the Piedmont and highland areas of the region, but secretive habits result in few records.

HABITS: Circadian with nocturnal activity peak; active throughout the year under rotten logs and wet leaf litter; sometimes caches food.

DIET: Soil and litter invertebrates, especially caterpillars, beetles, grasshoppers, and spiders.

REPRODUCTION: Breeds Apr–Aug; single litter/year; gestation—2–3 weeks; litter size—5–6; sexual maturity—next breeding season after birth.

CONSERVATION STATUS: S2—WV; secretive habits make status difficult to assess.

RANGE: Northern North America, south in mountains to GA and NM.

KEY REFERENCES: Long (1974).

Southeastern Shrew *Sorex longirostris*

DESCRIPTION: L = 7–11 cm (2.8–4.3 in); pelage is reddish-brown above, grayish below; long, pointed snout; tiny, beady eyes; small ears hidden by fur; long tail.

HABITAT AND DISTRIBUTION: Dense herbaceous ground cover of forest, bog, swamp, marsh, and grassland; often associated with disturbed sites; VA and southern MD.

HABITS: Circadian; active throughout the year; sometimes emit chipping vocalizations of unknown function; owls seem to be the main predators on this species.

DIET: Spiders, caterpillars, slugs, snails and other invertebrates.

REPRODUCTION: Breeds Mar–Oct; 2 litters/year; nest is a shallow depression lined with leaves and

grasses under a rotting log; gestation 2–3 weeks; sexual maturity may be reached late in first year of life.

CONSERVATION STATUS: S3—MD; limited distribution at the periphery of the range.

RANGE: Southeastern United States from MD west to MO south to LA and FL.

KEY REFERENCES: French (1980), Webster et al. (1985).

American Water Shrew *Sorex palustris*

DESCRIPTION: L = 14–16 cm (5.5–6.3 in); a large, long-tailed shrew; pelage is dark above, silvery below; feet are fringed with bristles and the hind feet are partially webbed (between third and fourth toes); long snout; tiny ears; small, beady eyes; tail is distinctly bicolored, dark above and light below.

HABITAT AND DISTRIBUTION: Rocky streams and cataracts in boreal forest; spottily distributed in highlands of VA, WV, MD, NJ, and PA.

HABITS: Circadian with crepuscular activity peaks; active year-round; fringed feet allow this shrew to run on top of the water surface, as well as to swim and dive in pursuit of prey.

DIET: Forages for aquatic invertebrates in or near rocky streams and pools, capturing prey located mostly by touch and smell.

REPRODUCTION: Breeds Feb–Aug; 2–3 litters/year; nest is placed in stream-bank hollows, tunnels,

muskrat or beaver lodges, or under logs or rocks, and built of moss; gestation—3 weeks; litter size—6; sexual maturity is reached in second year after birth.

CONSERVATION STATUS: S1—VA, MD, WV; highland boreal forest habitat in the Appalachians is threatened by pollution, climate change, logging, and development.

RANGE: Northern North America south in mountains to GA, NM, AZ, and CA.

KEY REFERENCES: Beneski and Stinson (1987)

Northern Short-tailed Shrew *Blarina brevicauda*

DESCRIPTION: L = 10–13 cm (3.9–5.1 in); a large, gray shrew with a short tail; cone-shaped head; tiny ears and eyes.

HABITAT AND DISTRIBUTION: Moist litter of forests and fields nearly throughout the region except parts of the Coastal Plain of southeastern VA (where replaced by *B. carolinensis*).

HABITS: Circadian but with a nocturnal peak; active throughout the year; saliva is poisonous, immobilizing prey for caching; the most commonly observed shrew in the Mid-Atlantic due to its active, aboveground, daytime foraging; many mammalian predators find this shrew distasteful.

DIET: Worms, millipedes, insects, and other soil and litter invertebrates as well as some plant material (e.g., nuts); caches food.

REPRODUCTION: Breeds Mar–Nov; 3–4 litters/year; nest, 10 cm dia. (4 in), is placed under a log, rock, or other debris and built of leaves, hair, and grasses; gestation—3 weeks; litter size 6–7; weaning at 3–4 weeks; sexual maturity is reached at 3 months.

RANGE: Eastern and central North America from southern Canada south to CO and GA.

KEY REFERENCES: George et al. (1986).

Southern Short-tailed Shrew *Blarina carolinensis*

DESCRIPTION: L = 9–11 cm (3.5–4.3 in); a large, dark gray shrew with a short tail; cone-shaped head; tiny ears and eyes; this species is a smaller, darker shrew than its northern relative, *B. brevicauda,* of which it was formerly considered a subspecies.

HABITAT AND DISTRIBUTION: Moist litter of forests and fields; parts of the Coastal Plain of southeastern VA.

HABITS: Circadian but with a nocturnal peak; active throughout the year; saliva is poisonous, immobilizing prey for caching; many mammalian predators find this shrew distasteful.

DIET: Worms, millipedes, insects, and other soil and litter invertebrates as well as some plant material (e.g., nuts); caches food.

REPRODUCTION: Breeds Mar–Nov; 3–4 litters/year; nest, 10 cm dia. (4 in), is placed under a log, rock, or other debris and built of leaves, hair, and grasses; gestation—3 weeks; litter size 6–7; weaning at 3–4 weeks; sexual maturity is reached at 3 months.

RANGE: Southeastern United States from MD west to IL and south to FL and TX.

KEY REFERENCES: McCay (2001).

Least Shrew *Cryptotis parva*

DESCRIPTION: L = 8–9 cm (3.1–3.5 in); a small, brown, short-tailed shrew; cone-shaped head; tiny ears and eyes.

HABITAT AND DISTRIBUTION: Fields, grassland, and second growth with dense herbaceous understory throughout the Mid-Atlantic.

HABITS: Circadian; active all year; sometimes uses colonial dormitories in winter.

DIET: Spiders, caterpillars, worms, isopods, and other soil and litter invertebrates.

REPRODUCTION: Breeds Mar–Nov; 3 litters/year; nest, 8–13 cm dia. (4–5 in), is placed under a log or rock and built of grass and leaves; gestation 3 weeks; litter size 2–7; weaning at 3 weeks; both parents care for young; sexual maturity is reached at 5 weeks.

CONSERVATION STATUS: S1—PA; S2—WV; small number of isolated populations in ephemeral, seral habitat makes the species vulnerable.

RANGE: Eastern, central, and southern North America from southern Canada south through eastern Mexico and Central America to Costa Rica.

KEY REFERENCES: Whitaker (1974).

FAMILY TALPIDAE

Hairy-tailed Mole *Parascalops breweri*

DESCRIPTION: L = 15–17 cm (5.9–6.7 in); a small mole with black, velvety pelage; head is flattened with tiny ears and eyes and naked pink nose; forefeet are large, naked, and almost circular in shape; short hairy tail.

HABITAT AND DISTRIBUTION: Highland forests and grasslands with moist, loamy soil; PA, WV, western VA, MD, and NJ.

HABITS: Circadian; active throughout the year; constructs networks of shallow tunnels for feeding, the presence of which can be observed as humped irregular ridges on the soil surface; a second tunnel system is built 25–45 cm (10–18 in) below the surface for breeding and wintering.

DIET: Beetle larvae, millipedes, worms, and other soil invertebrates; prey are detected mainly by smell and touch.

REPRODUCTION: Breeds Mar–Apr; single litter/year; nest, 15 cm dia. (6 in), is located in a burrow 30–40 cm (12–16 in) below ground and built of leaves and grasses; gestation 4 weeks; litter size 4–5; weaning at 4 weeks; sexual maturity is reached in spring of the year following birth.

RANGE: Northeastern North America from southern Canada to GA.

KEY REFERENCES: Hallett (1978).

Eastern Mole *Scalopus aquaticus*

DESCRIPTION: L = 14–21 cm (5.5–8.3 in); a large mole with black, gray, or brown, velvety pelage; head with tiny eyes (covered by a thin membrane) and ears; long snout; short, naked tail; large, mostly hairless forefeet with long claws; webbed toes.

HABITAT AND DISTRIBUTION: Loose soil in forest or open areas; VA, MD, NJ, DE, southern WV, and southeastern PA.

HABITS: Circadian; active year-round; constructs networks of shallow tunnels for feeding, the presence of which can be observed as humped irregular ridges on the soil surface; a second tunnel system is built 25–45 cm (10–18 in) below the surface for breeding, resting, and wintering; some mammalian predators find this species distasteful.

DIET: Feeds mostly on worms, insect larvae, and other soil invertebrates located by touch and smell; also takes some plant matter.

REPRODUCTION: Breeds Mar–Apr; single litter/year; nest is located in an underground burrow chamber, 10–20 cm dia. (4–8 in), and built of shredded leaves, roots, and grasses; gestation 6 weeks; litter size 2–5; weaning at 1 month; sexual maturity is reached in spring of the year following birth.

CONSERVATION STATUS: S3—WV.

RANGE: Eastern and central United States from MA west to WY and south to TX and FL; also the Mexican states of Tamaulipas, Coahuila, and Chihuahua along the TX border.

KEY REFERENCES: Yates and Schmidly (1978).

Star-nosed Mole *Condylura cristata*

DESCRIPTION: L = 18–20 cm (7.1–7.9 in); black or brown pelage; tiny eyes and ears; long whiskers; extraordinary, naked nose with 22 radiating appendages (Eimer organs) that serve as tactile and electrosensory prey detectors; forefeet with long, heavy claws; scaly, fleshy tail may serve as a rudder while swimming and for fat storage.

HABITAT AND DISTRIBUTION: Wet soils in riparian areas, swamps, marshes, and meadows; found throughout the Mid-Atlantic except western WV and southwestern PA.

HABITS: Circadian, occasionally foraging above ground at night; active year-round; constructs tunnel systems just below soil surface for feeding, some of which open directly into water; deeper tunnel systems are built for resting, breeding, and wintering; several moles may share the same tunnel system.

DIET: Feeds on aquatic and soil invertebrates, evidently detected by electrosensory organs in the star-shaped nose as well as by touch; sometimes forages under ice on stream bottoms in winter when soil is frozen.

REPRODUCTION: Pair bond may be maintained throughout the breeding season; breeds Mar–Aug; single litter/year; nest, 15 cm dia. (6 in), is built of leaves, plant stems, and grasses; gestation 6 weeks; litter size 3–7; weaning at 3 weeks; sexual maturity is reached in spring of the year following birth.

CONSERVATION STATUS: S2—WV.

RANGE: Eastern North America from southern Canada south to north-

ern FL along the east coast, and west to ND, WI, IL, and TN.

KEY REFERENCES: Peterson and Yates (1980), Gould et al. (1993).

ORDER CHIROPTERA; FAMILY VESPERTILIONIDAE

Southeastern Myotis *Myotis austroriparius*

DESCRIPTION: L = 8–10 cm (3.1–3.9 in); wooly, short, thick pelage is grayish or orangish-brown above, buffy below; long hairs on toes extend beyond claws.

HABITAT AND DISTRIBUTION: Roosts in caves, hollow trees, mines, buildings, and similar sites, almost always near permanent wetlands; known from a few sites in southeastern VA, mainly along the Blackwater River near the town of Zuni.

HABITS: Nocturnal; colonial (females and young); migratory; hibernates.

DIET: These bats forage over open water at night, flying low over the surface to capture flying insects, detecting them using echolocation.

REPRODUCTION: Mating occurs in Sep–Oct at hibernation caves; females store sperm, and fertilization occurs in spring after emergence from hibernation; 2 young are born in Apr–May, weaned in 5–6 weeks.

CONSERVATION STATUS: S1—VA; small populations located in a few sites.

RANGE: Southeastern United States from

southeastern VA west to southern IL, MO, AK, and OK, and south to FL and eastern TX.

KEY REFERENCES: Jones and Manning (1989), Linzey (1998), Virginia Department of Game and Inland Fisheries (2004).

Gray Myotis *Myotis grisescens*

DESCRIPTION: L = 8–10 cm (3.1–3.9 in); pelage is grayish-brown above, paler below; the only *Myotis* in which the wing membrane is inserted at the ankle instead of the side of the foot.

HABITAT AND DISTRIBUTION: Roosts in caves near water, and forages generally over water; found in 3 counties in southwestern VA (Lee, Scott, and Washington); 2 records from 1 county (Pendleton) in WV.

HABITS: Nocturnal; colonial; migratory; hibernates; the entire population of this species resides in a small number of caves scattered throughout its range; different caves may be used at different times of the year for hibernating, maternity, or roosting, and some individuals or populations migrate considerable distances between caves used for these different purposes; most of the few caves known for VA apparently serve as summer roosts for male bachelor colonies or roosts for migrating individuals; a single maternity colony is known for the state from Washington County.

DIET: These bats forage in loose flocks over open water at night to capture flying insects, detecting them using echolocation.

REPRODUCTION: Mating occurs in Sep–Oct at hibernation caves; females store sperm, and fertilization occurs in spring after emergence from

hibernation; one young is born in May or early Jun, and weaned in Jul.

CONSERVATION STATUS: S1—VA; "Endangered"— USFWS; small populations located in a few caves are vulnerable to disturbance.

RANGE: Eastern United States from southwestern VA west to KS south to the FL panhandle and coastal MS.

KEY REFERENCES: Decher and Choate (1995), Linzey (1998).

Eastern Small-footed Myotis *Myotis leibii*

DESCRIPTION: L = 7–8 cm (2.8–3.1 in); a small bat with brown pelage tinged with yellow above, whitish or buffy below; dark face gives a mask-like appearance; small feet.

HABITAT AND DISTRIBUTION: Summer breeding habitat is highland coniferous and mixed forest; small maternity colonies have been found in buildings and hollow trees; migrating and hibernating roosts are in caves and mine shafts; scattered records from nearly throughout the region except the Piedmont and Coastal Plain of VA.

HABITS: Nocturnal; colonial; migrates, collecting in cave roosts with other species; hibernates (Nov–Mar) in relatively small colonies (< 200); foraging flight is distinctive, involving a slow fluttering in elliptical patterns over the foraging substrate, usually tree canopy or water.

DIET: Flying insects detected by echolocation.

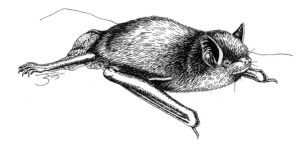

REPRODUCTION: Details are poorly known, but assumed to be similar to other *Myotis*; mating is thought to occur in fall before hibernation, with female storing sperm until spring when fertilization occurs; gestation is perhaps 2 months; females associate in small maternity colonies in hollow trees or buildings, and produce a single offspring, probably in July.

CONSERVATION STATUS: S1—WV, PA, VA, MD, NJ; lack of information on the species and vulnerability of hibernacula are the principal causes of concern.

RANGE: Eastern North America from ME and southern Ontario south to GA and east OK.

KEY REFERENCES: Best and Jennings (1997).

Little Brown Myotis *Myotis lucifugus*

DESCRIPTION: L = 8–9 cm (3.1–3.5 in); the pelage of the Little Brown Myotis (colloquially known as the little brown bat) is glossy brown above, paler below; ears, wings, and tail membranes are dark and nearly hairless.

HABITAT AND DISTRIBUTION: These bats frequent a variety of habitat types, often near water; attics, crevices, and hollow trees serve as maternity roosts; hibernation roosts are located in caves and mine shafts; found throughout the Mid-Atlantic.

HABITS: Nocturnal, with dusk and predawn activity peaks; colonial; migrates to hibernation sites; hibernates Oct–Apr; homing up to 430 km

(270 mi) has been demonstrated by displaced individuals.

DIET: Flying insects detected by echolocation.

REPRODUCTION: Mating occurs prior to hibernation in fall; females store sperm, and fertilization occurs in spring after emergence from hibernation; gestation is 2 months; single offspring is born in Jun–Jul and weaned in 1 month; sexual maturity is reached in 9–12 months.

RANGE: North America from boreal forest of AK and Canada south to FL, NM, and CA; southwestern and Mexican populations are now considered separate species.

KEY REFERENCES: Fenton and Barclay (1980).

Northern Long-eared Myotis *Myotis septentrionalis*

DESCRIPTION: L = 8–9 cm (3.1–3.5 in); similar in appearance to the Little Brown Myotis, but with longer ear (extends beyond tip of nose when laid forward); formerly considered a subspecies of *M. keenii*, eastern accounts of which pertain to *M. septentrionalis*.

HABITAT AND DISTRIBUTION: Frequents woodlands in summer; maternity roosts are in hollow trees, under loose bark, and in buildings; hibernation roosts are caves and mine shafts; found throughout the Mid-Atlantic.

HABITS: Nocturnal; colonial; forages within forest,

above shrub layer and in openings, and over water; few data on migratory movements, as most populations appear to hibernate near breeding areas; hibernates in mixed colonies with other bat species from Oct–Apr.

DIET: Flying insects detected by echolocation.

REPRODUCTION: Most mating occurs prior to hibernation in fall, although there is some evidence of posthibernation copulation; females store sperm, and fertilization occurs in spring after emergence from hibernation; gestation is 2 months; single offspring is born in Jun–Jul and weaned in 1 month; sexual maturity is reached in 9–12 months.

CONSERVATION STATUS: S3—PA, WV, VA; lack of information on the species and vulnerability of hibernacula are the principal causes of concern.

RANGE: Northern and eastern North America from Nova Scotia to eastern British Columbia and south in the eastern United States to GA and AK.

KEY REFERENCES: Caceres and Barclay (2000).

Indiana Bat *Myotis sodalis*

DESCRIPTION: L = 7–9 cm (2.8–3.5 in); similar in appearance to the Little Brown Myotis, although its pelage lacks the sheen of that species.

HABITAT AND DISTRIBUTION: Frequents woodlands in summer, foraging at canopy level or over water, often near streams; maternity roosts are in hollow

trees or under loose bark; hibernation roosts are limestone caves; found at a few localities in western VA, eastern WV, southwestern PA, western MD, and northwestern NJ.

HABITS: Nocturnal; colonial; migratory; hibernates in dense clusters in caves from Oct–Apr.

DIET: Flying insects detected by echolocation.

REPRODUCTION: Mating occurs prior to hibernation in fall; females store sperm, and fertilization occurs in spring after emergence from hibernation; gestation is 2 months; single offspring is born in Jun–Jul and weaned in 1 month; sexual maturity is reached in 6–12 months.

CONSERVATION STATUS: S1—PA, WV, VA, MD, NJ; "Endangered"—USFWS; rarity, declining populations, lack of information on the species, and vulnerability of hibernacula are the principal causes of concern.

RANGE: Eastern and central United States from VT west to NE south to the FL panhandle and east TX.

KEY REFERENCES: Thomson (1982).

Eastern Red Bat *Lasiurus borealis*

DESCRIPTION: L = 9–11 cm (3.5–4.3 in); a handsome, medium-sized bat with reddish pelage, more or less frosted with silver; females are more frosted than males; buffy shoulder patches; tail membrane is furred (not sparsely haired as in *Myotis*); members of the Red Bat superspecies are found throughout the Western Hemisphere from Canada to Argentina; recently, western and tropical taxa of the group have been recognized as separate species.

HABITAT AND DISTRIBUTION: Woodlands during the breeding season, foraging above the canopy, over water, or in open areas, and roosting in dense foliage of tree crowns; found throughout the region.

HABITS: Nocturnal; solitary except during migration and breeding; migratory, departing for wintering areas south of 40% N in Aug–Sep.; probably hibernates during cold periods in hollow trees.

DIET: Flying insects, especially moths, detected by echolocation.

REPRODUCTION: Mating usually occurs in fall; females store sperm, and fertilization occurs in spring; gestation is 2–3 months; 1–4 offspring are born in Jun–Jul and weaned in 5–6 weeks; sexual maturity is reached the year following birth.

RANGE: Eastern North America from Nova Scotia to southern Saskatchewan south to FL, TX, and northeastern Mexico; northern populations migrate to more southern portions of the breeding range in winter.

KEY REFERENCES: Shump and Shump (1982a).

Hoary Bat *Lasiurus cinereus*

DESCRIPTION: L = 13–15 cm (5.1–5.9 in); a large, beautiful bat with a broad wing span, 34 cm (13 in); pelage is dark frosted with silvery white

above, whitish below; buffy collar; tail membrane is furred; ears with a dark rim.

HABITAT AND DISTRIBUTION: Preferred breeding habitat is coniferous and mixed forest, roosting in dense foliage of conifer crowns and foraging over woodland or open areas; found throughout the Mid-Atlantic in summer (Apr–Sep).

HABITS: Nocturnal, commencing foraging well after dark; solitary except during breeding and migration; migratory, with most individuals departing northern portions of the range in fall; may hibernate during winter cold spells; flight is much more swift and direct than that of other bats of the region.

DIET: Flying insects, perhaps especially moths, detected by echolocation.

REPRODUCTION: Mating usually occurs at wintering sites, where many males may remain throughout the year; females store sperm, and fertilization occurs in spring; gestation is 3 months; 2 offspring are born in Jun–Jul and weaned in 5–6 weeks; sexual maturity may be reached within a few months after birth.

CONSERVATION STATUS: S3—WV; rarity and lack of information are main causes of concern.

RANGE: Found nearly throughout the Western Hemisphere except northwestern Canada, AK, Central America, and most of northern South America; in the southeastern United States it is present mostly as a transient or winter resident (although there are some breeding records).

KEY REFERENCES: Shump and Shump (1982b).

Silver-haired Bat *Lasionycteris noctivagans*

DESCRIPTION: L = 9–11 cm (3.5–4.3 in); a medium-sized bat with beautiful long, dark fur frosted with silver on the back, paler below; similar to the much larger Hoary Bat, but lacks the buffy collar and heavily furred tail membrane (lightly

furred in *L. noctivagans*), and has naked, black ears (partially furred in the Hoary Bat).

HABITAT AND DISTRIBUTION: Coniferous and mixed forest in summer (May–Sep) (PA and probably NJ and WV), where it roosts singly in hollow trees, dense foliage, or under loose bark, and forages over water; migrates south in winter (Sep–Apr); the species migrates through and winters in VA, when it can be found throughout the state, roosting singly or in small groups in hollow trees, rocky crevices, or buildings; there are no summer records for this species from VA.

HABITS: Nocturnal, foraging with a characteristic slow, erratic flight of glides, twists, and sharp turns; migratory; hibernates at times during cold periods.

DIET: Feeds on flying insects captured using echolocation.

REPRODUCTION: Mating usually occurs during fall migration; females store sperm, and fertilization occurs in spring; gestation is 2 months; 2 offspring are born in Jun–Jul and weaned in 4–5 weeks; sexual maturity may be reached within a few months after birth.

CONSERVATION STATUS: S2—WV; rarity and lack of information are main causes of concern.

RANGE: North America from southeastern AK and southern Canada south to GA, TX, and CA; also northeastern Mexico.

KEY REFERENCES: Kunz (1982).

Eastern Pipistrelle *Pipistrellus subflavus*

DESCRIPTION: L = 8–9 cm (3.1–3.5 in); a small bat with fluffy, tricolored hair—gray at the base, yellowish-brown in the middle, and dark at the tip; skin of the forearms is reddish; ears are light brown, not dark as in *Myotis*.

HABITAT AND DISTRIBUTION: Woodlands and fields in summer (May–Sep), foraging with slow, erratic, butterfly like movements in early evening above the canopy or over water; maternity roosts are in hollow trees, buildings, and rock crevices; males roost singly in trees; hibernation roosts are in caves and mine shafts; found throughout the region.

HABITS: Nocturnal, but with activity periods beginning earlier in the evening than for other Mid-Atlantic bats; hibernates, often in colonies with other bat species.

DIET: Feeds on flying insects captured using echolocation.

REPRODUCTION: Mating usually occurs in fall (older individuals) or spring (first-year females); females store sperm, and fertilization occurs in spring; gestation period is not known; 2 offspring are born in Jun–Jul and weaned in 4 weeks; sexual maturity is reached in the first spring after birth.

RANGE: Eastern North America from Nova Scotia west to MN and south to FL and east TX; also eastern Mexico south to Guatemala.

KEY REFERENCES: Fujita and Kunz (1984).

Big Brown Bat *Eptesicus fuscus*

DESCRIPTION: L = 10–13 cm (3.9–5.1 in); a large bat with long, two-tone, glossy fur, black at the base and brown at the tip; black wing and tail membranes are mostly hairless.

HABITAT AND DISTRIBUTION: During summer, these bats forage over a variety of habitats, including fields, agriculture, residential sites, and woodlands; maternity roost sites are often in buildings; bachelor roosts are generally solitary and located in hollow trees, crevices, or buildings; hibernation roosts are caves, mine shafts, and, occasionally, buildings; found throughout the Mid-Atlantic region.

HABITS: Nocturnal, departing roosting sites shortly before dark; foraging flight is relatively slow and straight at 5–10 m above the substrate; hibernates.

DIET: Flying insects, especially beetles, detected by echolocation.

REPRODUCTION: Mating takes place in fall and winter at hibernacula; females store sperm, and fertilization occurs in spring; gestation is 2 months; 2 offspring are born in May–Jul and weaned in 4–5 weeks; sexual maturity may be reached within a few months after birth.

RANGE: North America from central Canada south to Panama; also the Caribbean and parts of northern South America.

KEY REFERENCES: Kurta and Baker (1990).

Evening Bat *Nycticeius humeralis*

DESCRIPTION: L = 9–10 cm (3.5–3.9 in); a medium-sized bat with dark brown pelage; naked wing and tail membranes are dark as well; tragus (skin flap at the base of the ear) is blunt and curved forward, not narrow and pointed as in *Myotis.*

HABITAT AND DISTRIBUTION: In summer, this species frequents woodlands and adjacent open areas and waterways at lower elevations of the Piedmont and Coastal Plain of VA, MD, and southeastern PA; also in the southwest corner of PA and adjacent WV (rare); maternity roosts are in buildings or trees, especially those with Spanish moss; bachelor roosts are in trees; most Evening Bats apparently leave the Mid-Atlantic in fall on southbound migration, although the fall and winter portion of the life cycle are poorly known; winter roosts and hibernacula are in buildings or hollow trees.

HABITS: Nocturnal; migratory; hibernates.

DIET: Feeds on flying insects detected by echolocation.

REPRODUCTION: Mating evidently occurs in fall; females store sperm, and fertilization occurs in spring; 2 offspring are born in May–Jun and weaned at 6–9 weeks; sexual maturity may be reached within a few months after birth.

CONSERVATION STATUS: S1—WV; rarity and lim-

ited habitat at the periphery of the range are the principal causes of concern.

RANGE: Eastern and central United States from PA west to NE and south to FL and TX; also northeastern Mexico.

KEY REFERENCES: Watkins (1972).

Rafinesque's Big-eared Bat
Corynorhinus rafinesquii

DESCRIPTION: L = 9–11 cm (3.5–4.3 in); a medium-sized gray bat of striking appearance with very large ears, squashed nose, and a glandular mass between each eye and the nose; similar to Townsend's Big-eared Bat, but with white rather than buffy underparts; was placed in the genus *Plecotus*.

HABITAT AND DISTRIBUTION: In summer, this species frequents woodlands; maternity roosts are in abandoned buildings, trees, and bridges; winter roosts and hibernacula probably include abandoned buildings and hollow trees; there are only 5 records for VA, all from the southeast corner of the state; the species is rare in southern WV.

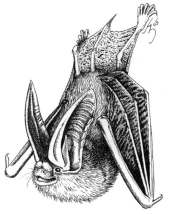

HABITS: Nocturnal, leaving roosts to forage well after dark; hibernates, at least during some periods in some portions of its range; rolls ears backwards like a scroll when roosting to limit heat loss.

DIET: Flying insects, especially moths, detected by echolocation.

REPRODUCTION: Mating evidently occurs in fall and winter; females store sperm, and fertilization occurs in spring; gestation lasts 2–3 months; single offspring is born in May–Jun and weaned at 2 months; sexual maturity is reached in year following birth.

CONSERVATION STATUS: S1—WV; S2—VA; rarity and limited habitat at the periphery of the range are the principal causes of concern.

RANGE: Southeastern United States from VA west to MO south to FL and east TX.

KEY REFERENCES: Jones (1977), Linzey (1998).

Townsend's Big-eared Bat
Corynorhinus townsendii

DESCRIPTION: L = 9–11 cm (3.5–4.3 in); a medium-sized tan bat of striking appearance with very large ears, nose with elongated nostrils, and a glandular mass between each eye and the nose; similar to Rafinesque's Big-eared Bat, but with buffy rather than white underparts; was placed in the genus *Plecotus*.

HABITAT AND DISTRIBUTION: In summer, this species frequents woodlands; maternity roosts are mainly in limestone caves; winter roosts and hibernacula are in cooler caves; in the Mid-Atlantic region they are found in the mountains of VA, and nearly throughout WV (highest known populations of the species).

HABITS: Nocturnal, leaving roosts to forage well after dark; hibernates, at least during some periods in some portions of its range; rolls ears backwards like a scroll when roosting to limit heat loss.

DIET: Flying insects, especially moths, detected by echolocation.

REPRODUCTION: Mating evidently occurs in fall and winter; females store sperm, and fertilization occurs in spring; gestation lasts 2–3.5 months; single offspring is born in May–Jun and weaned at 2 months; sexual maturity may be reached in first summer, at least by females.

CONSERVATION STATUS: S1—WV; S2—VA; "Endangered"—USFWS; human disturbance of roosting caves is the principal cause of concern.

RANGE: Western North America from southeastern AK and British Columbia south to southern Mexico; isolated populations in northwestern AK and  adjacent states; and in eastern KY, TN, NC, WV, and western VA.

KEY REFERENCES: Kunz and Martin (1982), Linzey (1998).

ORDER LAGOMORPHA; FAMILY LEPORIDAE

Eastern Cottontail *Sylvilagus floridanus*

DESCRIPTION: L = 38–49 cm (15–19 in); A medium-sized rabbit, brown or gray above and white below; short, fluffy tail, brown above and white below; buffy throat; rusty patch on nape; white eyering; dark tips on long, tan ears; furry feet.

HABITAT AND DISTRIBUTION: Found across a range of field, agricultural, residential, and woodland habitats throughout the Mid-Atlantic region wherever preferred herbaceous vegetation for food is available.

HABITS: Circadian, but with crepuscular feeding peaks; active throughout the year.

DIET: Feeds on a wide variety of plants, including herbs, grasses, and crops in summer and seeds, twigs, bark, and buds in winter; coprophagous.

REPRODUCTION: Most breeding in the Mid-Atlantic occurs from Feb–Aug; several litters/year can be produced; female digs a shallow burrow and lines it with grass, plant stems, and fur; gestation is 1 month; litter size 3–6; weaning at 5 weeks; sexual

maturity may be reached during the first year for females, later for males.

RANGE: Eastern and central North America from southern Canada to FL, TX, NM, and AZ; also parts of Mexico, Central, and northern South America.

KEY REFERENCES: Chapman et al. (1980).

Appalachian Cottontail *Sylvilagus obscurus*

DESCRIPTION: L = 38–41 cm (15–16 in); A medium-sized rabbit, brown or gray and black above, giving the animal a grizzled appearance; white below; short, fluffy tail, brown above and white below; dark anterior on tan, partially furred ears; furry feet; black patch between the ears; similar to the Eastern Cottontail but smaller, darker, and with shorter ears; formerly recognized as a subspecies of the New England Cottontail

S. transitionalis found west of the Hudson River; some authors continue this classification.

HABITAT AND DISTRIBUTION: Forests with dense understory in highlands of WV, VA, western MD, PA, and NJ.

HABITS: Circadian, but with crepuscular feeding peaks; active throughout the year.

DIET: Herbs, shrub leaves, grasses, and fruits in summer; seeds, twigs, bark, and buds in winter; coprophagous.

REPRODUCTION: Breeds Mar–Jul; several litters/year can be produced; female digs a shallow burrow and lines it with grass, plant stems, and fur; gestation is 1 month; litter size 3–5; weaning at 5 weeks; sexual maturity is reached in the second year for most individuals.

CONSERVATION STATUS: S1—MD; S3—WV; rarity, limited forest habitat threatened by clearing at the periphery of the range, and interbreeding with the Eastern Cottontail are the principal causes of concern.

RANGE: Appalachian Mountains from NY to GA.

KEY REFERENCES: Chapman (1975), Linzey (1998), Chapman et al. (1992).

Marsh Rabbit *Sylvilagus palustris*

DESCRIPTION: L = 35–44 cm (14–17 in); A medium-sized rabbit, dark blackish or reddish-brown above with a reddish nape; feet look small because they lack heavy fur of other *Sylvilagus*;

long claws; ears are relatively broad; no white on tail.

HABITAT AND DISTRIBUTION: Swamps, bottomland forest, marshes; found only in the southeastern corner of VA in our region.

HABITS: Nocturnal; active throughout the year; swims readily.

DIET: Aquatic and emergent vegetation; roots, bulbs, shoots, twigs, seeds, fruits, and bark of riparian plants.

REPRODUCTION: Breeds Feb–May; several litters/year can be produced; female makes a nest of grasses and fur under a log or in a dense thicket (not a burrow); gestation is 30–37 days; litter size 3–5; weaning at 5 weeks; sexual maturity is reached in the second year for most individuals.

CONSERVATION STATUS: S3—VA; limited wetland habitat threatened by clearing at the periphery of the range.

RANGE: Southeastern United States from southeastern VA to FL and southeastern AL.

KEY REFERENCES: Linzey (1998).

Snowshoe Hare *Lepus americanus*

DESCRIPTION: L = 47–52 cm (19–21 in); a large rabbit; brown and black above and white below in summer; entirely white except for black ear tips in winter.

HABITAT AND DISTRIBUTION: Highland boreal forest at scattered localities in the mountains of eastern WV and PA; formerly found in spruce-fir habitats on the highest mountains in VA, the species is now restricted solely to a population in northwestern Highland County; recent introductions in VA are of unknown success; probably extirpated in MD and NJ.

HABITS: Nocturnal with crepuscular peaks in foraging activity; active throughout the year.

DIET: Herbs, shrub leaves, grasses, and fruits in summer; seeds, twigs, bark, and buds in winter; coprophagous; also eats carrion when available.

REPRODUCTION: Breeds Feb–Aug; 1–4 litters/year; rock piles, upturned trees, hollow logs, and dense thickets are used as dens; gestation is 5 weeks; litter size averages 3; weaning at 4–6 weeks; sexual maturity is reached in second year.

CONSERVATION STATUS: S1—VA; S3—PA; SX—NJ; SH—MD; rarity and limited habitat degraded by acid rain, clearing, burning, and deer browsing at the periphery of the range are the principal causes of concern.

RANGE: Boreal North America from tree line south to the northern United States, and south in western mountains to CA, UT, and NM and in the Appalachians to TN.

KEY REFERENCES: Merritt (1987).

Black-tailed Jackrabbit *Lepus californicus*

DESCRIPTION: L = 48–60 cm (19–24 in); a large, rangy rabbit mottled gray and black with extremely long, black-tipped ears, long legs, and black-striped tail.

HABITAT AND DISTRIBUTION: Grassland, desert, and scrub habitats of western North America; introduced onto VA barrier islands where populations persist evidently on Cobb, Little Cobb, Rogue, Hog, and Castle Ridge islands; records also for MD and NJ.

HABITS: Nocturnal with crepuscular peaks in activity; active throughout the year.

DIET: Grasses, herbs, shrubs, fruits, seeds, crops; coprophagous.

REPRODUCTION: Breeds Dec–Sep; 1–4 litters/year; gestation 6 weeks; litter size usually 2–4; weaning 3–12 weeks.

RANGE: Western North America from WA east to SD and south to central Mexico.

KEY REFERENCES: Best (1996), Linzey (1998).

ORDER RODENTIA; FAMILY SCIURIDAE

Eastern Chipmunk *Tamias striatus*

DESCRIPTION: L = 22–27 cm (9–11 in); a small squirrel with tan dorsal pelage broadly striped with black and white; whitish below; long, hairy tail; quite vocal, giving loud chips and trills; has cheek pouches for food storage during foraging.

HABITAT AND DISTRIBUTION: Open deciduous and mixed forest, second growth, and residential areas throughout the Mid-Atlantic.

HABITS: Diurnal; terrestrial; fossorial, digging tunnel systems and burrows for predator escape, resting,

reproduction, food storage, and hibernating; arboreal, searching shrubs and saplings for nuts, fruits, and bird eggs; hibernates.

DIET: Nuts, seeds, crops, fruits, bird eggs, small invertebrates and vertebrates.

REPRODUCTION: Two breeding peaks: Mar–Apr and Jul–Aug; 1–2 litters/year; nest is located in a burrow lined with leaves, grass, and plant stems; gestation 1 month; litter size usually 3–5; weaning at 6 weeks; sexual maturity is reached at 3 months, but most individuals probably do not breed until the second year.

RANGE: Eastern North America from Quebec west to Manitoba and south to GA and eastern OK.

KEY REFERENCES: Snyder (1982).

Woodchuck *Marmota monax*

DESCRIPTION: L = 53–65 cm (21–26 in); the woodchuck or groundhog is a large, heavy bodied ground squirrel; light brown body variously grizzled with white; feet and tail are dark; vocalizations include a shrill whistle.

HABITAT AND DISTRIBUTION: Fields, open woodlands, agricultural and residential areas throughout most of the Mid-Atlantic (absent from southeastern VA and the southern Delmarva Peninsula).

HABITS: Diurnal; terrestrial; fossorial, digging extensive systems of tunnels and burrows for predator avoidance, resting, food storage, reproduction, latrines, and hibernation; arboreal, climbing into trees for fruits, or, on occasion, to escape predators.

DIET: Grass, herbs, crops, fruits.

REPRODUCTION: Breeds Mar–Apr; 1 litter/year; nest is built in a burrow lined with leaves and grass; gestation 1 month; litter size averages 4; weaning at 4–6 weeks; sexual maturity is reached the year following birth.

RANGE: Northern North America from AK east to Labrador, and in eastern North America south to OK and GA.

KEY REFERENCES: Kwiecinski (1998).

Eastern Gray Squirrel *Sciurus carolinensis*

DESCRIPTION: L = 42–54 cm (17–21 in); a medium-sized tree squirrel, gray above and white below with a long, bushy tail; partial white eyering; melanism and albinism are not rare in this species, with some populations being entirely black, entirely white, or various combinations; vocalizations include various chirs and barks.

HABITAT AND DISTRIBUTION: Deciduous and mixed woodlands and residential areas throughout the Mid-Atlantic.

HABITS: Diurnal with activity peaks in morning and afternoon; arboreal; terrestrial; builds nests in tree cavities for resting, protection from the weather, and reproduction; also builds nests of leaves and twigs in tree crowns.

DIET: Nuts, especially acorns, hickory nuts, walnuts, beechnuts, and hazelnuts; also buds, fruits, seeds, fungi, and insects; caches food by burying it for relocation by smell and memory in winter; active year-round.

REPRODUCTION: Breeds in 2 peaks: Dec–Feb and May–Jun; 1–2 litters/year; nest is built in a tree cavity lined with leaves, moss, bark shreds, grass, and plant stems; gestation 6 weeks; litter size 2–3; weaning 10–12 weeks, but young of late litters may remain with the mother through the winter; sexual maturity is reached in 5 months.

RANGE: Eastern North America from Nova Scotia to southern Manitoba south to FL and eastern TX.

KEY REFERENCES: Koprowski (1994a).

Eastern Fox Squirrel *Sciurus niger*

DESCRIPTION: L = 51–59 cm (20–23 in); a large tree squirrel; rusty brown above and tan below with a long, bushy, rusty tail; variable in coloration with some populations grayer or blacker, especially along the eastern Coastal Plain (*S. n. niger*); the Delmarva Fox Squirrel (*S. n. cinereus*), found on the eastern shore of MD and VA, is usually gray with a black face; the coastal populations of the Carolinas and southeastern VA are gray with black on the head and feet, and white belly, nose, and ear tips; melanism and albinism are not rare.

HABITAT AND DISTRIBUTION: Open woodlands, savanna, parklands, and residential areas; coastal oak scrub, woodlands, and pine-oak savanna; western populations (*S. n. rufiventer*) are found in western and northern PA and WV; Appalachian and adjacent Piedmont populations (*S. n. vulpinus*) are found in southern PA, eastern WV, western MD, and western VA; Coastal Plain populations (*S. n. niger*) supposedly were common in southeastern VA, but are now restricted to single localities in Amelia and Surry counties; the Delmarva Fox Squirrel (*S. n. cinereus*) is restricted to a few sites on the Delmarva Peninsula and adjacent barrier islands.

HABITS: Diurnal; arboreal; terrestrial; active year-round; caches food by burying for re-location by smell and memory in winter; builds nests in tree cavities for resting, protection from the weather, and reproduction; also builds nests of leaves and twigs in tree crowns.

DIET: Acorns, pine seeds, hickory nuts, beechnuts, hazelnuts, walnuts, fruits, fungi, bark, buds, shoots, crops.

REPRODUCTION: Breeds in 2 peaks: Jan–Apr (main breeding period) and Jul–Aug; 1–2 litters/year; nest is built in a tree cavity lined with leaves, moss, bark shreds, grass, and plant stems; gestation 6 weeks; litter size 2–3; weaning 10–13 weeks, but young of late litters may remain with the mother through the winter; sexual maturity is reached in year following birth.

CONSERVATION STATUS: SX—NJ; S1—VA;—the Delmarva Fox Squirrel is listed as "Endangered" by the USFWS and VA; declines caused by destruction of coastal woodlands.

RANGE: Historically nearly throughout eastern North America from NY west to WY south to FL and western TX; also Coahuila and Tamaulipas in northern

Mexico; however, populations in parts of the Piedmont and Coastal Plain of NC, VA, MD, and NJ have declined or disappeared.

KEY REFERENCES: Koprowski (1994b).

Red Squirrel *Tamiasciurus hudsonicus*

DESCRIPTION: L = 24–35 cm (10–14 in); a small tree squirrel, reddish above, whitish below, with a long, bushy tail and partial white eyering; winter pelage is mottled gray with a rusty band down the back to the tip of the tail, grayish belly, and tufted ears; in summer the back is more uniformly gray, there are no ear tufts, the belly is white, and there is a dark line along the side; vocalizations include prolonged, scolding chirs and various squeaks and barks.

HABITAT AND DISTRIBUTION: Coniferous, mixed, and, occasionally, deciduous forest nearly throughout the region (absent from southern VA, WV, and MD).

HABITS: Diurnal with early morning and late afternoon activity peaks; solitary, except briefly for mating; arboreal; terrestrial, constructing one or more cache middens in stumps, hollow logs, tree cavities, or abandoned burrows where they store nuts for winter use; these middens are vigorously defended against other squirrels or any other potential robber; active year-round; builds nests in cavities, tree crowns, abandoned hawk nests, or logs for resting, protection from the weather and predators, and reproduction.

DIET: Acorns, hickory nuts, walnuts, fruits, fungi, beechnuts, birds eggs, conifer seeds; in mixed forest where both Red and Gray squirrels occur, Red Squirrels appear restricted to use of conifer areas.

REPRODUCTION: Breeds Mar–Apr and Jun–Jul; 1–2 litters/year; nest is usually located in a tree cavity lined with grasses, leaves, plant stems, bark shards, but can be placed in tree crowns, stumps, logs, or abandoned burrows; gestation 35–38 days; litter size 4–5; weaning at 9–11 weeks; sexual maturity is reached at 1 year.

CONSERVATION STATUS: S3—DE; limited habitat.

RANGE: Boreal and north temperate North America from tree line south in mountains in the west to AZ and NM, and in the east to GA.

KEY REFERENCES: Steele (1998).

Northern Flying Squirrel *Glaucomys sabrinus*

DESCRIPTION: L = 25–28 cm (10–11 in); a small squirrel, glossy reddish-brown above; creamy white below; large, dark eyes; broad, furred skin fold between arms and legs used when extended for gliding between trees; flattened, bushy tail serves as a rudder.

HABITAT AND DISTRIBUTION: Boreal coniferous and mixed forest in highlands of northern and west central PA, northwestern NJ, eastern WV, and 5 localities in 3 counties in western VA (Smyth, Grayson, and Highland).

HABITS: Nocturnal with an activity peak shortly after dark; active year-round; several individuals share winter nests, evidently to maximize warmth; gliding is used to move between trees, with average gliding distance about 20 m (70 ft) in an Alaskan population; builds nests for resting, protected feeding, and reproduction, mainly in abandoned woodpecker holes lined with bark shreds, leaves, pine needles, fur, grasses, moss, and feathers; also builds nests of leaves and sticks in abandoned hawk nests, in tree branches, or even in burrows; usually more than one nest is used per individual, with a single primary site and several alternate sites; caches food.

DIET: Conifer seeds, acorns, cherry pits, beechnuts, lichens, fungi, sap, fruits, bird eggs and young.

REPRODUCTION: Breeds Feb–May and Jul–Aug; 1–2 litters/year; nest is usually built in a tree cavity; gestation 37–42 days; litter size is usually 4–5; weaning at 2 months; family groups may remain together through the winter; sexual maturity is reached at 6–12 months.

CONSERVATION STATUS: S1—VA; S2—WV; relic populations in limited boreal forest habitat threatened by clearing, fire, acid rain, and, perhaps, competition with the Southern Flying Squirrel.

RANGE: Boreal North America south of tree line south in mountains in the west to CA and UT, and in the east at scattered localities in the Appalachians south to GA.

KEY REFERENCES: Wells-Gosling and Haney (1984), Linzey (1998).

Southern Flying Squirrel *Glaucomys volans*

DESCRIPTION: L = 21–25 cm (8–10 in); a small squirrel, glossy gray above; white below; large, dark eyes; broad, furred skin fold (patagium) between arms and legs used when extended for gliding between trees; flattened, bushy tail serves as a rudder; similar to *G. sabrinus,* which is more reddish on the back, has two-tone belly fur (gray at base, white at tip), and is larger.

HABITAT AND DISTRIBUTION: Deciduous and mixed forest throughout the Mid-Atlantic.

HABITS: Nocturnal with an activity peak shortly after dark; active year-round;; gliding is used to move

between trees, with glides averaging 6–9 m, although exceptionally > 30 m; builds nests for resting, protected feeding, and reproduction in abandoned woodpecker holes lined with bark shreds, leaves, pine needles, fur, grasses, moss, and feathers; also builds nests of leaves and sticks in abandoned hawk nests, in tree branches, or even in burrows; usually more than one nest is used per individual; several individuals may share the same winter nest, evidently to maximize warmth; caches food in tree cavities.

DIET: Acorns, cherry pits, beechnuts, lichens, fungi, sap, fruits, bird eggs and young, buds, insects, mice, carrion.

REPRODUCTION: Breeds Apr–May and Aug–Sep; 1–2 litters/year; nest is usually built in a tree cavity; gestation 40 days; litter size is usually 3–4; weaning at 2 months; family groups may remain together through the winter; sexual maturity is reached at 6–12 months.

RANGE: Eastern North America from Nova Scotia west to MN and south to FL and eastern TX; also parts of Mexico, Guatemala, and Honduras.

KEY REFERENCES: Dolan and Carter (1977).

FAMILY CASTORIDAE

American Beaver *Castor canadensis*

DESCRIPTION: L = 0.8–1.4 m (32–55 in); a large, heavy bodied rodent; dark brown body; large, scaly, paddle-shaped tail; webbed hind feet.

HABITAT AND DISTRIBUTION: Small to medium-sized watersheds, ponds, and lakes throughout the Mid-Atlantic.

HABITS: Mostly nocturnal; semi-aquatic; active year-round; constructs dams to create impoundments for protective lodges; builds lodges in standing water to serve as a site protected from predators and the weather for resting, feeding, and reproduction.

DIET: Feeds mostly on bark and cambium from shrubs, branches, and small trees, harvested by gnawing at the base, usually within 100 m (330 ft) of water; favored tree species include poplar (*Populus* sp.), beech (*Fagus* sp.), maple (*Acer* sp.), birch (*Betula* sp.), and dogwood (*Cornus* sp.); stores branches in underwater caches in autumn for winter use.

REPRODUCTION: Monogamous; copulation occurs Jan–Mar; 1 litter/year; young are born in the lodge; gestation 105–107 days; litter size averages 3–4; weaning at 6 weeks, but young can remain with parents for up to 2 years before dispersal from the natal lodge; sexual maturity at 21 months for males, 3 years for females.

CONSERVATION STATUS: S3—DE; formerly extirpated from much of its range, the beaver has recovered in most areas, and reached nuisance status in many as a result of changes in habitat and game laws. They can cause significant alterations to watersheds and damage to trees.

RANGE: North America from south of tundra to northern Mexico.

KEY REFERENCES: Jenkins and Busher (1979), Linzey (1998).

FAMILY MURIDAE

Marsh Rice Rat *Oryzomys palustris*

DESCRIPTION: L = 22–26 cm (9–10 in); a medium-sized rat; thick, grayish-brown fur above, grayish white below; small, furred ears and long, sparsely haired tail.

HABITAT AND DISTRIBUTION: Fresh, brackish, and saltwater wetlands along the Coastal Plain of NJ, MD, and VA; may have occurred in marshes along the Delaware River in PA.

HABITS: Circadian but more active at night; active year-round; solitary except during mating; builds globular nests, 25–45 cm dia. (10–18 in), of woven grasses, leaves, and sedges in marsh vegetation 1 m (3 ft) or so above water for resting, hiding, and reproduction.

DIET: Seeds and marsh vegetation; also takes small invertebrates and vertebrates (e.g., bird eggs and young) as available.

REPRODUCTION: Promiscuous; breeds Mar–Nov; several litters/year; gestation 21–28 days; litter size usually 4–6; weaning at 2 weeks; sexual maturity is reached in 2 months.

CONSERVATION STATUS: SX—PA; S3—NJ, DE; destruction of coastal wetland habitat.

RANGE: Eastern United States from NJ west to KS south to FL and southern TX.

KEY REFERENCES: Wolfe (1982).

Eastern Harvest Mouse
Reithrodontomys humulis

DESCRIPTION: L = 11–15 cm (4.3–5.9 in); a small mouse; back is brown down the center and more grayish on the sides; underparts grayish-white; prominent, rounded ears are furred; sparsely haired, long, bicolored tail (gray above, whitish below).

HABITAT AND DISTRIBUTION: Old fields, hedgerows, and grasslands with scattered shrubs in VA, southern and western WV, and southwestern MD.

HABITS: Mostly nocturnal; active year-round; solitary; builds nest, 7–10 cm dia. (3–4 in), of woven grasses and leaves lined with fine grasses and plant stems in shrubs, grass clumps, or on the ground for resting or reproduction.

DIET: Seeds, fruits, grasses; caches food.

REPRODUCTION: Promiscuous; breeds nearly year-round; several litters/year; gestation 3 weeks; litter size 2–4; weaning 2–4 weeks; sexual maturity 2–4 months.

CONSERVATION STATUS: SX—MD; S1—WV; limited habitat at periphery of range.

RANGE: Eastern United States from MD west to OH, KY, and OK, and south to FL and east TX.

KEY REFERENCES: Stalling (1997).

Cotton Mouse *Peromyscus gossypinus*

DESCRIPTION: L = 18–20 cm (7.1–7.9 in); a medium-sized mouse; grayish-brown back; tawny brown sides; white belly and feet; sparsely haired, indistinctly bicolored tail is half body length; very similar to the White-footed Mouse but with somewhat longer hind foot, 22–26 mm (8.7–10.2 in) versus 19–24 mm (7.5–9.4 in).

HABITAT AND DISTRIBUTION: Swamps, wet bottomlands, riparian forest, and cypress bays in southeastern VA.

HABITS: Nocturnal; solitary; active year-round; terrestrial; arboreal, often foraging in shrubs and trees; builds nest for resting or breeding in tree cavities, logs, rock piles, and abandoned buildings of leaves, grasses, plant stems, and bark shreds.

DIET: Seeds, nuts, small invertebrates, especially beetles and caterpillars, and fruits.

REPRODUCTION: Promiscuous; breeds Mar–Oct; several litters/year; gestation 23 days; litter size 3–5; weaning at 3–4 weeks; sexual maturity is reached at 1–2 months.

CONSERVATION STATUS: S3—VA; limited habitat at periphery of range.

RANGE: Eastern United States from VA west to IL and OK, and south to FL and east TX.

KEY REFERENCES: Wolfe and Linzey (1977).

White-footed Mouse *Peromyscus leucopus*

DESCRIPTION: L = 15–20 cm (5.9–7.9 in); a medium-sized mouse with moderately large ears, large eyes and a pointed snout; grayish-brown back; tawny brown sides; white belly and feet; very similar to the Deer Mouse, but usually with an indistinctly bicolored tail that is less than its body length, while that of the Deer Mouse is usually longer than its body length, distinctly bicolored, and has a white terminal tuft.

HABITAT AND DISTRIBUTION: Woodlands, fields, agricultural and residential areas (including inside houses) throughout the Mid-Atlantic.

HABITS: Nocturnal; active year-round; undergoes torpor at times in winter; sometimes associates in "huddles" in winter; subnivean foraging; nests for resting, hiding, or reproduction are built in tree cavities, logs, burrows, buildings, and abandoned bird nests of leaves, grasses, bark shreds, and moss.

DIET: Arthropods, seeds, fruits; caches food.

REPRODUCTION: Promiscuous; breeds Mar–Dec; several litters/year; gestation 22–25 days; litter size 4–5; weaning 3 weeks; sexual maturity 5–7 weeks.

RANGE: Eastern and central North America from Nova Scotia west to southern Alberta south to GA and AZ; also eastern and central Mexico.

KEY REFERENCES: Lackey et al. (1985).

Deer Mouse *Peromyscus maniculatus*

DESCRIPTION: L = 15–20 cm (5.9–7.9 in); a medium-sized mouse with moderately large ears, large eyes, and pointed snout; grayish-brown back; brown sides; white belly and feet; juveniles are gray until 4–6 weeks old; very similar to the White-footed Mouse, which usually has an indistinctly bicolored tail that is less than its body length, while that of the Deer Mouse is usually longer than its body length, distinctly bicolored, and often has a white terminal tuft.

HABITAT AND DISTRIBUTION: Woodlands (*P. m. nubiterrae, P. m. gracilis*) and fields (*P. m. bairdii*) of WV, PA, western VA, MD, and NJ; mostly absent from the Piedmont and Coastal Plain.

HABITS: Nocturnal; active year-round; undergoes torpor at times in winter; sometimes associates in "huddles" in winter; subnivean foraging; nests 15–20 cm dia. (6–8 in), for resting, hiding, or reproduction are built in tree cavities, logs, burrows, buildings, and abandoned bird nests of leaves, grasses, bark shreds, hair, feathers, and moss.

DIET: Arthropods, seeds, fruits, fungi; caches food;

REPRODUCTION: Promiscuous; breeds Mar–Oct; several litters/year; gestation 23 days; litter size 4–5; weaning 3 weeks; sexual maturity is reached at 2 months.

RANGE: Eastern and central North America from Nova Scotia west to southern Alberta south to GA and AZ; also eastern and central Mexico.

KEY REFERENCES: Linzey (1998), DeGraaf and Yamasaki (2001).

Golden Mouse *Ochrotomys nuttalli*

DESCRIPTION: L = 14–20 cm (5.5–7.9 in); a beautiful medium-sized mouse; body, head, and ears covered with russet fur; small feet are white or cinnamon; long, semiprehensile tail is tawny above and white below.

HABITAT AND DISTRIBUTION: Dense understory and thickets of woodlands and old fields in western and southern VA and eastern and southern WV.

HABITS: Nocturnal; gregarious; semi-arboreal, often foraging and nesting in shrubs and trees; builds nest, 14 cm dia. (6 in), for resting, feeding, and reproduction in a vine tangle, the crotch of a shrub or sapling, or in a hollow log, using leaves and pine needles lined with bark shards, grasses, and feathers; seeds are often stored in the nest; Golden Mice also build feeding platforms where they bring food in cheek pouches to store and eat.

DIET: Seeds, fruits, nuts, insects.

REPRODUCTION: Promiscuous?; breeds Mar–Oct; several litters/year; gestation 25–30 days; litter size 2–3; weaning 3 weeks.

CONSERVATION STATUS: S2—WV; limited habitat at periphery of range.

RANGE: Southeastern

United States from VA west to MO and south to FL and east TX.

KEY REFERENCES: Linzey and Packard (1977), Linzey (1998).

Hispid Cotton Rat *Sigmodon hispidus*

DESCRIPTION: L = 23–33 cm (9–13 in); a chunky, medium-sized rat having a grizzled appearance, with mixed brown, gray, and black pelage above, grayish-white below; mostly naked, scaly tail is shorter than the body; ears almost hidden by fur.

HABITAT AND DISTRIBUTION: Grassland, pasture, old fields, agricultural areas, and marshes in southern VA.

HABITS: Circadian with crepuscular peaks; active year-round; terrestrial, building tunnels and runway systems through and under dense grass; nest for resting, hiding, and reproduction is placed under a log, rock, board, or other debris and built of grasses and plant stems.

DIET: Grasses, leaves, stems, roots, buds, nuts, seeds, fruits, arthropods.

REPRODUCTION: Promiscuous; breeds throughout the year; several litters/year; gestation 27 days; litter size averages 5–7; weaning 15–25 days; sexual maturity is reached at 2–3 months.

RANGE: Southeastern United States from VA west to eastern CO and south to FL and southeastern AZ; also Mexico, Central America, and parts of northern South America.

KEY REFERENCES: Cameron and Spencer (1981).

Appalachian Woodrat *Neotoma magister*

DESCRIPTION: L = 35–43 cm (14–17 in); a fairly large rat; brownish-gray above with black guard hairs; whitish below; large eyes; long whiskers; large, nearly naked ears; long, hairy tail; formerly considered to be a subspecies of the Eastern Woodrat (*N. floridana*).

HABITAT AND DISTRIBUTION: Forest, often associated with rocky slopes, crevices, boulder fields, cliffs, caves, talus, and stream borders in the highlands of PA, WV, western VA, MD, and NJ.

HABITS: Nocturnal; solitary; active year-round; nests for resting, food storage, and reproduction are large mounds 0.5–3 m in dia. (3–10 ft), of sticks, leaves, twigs, and other debris with an inner nest of bark shreds and grasses; woodrats have the curious habitat of collecting odd materials for their mounds, including pine cones, bones, feathers, paper, coins, shotgun shells, or almost any other small item, often leaving another item "in exchange" where it was gathered—hence the "pack rat" designation of Western fame; *N. floridana* in TX has been found to be resistant to rattlesnake venom; woodrats may chatter their teeth or "drum" with their hind legs when disturbed.

DIET: Nuts, leaves, fruits, arthropods, fungi; caches food on or in its house.

REPRODUCTION: Promiscuous; breeds Mar–Sep; 2–3 litters/year; gestation 32–38 days; litter size 2–4; weaning at 4 weeks; sexual maturity may be reached late in the first year, but second year is more usual.

CONSERVATION STATUS: S1—MD, NJ; S3—PA, WV, VA; declines have been recorded from throughout the range for reasons perhaps related to loss of mature forest.

RANGE: Appalachian region from PA and NJ to GA; formerly to NY and CT.

KEY REFERENCES: Wiley (1980), Linzey (1998).

Norway Rat *Rattus norvegicus*

DESCRIPTION: L = 32–46 cm (13–18 in); a large rat; brown on the back; paler below; naked ears; scaly tail is dark above, lighter below, and shorter than the length of the body.

HABITAT AND DISTRIBUTION: An introduced species found in urban, suburban, residential, and agricultural areas throughout the Mid-Atlantic; seldom found far from human habitation.

HABITS: Nocturnal; colonial with some evidence of cooperative breeding; terrestrial; fossorial, constructing systems of tunnels and burrows under debris, litter, and rocks.

DIET: Omnivorous, feeding on crops, garbage, fruits, and similar gleanings from human refuse.

REPRODUCTION: Promiscuous; breeds year-round; several litters/year; nest of leaves, string, paper, and other detritus are built under debris; gestation 22–24 days; litter size 7–8; weaning 3–4 weeks; sexual maturity is reached at 3–4 months.

CONSERVATION STATUS: Norway Rats cause serious economic damage and serve as hosts for a variety of diseases harmful to humans, including bubonic plague and typhus. The laboratory rat is derived from this species.

RANGE: Native to southeast-
ern Siberia and China
but introduced nearly
worldwide; found
throughout North
America from AK and
southern Canada to
Panama.

KEY REFERENCES: Armitage (2004).

Black Rat *Rattus rattus*

DESCRIPTION: L = 33–46 cm (13–18 in); a large rat, brown or black above, paler below; naked ears and long, naked tail (not bicolored as in Norway Rat, and longer than the body).

HABITAT AND DISTRIBUTION: An introduced species found around human habitation mainly in coastal areas or lowlands, < 250 m (800 ft) throughout the Mid-Atlantic, although declining or displaced in some areas by Norway Rats. Linzey (1998) believes this species to be absent from most of VA at present.

HABITS: Nocturnal; terrestrial; arboreal; active year-round.

DIET: Mostly vegetarian, especially grains, but will eat a variety of human refuse.

REPRODUCTION: Polygynous; breeds throughout the year; several litters/year; nest is built of sticks, leaves, and detritus, under debris, in abandoned burrows of other mammals, or in trees; gestation 21–29 days; litter size averages 8; weaning 3–4 weeks; sexual maturity 3–5 months.

CONSERVATION STATUS: Black Rats cause serious eco-

nomic damage, and
serve as hosts for a vari-
ety of diseases harmful
to humans, including
bubonic plague and
typhus.

RANGE: Native to India, but
introduced worldwide to coastal areas.

KEY REFERENCES: Gillespie (2004).

House Mouse *Mus musculus*

DESCRIPTION: L = 15–18 cm (6–7 in); a small mouse; pelage is almost uniformly grayish-brown, somewhat paler below; naked ears; long, sparsely haired tail (longer than body).

HABITAT AND DISTRIBUTION: Found in association with human habitation throughout the Mid-Atlantic.

HABITS: Nocturnal; colonial; mostly terrestrial, but climbs readily; fossorial, constructing tunnel and burrow systems; active year-round; caches food in nest chambers.

DIET: Seeds, fruits, leaves, roots, arthropods, most types of human garbage.

REPRODUCTION: Polygynous; breeds throughout the year; several litters/year; nest is built from shredded cloth, paper, string, leaves, or similar material and placed in a burrow, between walls, or under floors, boards, logs, or rocks; gestation 21 days; litter size averages 5; weaning 3 weeks; sexual maturity 5–7 weeks.

CONSERVATION STATUS: House mice can be serious pests on crops and stored grains. The laboratory mouse is derived from this species.

RANGE: Originally perhaps from south-temperate Eurasia, but now distributed nearly worldwide in association with humans.

KEY REFERENCES: Ballenger (1999).

Southern Red-backed Vole *Clethrionomys gapperi*

DESCRIPTION: L = 12–16 cm (5–6 in); a small vole (i.e., a type of mouse with relatively blunt, furry face, small eyes, small, rounded ears, and short, hairy tail); reddish-brown pelage on the back shading to gray on the sides and paler gray or creamy white below.

HABITAT AND DISTRIBUTION: Leaf litter, rotted logs, tree roots, moss, rocks, and boulder fields of coniferous, mixed, and deciduous forest in highlands of PA, WV, western VA, MD, and NJ.

HABITS: Circadian with crepuscular peaks; active year-round; fossorial, using tunnels, burrows, and runways of other small mammals; forages under snow in winter.

DIET: Seeds, nuts, fruits, fungi, arthropods; sometimes caches food in autumn.

REPRODUCTION: Promiscuous; breeds Feb–Oct; 1–6 litters/year; globular nest, 7–10 cm dia. (3–4 in), is placed in a burrow, natural cavity, or abandoned nest of another small mammal, and built of grass, plant stems, leaves, and moss; gestation 17–19 days; litter size averages 5–6; weaning 17 days; sexual maturity is reached at 3 months.

RANGE: Northern North America from Labrador west

to British Columbia and south in western mountains to AZ and NM, and eastern mountains to GA.

KEY REFERENCES: Merritt (1981).

Rock Vole *Microtus chrotorrhinus*

DESCRIPTION: L = 14–17 cm (5.5–6.7 in); a small vole; fluffy yellowish-brown pelage above, grayish below; yellowish snout.

HABITAT AND DISTRIBUTION: Leaf litter, rotten logs, mossy talus, rocky slopes, and boulder fields in

highland coniferous and mixed forest; found in the Mid-Atlantic in isolated populations in northeastern PA (Luzerne, Sullivan, Wayne, and Wyoming counties), VA (Bath County), and eastern WV.

HABITS: Circadian with morning foraging peak; active year-round.

DIET: Fruits, nuts, foliage, moss, fungi, seeds, arthropods; caches food in burrows.

REPRODUCTION: Promiscuous; breeds Mar–Oct; 1–3 litters/year; nest is placed under a rock or log and built of moss with a lining of grasses; gestation 19–21 days; litter size averages 3–4; weaning 3–4 weeks; sexual maturity 3–4 months.

CONSERVATION STATUS: S1—VA, MD; S2—WV, PA; relic populations in

limited boreal forest habitat threatened by clearing, fire, acid rain.

RANGE: Northeastern North America from Labrador west to southwestern Ontario and south in the Appalachians to SC.

KEY REFERENCES: Kirkland and Jannett (1982).

Meadow Vole *Microtus pennsylvanicus*

DESCRIPTION: L = 13–19 cm (5–7.5 in); a small to medium-sized vole; fluffy, dark brown pelage above, grayish below; tail is relatively long for a vole (twice length of hind foot).

HABITAT AND DISTRIBUTION: Dense, grassy understory of meadows, forest openings, marshes, old fields, and agricultural areas throughout the Mid-Atlantic.

HABITS: Circadian with crepuscular peaks; active year-round; builds tunnel and burrow systems through grass and leaf litter, and under snow.

DIET: Leaves, plant stems, fungi, seeds, roots; caches food for winter.

REPRODUCTION: Promiscuous; breeds Apr–Oct; several litters/year; globular nest, 13–20 cm dia. (5–8 in), is placed in a burrow or at the base of a grass tussock and built of woven grasses lined with fine grass; gestation 3 weeks; litter size averages 4–5; weaning at 2 weeks; sexual maturity is reached at 1 month.

RANGE: Northern North America from northern Canada and AK south to WA, UT, NM, OK, and GA.

KEY REFERENCES: Reich (1981).

Woodland Vole *Microtus pinetorum*

DESCRIPTION: L = 11–14 cm (4.3–5.5 in); a small vole with shiny, brown fur above, paler below; short tail (even for a vole—about the length of the hind foot); enlarged front claws (for burrowing).

HABITAT AND DISTRIBUTION: Dense herbaceous understory, loose soil, and leaf litter of woodlands and old fields throughout the Mid-Atlantic.

HABITS: Circadian; fossorial, creating burrow systems in loose soil and under snow; active year-round.

DIET: Roots, bulbs, seeds, fruits; caches food in burrows.

REPRODUCTION: Promiscuous; breeds Feb–Nov; 1–4 litters/year; globular nest, 15–18 cm dia. (6–7 in), is placed under debris, in a burrow, or against a tree trunk, and built of leaves, roots, and grasses; gestation 24 days; litter size 3–6; weaning 3–4 weeks; sexual maturity is reached at 2 months.

CONSERVATION STATUS: Causes extensive damage to orchards in some areas.

RANGE: Eastern North America from southeastern Quebec west to NE and south to FL and central TX.

KEY REFERENCES: Smolen (1981).

Common Muskrat *Ondatra zibethicus*

DESCRIPTION: L = 0.5–0.6 m (20–24 in); a large rodent with rich, thick, chestnut brown pelage; blunt face; small ears mostly hidden by fur; par-

tially webbed hind feet; looks like a small beaver, but with a long, rat-like, laterally compressed, scaly tail.

HABITAT AND DISTRIBUTION: Open freshwater or brackish wetlands throughout the Mid-Atlantic.

HABITS: Mostly nocturnal; active year-round; muskrats build a variety of dens, runways, canals, and lodges for resting, feeding, and reproduction including 1) dens dug into banks, usually with entrances below water level, 2) small mounds of mud and plant material in shallow water with underwater entrances that serve as feeding sites protected from predators, and 3) larger mounds of mud and plant material, 2–3 m (7–10 ft) wide and 0.5–1 m (1.6–3 ft) high, with two or more underwater entrances that serve as nesting sites; slaps its tail, as beavers do, to communicate to other muskrats; has anal scent glands for marking territory and attracting mates.

DIET: Primarily aquatic plants (e.g., cattails, bulrushes, pondweed), but also some aquatic invertebrates and vertebrates; caches food for winter use.

REPRODUCTION: Monogamous in some populations, polygynous in others; breeds Mar–Oct; 2 litters/year; gestation 25–30 days; litter size averages 6; weaning 3–4 weeks; young are normally forced to disperse after weaning, but some late summer litters are allowed to remain on the female's territory through the winter; sexual maturity is reached the year following birth.

RANGE: North America from northern AK and Canada south to CA, TX, and GA.

KEY REFERENCES: Willner et al. (1980).

Southern Bog Lemming *Synaptomys cooperi*

DESCRIPTION: L = 12–15 cm (4.7–5.9 in); a small vole; fluffy dark brown fur above, grayish below; small, beady eyes; ears hidden by fur; short tail (slightly longer than hind foot); broad, grooved incisors.

HABITAT AND DISTRIBUTION: Leaf litter of fields, woodland openings, marshes, pastures, and bogs in WV, PA, NJ, MD, DE, western and southeastern VA.

HABITS: Mostly nocturnal; active year-round; builds tunnel systems in matted grass, litter, and just below soil surface.

DIET: Grasses, foliage, seeds, fruits, mosses, fungi.

REPRODUCTION: Promiscuous; breeds year-round with a peak from Mar–Oct; several litters/year; nest, 10–20 cm dia. (4–8 in), is placed in a burrow (winter), under a log or rock, or in understory vegetation (summer) and built of grass and lined with fine grasses, hair, and feathers; gestation 21–23 days; litter size averages 2–5; weaning at 3 weeks; sexual maturity is reached at 2 months.

CONSERVATION STATUS: S2—NJ, WV; S3—MD; small, poorly known populations in limited range perhaps affected by loss of old field habitat.

RANGE: Northeastern North America from northern Quebec west to southeastern Manitoba south to KS and GA.

KEY REFERENCES: Linzey (1983).

FAMILY ZAPODIDAE

Meadow Jumping Mouse *Zapus hudsonius*

DESCRIPTION: L = 19–22 cm (7.5–8.7 in); a beautiful little mouse, somewhat reminiscent in appearance and habits of the kangaroo mice of the West; pelage is mottled dark gray and brown on the top of the head and back, tan on the sides, and white below; long hind feet; very long tail is bicolored with a dark tuft at the tip (white tuft in the similar Woodland Jumping Mouse); sparsely haired ears are dark with a pale rim.

HABITAT AND DISTRIBUTION: Old fields, wet meadows, swampy thickets, and grasslands bordering ponds, lakes, and streams throughout the Mid-Atlantic.

HABITS: Mostly nocturnal; hibernates (Oct–May) in a globular nest built of leaves and grass and placed in a burrow or under a log or other debris; solitary.

DIET: Mainly arthropods on emergence from hibernation in spring; seeds, grasses, fungi, fruits, nuts, and arthropods in summer and fall.

REPRODUCTION: Promiscuous; breeds May–Sep; 2 litters/year; nest is a ball of leaves and grass placed in a grass tussock, burrow, or log; gestation 17–20 days; litter size averages 4–6 ; weaning at 4 weeks; sexual maturity is reached for most in the second year.

CONSERVATION STATUS: S3—WV; small populations perhaps affected by loss of old field habitat.

RANGE: Boreal Canada and AK; eastern and central United States from ME to MT south to AL and CO.

KEY REFERENCES: Whitaker (1972), Linzey (1998).

Woodland Jumping Mouse *Napaeozapus insignis*

DESCRIPTION: L = 19–22 cm (7.5–8.7 in); similar in appearance to the Meadow Jumping Mouse described above; pelage is mottled dark gray and brown on the top of the head and back, tan on the sides, and white below; long hind feet; very long tail is bicolored with a white tuft at the tip (dark tuft in the similar Meadow Jumping Mouse); sparsely haired ears are larger than those of *Zapus hudsonius*.

HABITAT AND DISTRIBUTION: Cool, damp understory of coniferous and mixed highland forest, bogs, rocky streams, and swamps of WV, northern and western PA, western VA, and northwestern NJ.

HABITS: Mostly nocturnal; hibernates (Oct–May) in a globular nest built of leaves and grass and placed in a burrow or under a log or other debris; solitary.

DIET: Mainly arthropods on emergence from hibernation in spring; seeds, grasses, fruits, fungi, nuts, and arthropods in summer and fall.

REPRODUCTION: Promiscuous; breeds May–Sep; 1–2 litters/year; nest is a ball of leaves and grass placed in a grass tussock, burrow, or log; gestation 29 days; litter size averages 4–5; weaning at 5 weeks; sexual maturity is reached for most in the second year.

RANGE: Northeastern North America from Labrador west to southeastern Manitoba south to MN, WI, and OH, and in mountains to GA.

KEY REFERENCES: Whitaker and Wrigley (1972).

FAMILY ERETHIZONTIDAE

North American Porcupine *Erethizon dorsatum*

DESCRIPTION: L = 0.5–0.8 m (20–32 in); a large rodent of distinctive appearance; the usual view of a porcupine is that of a dark, beachball sized lump with white spines well up on the trunk or large branch of a tree—or flattened on the road; black, blunt, furry face; black feet with strong claws and thick pads; pelage on the back is of 3 types: 1) thick, black fur, 2) long, white-tipped guard hairs, and 3) long, thick, hollow quills, white with dark tips; short thick tail.

HABITAT AND DISTRIBUTION: Highland coniferous and mixed forest of central and northern PA; formerly rare in WV, VA, and western MD, but now presumed extirpated; there have been 4 reports of

porcupines in VA since 1950, unsupported by specimens, from Giles County, Bath County, Wythe County, and Henrico County (!?).

HABITS: Mostly nocturnal, usually spends the day huddled high in a tree against the trunk; active year-round; mostly arboreal; normally solitary, but may associate in communal dens in winter; dens for resting, protection from predators and the weather, or reproduction are located in a hollow log, tree cavity, small cave, rock piles, or abandoned building.

DIET: A variety of plant products in summer, including leaves, roots, buds, and fruits; mostly bark, especially the cambium layer, in winter (especially hemlock and sugar maple); sometimes gleans roadsides for salt.

REPRODUCTION: Polygynous; mating occurs in fall; 1 litter/year; maternity den is normally placed in a hollow log or cave lined with leaves and sticks; gestation 7 months with a single young born in May–Jun; weaning at 3 months, but young remain with the mother until about 6 months; sexual maturity is reached at about 1 year.

CONSERVATION STATUS: SX—VA; S1—MD; loss of mature highland forest.

RANGE: Northern and western North America from northern Canada and AK south in western mountains to CA, AZ, and TX; south to MN, MI, and southern

Ontario in central North America, and south in the Appalachians to MD (formerly VA, WV, and TN) in the east; also northern Mexico (Sonora, Chihuahua, Coahuila).

KEY REFERENCES: Woods (1973).

FAMILY MYOCASTORIDAE

Nutria *Myocastor coypus*

DESCRIPTION: L = 0.7–1.1 m (28–43 in); a large brown-pelaged rodent, similar in appearance to a beaver, but with a long, round, scaly tail and

heavy whiskers (like a walrus mustache); hind feet are partially webbed.

HABITAT AND DISTRIBUTION: Freshwater, brackish, and saltwater marshes along the coast of VA, MD, DE, and NJ; some inland populations in VA and MD as well.

HABITS: Mostly nocturnal; active year-round; burrow systems are built along well vegetated banks of water courses for resting, feeding, and reproduction; platforms of floating vegetation are also built for feeding; uses glandular secretions from the side of its mouth to waterproof fur; teats of the female are located high on its side to allow nursing by young while the mother is swimming.

DIET: Aquatic vegetation including cattails, sedges, marsh grass, and pondweed, which are harvested and then carried to a feeding platform for consumption; can be destructive of native aquatic plant communities because Nutria harvest methods kill the plant.

REPRODUCTION: Polygynous, with a single male sharing a burrow system with 2–3 females; breeds throughout the year; 2 litters/year; maternity nest is built in a burrow along the bank; gestation is 4 months; litter size 2–6; weaning 8 weeks; sexual maturity is reached at 5 months.

RANGE: Originally, South America from central Bolivia southward in lowlands; first introduced into CA for fur farming in 1899, and

later (1930s) in LA and elsewhere as a fur source for trappers and to control clogging of waterways by water hyacinth (another introduced species); releases, escapes, introductions, and natural dispersal have resulted in establishment of populations in many parts of the Coastal Plain of the eastern United States from NJ to TX; also around the Great Lakes region and in the Pacific Northwest.

KEY REFERENCES: Woods et al. (1992).

ORDER CARNIVORA; FAMILY CANIDAE

Feral Dog *Canis familiaris*

DESCRIPTION: L = 0.5–2.0 m (20–78 in); dogs come in a wide variety of shapes, sizes, and colors; data indicate that dogs from North America and Eurasia are derived from the Gray Wolf (*Canis lupus*), although the evidence is not conclusive.

HABITAT AND DISTRIBUTION: Woodlands and open areas throughout the Mid-Atlantic; it can be difficult to determine whether a particular animal living on its own is part of a breeding, wild population (i.e., "feral") or simply one that is lost or escaped. Nevertheless, reports from a number of managers of protected areas document the presence of Feral Dog packs (Drost and Feller 2005). Interbreeding with Coyotes or Gray Wolves can also occur to produce wild hybrid populations.

HABITS: Circadian; active year-round; social, associating in packs.

DIET: Mostly small to medium-sized mammals (including other canids, cats, fawns, and young livestock); sometimes hunts in packs; feeds on fruits when available.

REPRODUCTION: Pack organization—alpha male and female in the pack control reproduction of subordinate members; 1–2 litters/year; den is placed in a small cave, hollow log, burrow, or dense thicket; gestation is 9 weeks; litter size averages 3–9; weaning at 3 months; pack members assist in care of young, regurgitating food for both the female and young at the den site for the first few weeks after birth; sexual maturity is reached at 1–2 years.

RANGE: World-wide in the vicinity of human habitation.

KEY REFERENCES: Bhagat and Dewey (2002), Drost and Fellers (2005).

Coyote *Canis latrans*

DESCRIPTION: L = 1.2–1.4 m (47–55 in); looks like a small, rangy wolf; long, narrow snout; erect ears; long legs; thick gray pelage, often with a darker, middorsal band; thick bushy tail; vocalizes in group choruses of yips and barks in the evening.

HABITAT AND DISTRIBUTION: Woodlands and open areas throughout the Mid-Atlantic; records for the species in the region date back only for the last half century or so.

HABITS: Nocturnal with crepuscular peaks; active year-round; social, associating in family groups.

DIET: Mostly small to medium-sized mammals (including other canids, cats, fawns, and young livestock); sometimes hunts in pairs or family groups.

REPRODUCTION: Monogamous; mates in Jan–Mar; 1 litter/year; den is placed in a small cave, hollow log, or burrow, often in a dense thicket; gestation is 2 months; litter size averages 4–7; weaning at 3 months; both parents care for the young, with the male regurgitating food for both the female and young at the den site for the first few weeks after birth, and then both parents returning to

the den to regurgitate food for the young later; sexual maturity is reached at 1–2 years.

RANGE: North America from northern Canada and AK to Costa Rica; original distribution probably did not include eastern North America, but eradication of the Gray Wolf, and clearing of forest for farms, allowed range expansion.

KEY REFERENCES: Bekoff (1977).

Red Fox *Vulpes vulpes*

DESCRIPTION: L = 0.9–1.1 m (35–43 in); a small canid with thick, reddish fur above, whitish below; white throat and jaw line; long, narrow snout; black legs; erect ears tipped in black on the back; long, bushy, white-tipped tail; advertises territory with scats and screechy barks.

HABITAT AND DISTRIBUTION: Old fields, hedgerows, farm land, residential areas, and open woodland nearly throughout the Mid-Atlantic (absent from southeastern VA); archaeological evidence from pre-Columbian Indian campsites, fossil remains in caves, testimony of early settlers, and statements of indigenous people at time of settlement provide no evidence of the Red Fox in the region at time of settlement, although there is considerable evidence for the Common Gray Fox, indicating that this open-country species may not have been present in the forests of eastern North America prior to European colonization.

HABITS: Nocturnal with crepuscular peaks; active

year-round; solitary hunter except during the breeding period when mates may hunt jointly.

DIET: Small mammals, birds, arthropods, fruits, carrion.

REPRODUCTION: Monogamous; mates in winter; 1 litter/year; den is placed in an excavated burrow, hollow log, small cave, or abandoned building; gestation 51–56 days; litter size averages 4–5; weaning 8–10 weeks; males provide food for the female, and assist in care of the offspring during the post-weaning period; the family unit breaks up in the fall; sexual maturity is reached the winter following birth.

RANGE: North America from northern Canada and AK south to FL, TX, northern NM, AZ, and CA; also Eurasia, North Africa, and Australia (introduced).

KEY REFERENCES: Larivière and Pasitschniak-Arts (1996).

Common Gray Fox *Urocyon cineroargenteus*

DESCRIPTION: L = 0.9–1.0 m (35–39 in); a small, handsome canid; grizzled black, gray, and white on the back; reddish-tan on the neck, flanks, and chest; white throat; gray face with dark muzzle; long, bushy tail, gray above and reddish below with a dark tip; erect ears with white inner fur and reddish-tan and gray outer fur; territory is advertised with a loud, harsh bark, scats, and urine.

HABITAT AND DISTRIBUTION: Woodlands, old fields, swamps, and, less commonly, open areas throughout the region.

HABITS: Nocturnal with crepuscular peaks; active year-round; solitary except during raising of young; able to climb trees to escape predators.

DIET: Small mammals, birds, reptiles, amphibians, fruits, insects, grains.

REPRODUCTION: Monogamous; mates in Feb–Mar; 1 litter/year; den is placed in an excavated burrow,

hollow log, tree cavity, or small cave; gestation 50–60 days; litter size averages 4–6; weaning 8–12 weeks; males provide food for the female, and assist in care of the offspring during the post-weaning period; the family unit breaks up in the fall; sexual maturity is reached the winter following birth.

RANGE: North America south of boreal regions and the Great Plains to Panama and adjacent portions of northern South America.

KEY REFERENCES: Fritzell and Haroldson (1982).

FAMILY URSIDAE

American Black Bear *Ursus americanus*

DESCRIPTION: L = 1.3–1.8 m (51–71 in); a large, heavy, hairy, black mammal, standing as tall as a human on its hind legs.

HABITAT AND DISTRIBUTION: Distributed throughout the region at the time of European settlement, the species is now found in heavily wooded areas of WV, the mountains of western VA, parts of eastern VA, western MD, central and northern PA, and northern NJ; populations are expanding in some parts of the region.

HABITS: Mostly nocturnal with crepuscular peaks; hibernates in winter dens located in dense thick-

ets, excavated burrows, caves, stumps, and hollow logs; solitary.

DIET: Forages opportunistically on fruits, nuts, seeds, roots, fawns, fish, small mammals, insects, honey, bird eggs, and garbage.

REPRODUCTION: Polygynous; mating occurs in summer; implantation is delayed until late autumn; young are born in Jan–Feb while the mother is in her hibernation den; young females usually produce a single cub, while older females may give birth to 2 or 3; 1 litter every other year; weaning at 7 months; young may den with the mother during their first winter; sexual maturity is reached at 3–4 years.

CONSERVATION STATUS: S3—MD, NJ; small populations in limited range.

RANGE: Formerly found throughout most of North America from northern Canada and AK south to central Mexico, except desert regions of the

Southwest; currently distributed across most of Canada and AK, the western United States in forested areas south to CA, AZ, and western TX, AR, eastern MO, parts of the Great Lakes region, and in remaining areas of extensive forest in the eastern United States, where populations have been expanding in recent years.

KEY REFERENCES: Larivière (2001).

FAMILY PROCYONIDAE

Northern Raccoon *Procyon lotor*

DESCRIPTION: L = 0.7–0.9 m (28–35 in); with its rotund body shape, peppery pelage, black mask, ringed tail, and odd, humped-back gait, the raccoon is unmistakable.

HABITAT AND DISTRIBUTION: Woodland, old fields, farm land, parks, residential areas, often near water, throughout the Mid-Atlantic.

HABITS: Mostly nocturnal; undergoes periods of torpor during winter cold spells; dens for resting, hibernating, and breeding are located in tree cavities, crevices, hollow logs, or burrows; social, often remaining in family groups through the winter, sharing a common den; often "washes" food if water is available.

DIET: Depending on the season and availability, raccoons eat fish, crayfish, amphibians, bird eggs, fruits, nuts, grains (ready-to-pick corn is a favorite), arthropods, small mammals, and garbage.

REPRODUCTION: Promiscuous, but males may remain with pregnant females until young are born or for a short period thereafter; mating takes place Jan–Mar; 1 litter/year; gestation is 63 days; litter size averages 3–4; weaning at 16 weeks; cubs may remain with the mother through the winter; sexual maturity is reached at 1–2 years.

RANGE: North America

from southern Canada and southeastern AK south to Panama, except for some desert areas of the western United States; also introduced in France, Germany, and some former republics of the Soviet Union.

KEY REFERENCES: Lotze and Anderson (1979).

FAMILY MUSTELIDAE

Fisher *Martes pennanti*

DESCRIPTION: L = 0.8–1.0 m (32–39 in); a large version of the mink; rich, thick, dark brown pelage; small, rounded ears; long, bushy tail.

HABITAT AND DISTRIBUTION: Formerly distributed throughout the highland coniferous and mixed forest of the region, this species was completely extirpated until reintroduced into Tucker County, WV (1969–1978); this effort appears to have been self-sustaining, and some sightings have been recorded in the mountains of VA, presumably from this small population.

HABITS: Mostly crepuscular in summer and diurnal in winter; active year-round; terrestrial; arboreal; uses dens in tree cavities, rock crevices, or hollow logs for resting and breeding.

DIET: Small mammals (e.g., squirrels, rabbits, mice, shrews), birds, carrion, and fruit.

REPRODUCTION: Mating takes place in late winter or early spring, usually within a week of birth of the previous litter; implantation is delayed for 10–11 months and birth takes place within a month after that, in late winter; 1 litter/year; litter size averages 3; weaning at 3 months, but they remain with their mother until fall; young females mate

at 1 year (giving birth at age 2); males reach sexual maturity at 2 years.

CONSERVATION STATUS: S1—VA; S3—WV, MD, PA; SX—NJ; extirpated as a result of trapping and habitat destruction, but reintroduced in WV.

RANGE: Formerly found throughout boreal and north temperate North America, south in western mountains to CA and WY, and in eastern mountains to TN and NC.

KEY REFERENCES: Powell (1981).

Long-tailed Weasel *Mustela frenata*

DESCRIPTION: L = 29–45 cm (11–18 in); a large weasel; long, narrow body, brown above, cream colored or buffy below; short, brown legs; long, bushy, black-tipped tail; wedge-shaped head with erect, rounded ears, prominent whiskers, and white throat.

HABITAT AND DISTRIBUTION: Woodland, old fields, wetlands, and farm land throughout the Mid-Atlantic.

HABITS: Mostly nocturnal; active year-round, foraging under snow for prey in winter; solitary; dens in burrows built by other species, especially the Eastern Chipmunk; marks territory with pungent secretions from anal scent glands.

DIET: Small mammals are the main prey; takes some prey to its burrow for consumption.

REPRODUCTION: Mates in summer; fertilization follows copulation by 2–4 days; implantation delay is prolonged, lasting from 68–251 days

post-copulation; young are born in Apr–May, about 280 days after mating; 1 litter/year; nest, 22–30 cm dia. (9–12 in), is placed in the chamber of a burrow, lined with grasses and fur; litter size averages 4–7; weaning at 5 weeks; young disperse at 11–12 weeks; sexual maturity is reached by some females late in the year of birth, but by most males and females in the following year.

RANGE: North America from southern Canada to Panama exclusive of the desert southwestern United States and northwestern Mexico; also parts of northwestern South America.

KEY REFERENCES: Sheffield and Thomas (1997).

Ermine *Mustela erminea*

DESCRIPTION: L = 23–30 cm (9–12 in); a medium-sized weasel; brown above and white below in summer; entirely white in winter except for black-tipped tail; similar to the Long-tailed Weasel in summer pelage, but with white (not brown) fur on the inside of its hind feet and with a shorter tail (< 1/3 total length; that of *M. frenata* is > 1/3 total length).

HABITAT AND DISTRIBUTION: Woodlands and second growth, often bordering wetlands, in northwestern NJ and northern and eastern PA.

HABITS: Mostly nocturnal; active year-round, foraging under snow for prey in winter; solitary; dens in hollow logs, under rocks or large tree roots, stumps, or in burrows built by other species, espe-

cially the Eastern Chipmunk; marks territory with pungent secretions from anal scent glands.

DIET: Mostly small mammals; also birds and arthropods.

REPRODUCTION: Polygynous; mating occurs in summer; implantation is delayed several months; young are born in Apr–May, about 260 days after mating; 1 litter/year; nest is placed in a cavity, hollow log, stump, or burrow lined with leaves, grasses, fur, and feathers; litter size averages 6–7; some males may assist in raising of young, bringing prey to the nest; weaning at 8–10 weeks; young disperse at 11–12 weeks; sexual maturity is reached by some females late in the year of birth, but in most males and females by the following year.

RANGE: Northern North America south in the western mountains to CA, NV, UT, and NM, and in the east to PA and northwestern NJ; also northern Eurasia.

KEY REFERENCES: King (1983).

Least Weasel *Mustela nivalis*

DESCRIPTION: L = 18–21 cm (7–8 in); a small weasel; brown above and white below in summer; entirely or partially white in winter in northern parts of its range; similar to the larger Long-tailed Weasel and Ermine, but with a very short, entirely brown tail, lacking a black tip.

HABITAT AND DISTRIBUTION: Woodlands, second growth, old fields, hedgerows and farmlands in WV, western PA, MD, and VA.

HABITS: Nocturnal; active year-round, subnivean foraging in winter; solitary; dens in burrows built by other species, especially small rodents; marks territory with pungent secretions from anal scent glands.

DIET: Mostly small mammals, e.g., voles, mice, and shrews; also birds and arthropods.

REPRODUCTION: Polygynous; mating can occur almost any time of the year, but mainly spring

and late summer; gestation is 35–37 days; 2–3 litter/year; nest is usually placed in a small- mammal burrow lined with leaves, grasses, fur, and feathers; litter size averages 5; male may assist in raising young; weaning at 6–7 weeks; young disperse at 12–14 weeks; sexual maturity is reached by females late in the year of birth, but in the following year for males.

CONSERVATION STATUS: S2—MD; S3—PA, VA; secretive habits leave questions concerning population health; perhaps affected by loss of old field habitat.

RANGE: Northern North America south to southern British Columbia in the west, KS in the central United States, and in the Appalachians to south to GA in the east;  also Eurasia, and introduced in New Zealand, Malta, Crete, the Azores, and Sao Tome (off west Africa).

KEY REFERENCES: Sheffield and King (1994).

American Mink *Mustela vison*

DESCRIPTION: L = 46–68 cm (18–27 in); essentially a large weasel; long, narrow body covered with lustrous, thick, brown fur; white patches on chin and throat; short legs; long, bushy, tail; wedge-shaped head with erect, rounded ears, prominent whiskers, and white throat.

HABITAT AND DISTRIBUTION: Wetlands and wetland borders throughout the Mid-Atlantic.

HABITS: Nocturnal with crepuscular peaks; active year-round; forages both in and along waterways,

where individuals defend territories; solitary; dens in burrows built by itself or confiscated from other species, especially muskrats; marks territory with pungent secretions from anal scent glands.

DIET: Muskrats, crayfish, fish, frogs, rodents, waterfowl eggs and young.

REPRODUCTION: Promiscuous; mating occurs in Feb–Mar; implantation is delayed about a month; 1 litter/year; nest is placed in the chamber of a burrow, muskrat house, or similar site, lined with grasses, fur, and feathers; litter size averages 3–4; male may assist in care of young; weaning at 8–9 weeks; sexual maturity is reached in year following birth.

RANGE: North America from AK and northern Canada to OR in the west, CO and KS in the central United States, and east TX and FL in  the east; introduced in several areas in Eurasia.

KEY REFERENCES: Larivière (1999).

Northern River Otter *Lontra canadensis*

DESCRIPTION: L = 0.9–1.2 m (35–47 in); unique in appearance; tubular, sleek, brown body; long, muscular neck; small head with beady eyes, prominent whiskers, and small valvular ears and nose; webbed feet; long, furry tail; formerly placed in the genus *Lutra*.

HABITAT AND DISTRIBUTION: Forest rivers and streams in VA, MD, northern NJ, and northeastern PA. The species was extirpated from WV, but

249 otters were released through 1997 as part of a reintroduction program.

HABITS: Mostly nocturnal; active year-round; semi-aquatic; most foraging is done in water where it can swim at speeds of 11 km/h (7 mph), submerge for up to 8 min, and dive to depths of 14 m (50 ft); social; engages in "play" activities, e.g., sliding in snow and mud, and wrestling.

DIET: Aquatic vertebrates and invertebrates, especially fish, crayfish, and amphibians; some terrestrial vertebrates and invertebrates as well.

REPRODUCTION: Polygynous; mating occurs in Mar–Apr; implantation is delayed > 8 months; young are born Feb–Apr of the year following mating; 1 litter every other year; den is often placed in a burrow dug into the bank of a river or stream, often with 2 or more underwater entrances; sticks, grass, and leaves are used to line the den; litter size averages 2–3; weaning at 3 months; male may assist in care of offspring once young leave the den; family group remains together for a year or more before young disperse; sexual maturity is reached at 2 years, but most males < 6 years are not successful in breeding.

CONSERVATION STATUS: SH—MD; S1—WV; S3—PA; small population size vulnerable to extirpation resulting from wetland destruction and pollution.

RANGE: Formerly throughout North America north of Mexico except the desert southwest; now found in AK and most of Canada, the northwestern United States, parts of the Great Lakes region and Midwest, the southeast (VA to east TX), and the northeast

(NY, PA and New England); reintroduced in many areas.

KEY REFERENCES: Larivière and Walton (1998).

FAMILY MEPHITIDAE

Eastern Spotted Skunk *Spilogale putorius*

DESCRIPTION: L = 40–55 cm (16–22 in); obviously a skunk, but smaller than the common Striped Skunk, and with white spots and streaks rather than broad stripes; stench is also different.

HABITAT AND DISTRIBUTION: Woodlands, old fields, and farm lands, especially with rocky slopes or abandoned buildings for den sites; found in the region in WV, southwestern PA, western MD, and western VA.

HABITS: Nocturnal; undergoes periods of torpor in winter; builds dens in small caves, abandoned burrows of other mammals, hollow logs, and under buildings; sometimes winter dens are shared; does "handstand" on front feet to threaten, or deliver, spray from anal scent glands at potential predators; mostly terrestrial, but climbs readily for fruits in fall.

DIET: Arthropods, small mammals, birds, reptiles, amphibians, fruits, and fungi.

REPRODUCTION: Mating occurs in Mar–Apr with implantation 2 weeks later; 1 litter/year; gestation 50–65 days with young born in May–Jun; litter size averages 5–6 ; weaning at 54 days, and dispersal at 3–4 months; sexual maturity is reached at 9–10 months.

CONSERVATION STATUS: SH—PA; S1—MD; S2—WV; S3—VA; trapping for other species may cause extirpation of small populations.

RANGE: Central and southern United States from eastern MT to WI south to south TX, and in the east from southern PA to FL.

KEY REFERENCES: Kinlaw (1995).

Striped Skunk *Mephitis mephitis*

DESCRIPTION: L = 54–67 cm (21–26 in); familiar cat-sized mammal; usual color is black with broad white stripes down either side of the back, uniting at the neck, and with a white tuft on the very bushy, black tail; however, variations are common, from nearly all black to nearly all white.

HABITAT AND DISTRIBUTION: Woodlands, second growth, old fields, and farm lands throughout the Mid-Atlantic.

HABITS: Mostly nocturnal except in spring, when males may wander widely, day or night, in search of females in estrus; skunks do a handstand on their front feet when threatened by a predator, and can spray noxious fluid from their anal glands up to 3 m (10 ft) if sufficiently provoked; avoided by most experienced predators, but smell of the Great Horned Owl drawer at museums

indicates that this predator is either a slow learner or unaffected by the spray; undergoes periods of hypothermic torpor in winter; dens for resting in summer are usually above ground; wintering or reproduction dens are excavated under rocks, buildings, logs, or abandoned mammal burrows.

DIET: Insects (including bees and wasps from nests, to whose stings they seem immune), fruits, nuts, roots, grains, small mammals, bird eggs, small reptiles and amphibians, garbage.

REPRODUCTION: Promiscuous; mating occurs in Feb–Mar; 1 litter/year; gestation 59–77 days; litter size 6–8; weaning 6–8 weeks; some kits may remain with the mother through the winter; sexual maturity is reached in the year following birth.

RANGE: North America from central Canada south to central Mexico.

KEY REFERENCES: Wade-Smith and Verts (1978).

FAMILY FELIDAE

Feral Cat *Felis catus*

DESCRIPTION: L = 0.6–0.8 m (24–32 in); tabby can take a wide range of colors; a medium-sized felid with upright, pointed ears; large eyes with vertical pupils; long whiskers; long legs with retractable claws; long tail; derived from the European Wild Cat (*F. silvestris*).

HABITAT AND DISTRIBUTION: Open areas throughout the Mid-Atlantic.

HABITS: Circadian with nocturnal peaks; solitary; active year-round; dens are made in abandoned mammal burrows (e.g., rabbits), hollow log, tree cavity, cave, or abandoned building.

DIET: Mainly small mammals and birds.

REPRODUCTION: Promiscuous; mating takes place throughout the year; 1–2 litters/year; den is placed in a small cave, hollow log, abandoned building, or abandoned burrow; gestation is 65 days; litter size averages 1–4; weaning 35–40 days; sexual maturity is reached at 9 months.

CONSERVATION STATUS: The Feral Cat is one of the most destructive introductions humans have made, causing extinction of a number of island species.

RANGE: Nearly worldwide in the vicinity of human habitation.

KEY REFERENCES: Veitch (2005).

Bobcat *Lynx rufus*

DESCRIPTION: L = 0.7–1.2 m (28–47 in); looks like a large, rangy, long-legged, short-tailed house cat; tan or reddish-brown above streaked and spotted with black; whitish below; face is framed with a short ruff, and has large eyes with partial white eyering, erect, pointed ears (sometimes tufted), pink nose, long whiskers, and white throat; short tail is ringed with dark brown and tipped in white; can be quite vocal during the breeding season, producing a variety of screams and howls.

HABITAT AND DISTRIBUTION: Forest, swamp, old field, and open areas adjacent to woodland; VA, WV, isolated mountain areas of northern and central PA, western MD, and 4 NJ counties based on a recent scent post survey (Sussex, Warren, Morris, and Passaic).

HABITS: Mostly nocturnal; active year-round; solitary; dens for resting are selected in thickets, tree cavities, or hollow logs, and changed often; scrapes are made in forest litter on which they urinate to advertise territory.

DIET: Rabbits, rodents, and, occasionally, deer, are main prey items.

REPRODUCTION: Polygynous; mating occurs Jan–May; 1 litter/year; den is usually in a shallow cave or hollow log; gestation is 50–70 days; litter size averages 2–3; weaning at 2 months; dispersal in fall; sexual maturity is reached at 1–2 years.

CONSERVATION STATUS: S3—PA, NJ, MD; nearly extirpated in much of its eastern range, and still absent or precarious in some parts of the eastern United States and the Midwest.

RANGE: North America from southern Canada to southern Mexico; absent from some western desert regions and parts of the Midwest.

KEY REFERENCES: Larivière and Walton (1985).

ORDER PERISSODACTYLA; FAMILY EQUIDAE

Feral Horse *Equus caballus*

DESCRIPTION: L = 2–3 m (78–117 in); horses, along with cows, deer, pigs, and goats, are ungulates, i.e., they run on their ungules or toenails (hooves). Horses are among the odd-toed ungulates (Order Perissodactyla), while cows, deer, and pigs are even-toed (Order Artiodactyla). The horse's hoof is a modified third digit.

HABITAT AND DISTRIBUTION: Grasslands; currently 2 wild populations of horses exist in VA: along Wilburn Ridge on Mount Rogers (Smyth and Grayson counties) and on Assateague Island (Accomack County).

HABITS: Diurnal; active year-round; social, populations are divided into bachelor male herds and harems (stallion with families of females and offspring).

DIET: Grazes on herbaceous plants.

REPRODUCTION: Polygynous; males defend a herd containing females and young; mating occurs Mar–Sep; 1 foal/year; no nest or den is constructed—foals are precocial, able to move with the herd within a few hours of birth; gestation

330–340 days; weaning at 7 months; sexual maturity is reached at 2 years.

CONSERVATION STATUS: Feral Horses can cause extensive damage to grassland ecosystems, and require careful monitoring.

RANGE: Prairies of North America and Eurasia during the Pleistocene; North American populations became extinct, but were reintroduced  by Europeans during colonization, when feral herds once again became established across the continent, mostly in prairie areas; now restricted to protected herds, mostly in the western United States.

KEY REFERENCES: Linzey (1998), Drost and Fellers (2005).

ORDER ARTIODACTYLA; FAMILY SUIDAE

Feral Pig *Sus scrofa*

DESCRIPTION: L = 1–2 m (39–78 in); pigs vary widely in their size and coloration, depending on the amount of time the population has been feral; brown, reddish-brown, and black are common pelage colors.

HABITAT AND DISTRIBUTION: Woodlands and open areas; currently known only from Pulaski and Wythe counties in southwestern VA, barrier beaches, and perhaps Dismal Swamp, in Chesapeake County, VA.

HABITS: Circadian with crepuscular peaks (probably some feeding activity periods are controlled by tides for populations located on barrier beaches); active year-round; social, with females and young associating in herds; adult males are solitary.

DIET: Nuts, grains, tubers, bulbs, fungi, herbaceous plant leaves, fruits, bird eggs, carrion, manure; often seen along beaches feeding on dead vertebrates and invertebrates washed up by the tide.

REPRODUCTION: Polygynous; breeds in spring and summer mostly; 1 litter/year; den is a bed of

grasses in a thicket; gestation is 115 days; litter size averages 4–8; weaning 3–4 months; offspring remain with the female for up to 1 year; sexual maturity is reached at 8–10 months but females probably don't breed until 1.5 years of age, and males not until they are several years old and able to compete.

CONSERVATION STATUS: Feral Pigs can cause great damage to ecosystems because of their habit of rooting up ground vegetation; they also destroy breeding colonies of beach-nesting shorebirds, eating the eggs.

RANGE: Formerly Eurasia and North Africa, but now worldwide in the vicinity of human habitation.

KEY REFERENCES: Linzey (1998), Hruby and Dewey (2002).

FAMILY CERVIDAE

Elk *Cervus canadensis*

DESCRIPTION: L = 2–3 m (78–117 in); a large cervid with tan or light brown body pelage; long, dark-colored legs; short tail; white or beige rump; thick, dark brown fur on neck; long face and ears; male grows a large set of antlers each year, which drop off after the breeding season; formerly considered a subspecies of the European Red Deer (*C. elaphus*).

HABITAT AND DISTRIBUTION: Open woodlands and meadows in highlands of WV, western VA, western MD, and PA formerly; now restricted to reintroduced herds in McKean, Elk,

and Cameron counties in PA; some records from southwestern VA, presumably from introductions in KY.

HABITS: Diurnal; active year-round; social, females and offspring associating in herds; a dominant male will associate with a herd during rut (Sep–Oct), but males otherwise are solitary; migratory in some parts of its range, moving upslope to highland meadows in summer, and downslope into valleys in winter.

DIET: Grazes on herbaceous plants in summer; some browsing on shrubs in winter.

REPRODUCTION: Polygynous; dominant male defends a herd of females and yearling offspring from other males for breeding purposes during rut; mating occurs Sep–Oct; gestation 249–262 days with 1–2 young born in spring; females leave the herd to have their young; no den is made as young are precocial and able to move with the mother within hours of birth; however, females often hide them in thickets during the first few weeks of life; weaning at 2 months; sexual maturity is reached at 16 months.

CONSERVATION STATUS: SX—NJ, WV, MD, VA, DE; extirpated by hunting throughout its Mid-Atlantic range, but reintroduced in northwestern PA.

RANGE: Extirpated from large parts of its range, but formerly throughout much of Eurasia and North America; in North America, chiefly now in the western United States and Canada, but reintroduced in some areas in the Midwest and east; also introduced in Morroco, South America, New Zealand, and Australia.

KEY REFERENCES: Merritt (1987), Senseman (2002).

Sika Deer *Cervus nippon*

DESCRIPTION: L = 1–1.4 m (39–55 in); a medium-sized cervid; reddish or grayish-brown body pelage with white spotting; white rump and throat; short tail is brown above and white below;

males develop shaggy manes and grow antlers in fall, which are dropped following rut; quite vocal—herd members give a loud "chirp" when startled, males scream at night during rut, does and fawns communicate with neighs and bleats.

HABITAT AND DISTRIBUTION: Sika Deer prefer forested areas in their native range, but are found in coastal woodlands, thickets, and marshes in the Mid-Atlantic; introduced (1916, 1920s) on James Island (Dorchester County, MD) and Assateague Island (Accomack County, VA and Worcester County, MD).

HABITS: Mostly nocturnal; active year-round; females and offspring associate in small herds in late summer, fall, and winter; males are usually solitary but may associate in small bachelor herds outside the breeding season.

DIET: Browses on shrubs, e.g., wax myrtle, loblolly pine, and red maple; also grazes on marsh grasses.

REPRODUCTION: Polygamous; males establish territories and attempt to drive females into them for breeding; mating occurs Oct–Dec; gestation is 30 weeks; average litter size is 1; females establish an exclusive territory in spring for raising their newborn; fawns are able to walk within hours of birth, but are usually left hidden in thickets by the mother while she forages for the first few weeks after birth; most fawns are weaned in 5–6 months, but some continue nursing for 8–10 months; sexual maturity is normally reached during the second year, but some females may breed in the fall of their first year.

RANGE: Eastern Asia; introduced in Australia, New Zealand, Europe, and parts of the United States.

KEY REFERENCES: Feldhamer (1980), Linzey (1998).

White-tailed Deer *Odocoileus virginianus*

DESCRIPTION: L = 1.3–2.1 m (51–82 in); a medium-sized cervid with mostly light brown or grayish body pelage; belly, inside of legs, and rump are white; short tail is brown above, white below; head with prominent, upright ears is brown with white around the muzzle and white throat; males grow antlers in fall, which drop off after rut; usually silent, but males make loud blowing noises in rut, and fawns and mothers sometimes communicate with bleats.

HABITAT AND DISTRIBUTION: Woodlands and open areas throughout the Mid-Atlantic; deer populations in many parts of the region were low or extirpated in the early 1900s; after passage of protective harvest laws and reintroductions, populations have recovered, with some expanding to "pest" status, particularly in areas where croplands and woodlands intermix.

HABITS: Circadian with crepuscular peaks; active year-round; social, associating in herds of female and offspring during fall and winter; adult males associate in bachelor herds for much of the year, but become solitary and fight for access to females during rut.

DIET: Browsers for the most part, feeding on shrubs and seedlings; however, they will graze on certain plants, e.g., timothy, alfalfa, and other crops; also acorns, dead leaves, fruits, and fungi in season.

REPRODUCTION: Polygynous; mating occurs in fall (Oct–Dec) with a peak in Nov; gestation is 187–222 days with 1–2 young (depending on food supply) born in the spring (May–Jun); female leaves herd and establishes a territory before the young are born; no den is made as fawns are precocial and able to walk within hours of birth; however, the female establishes bedding areas in thickets where she leaves her fawns for the first few weeks after birth while she forages; weaning is at 10 weeks; family groups often join with others in late summer, and remain in herds until the following spring; some females are able to breed in the fall of their first year, but most probably do not breed until their second fall.

CONSERVATION STATUS: Formerly extirpated from most of its range in eastern North America by hunting, the species has repopulated and has gone from scarce to abundant in a matter of decades; high populations can damage ecosystems, threatening local populations of understory plants, birds, and small mammals; they also can threaten natural forest regeneration.

RANGE: Most of North America from central and southern Canada south to Panama; also northern South America; absent from much of the southwestern United States (most of CA, NV, UT, western CO, and northern AZ).

KEY REFERENCES: Smith (1991), McShea et al. (1997).

APPENDIX. CASUAL (C), ACCIDENTAL (A), OR EXTIRPATED (X) SPECIES

This list represents terrestrial vertebrates not expected to occur in the Mid-Atlantic region. Definitions for the various levels of scarcity are provided below:

Casual (C): A few records (3–40). The species does occur on occasion, but not every year.

Accidental (A): One or two records for the species. Not expected to recur.

Extirpated (X): The species formerly occurred in the region, but has been largely or completely extirpated.

AMPHIBIANS

Northern Zigzag Salamander *Plethodon dorsalis*—C

BIRDS

Western Grebe *Aechmophorus occidentalis*—C
Clark's Grebe *Aechmophorus clarkii*—C
Yellow-nosed Albatross *Thalassarche chlororhynchos*—A
Kermadec Petrel *Pterodroma neglecta*—A
Black-capped Petrel *Pterodroma hasitata*—A
White-faced Storm-Petrel *Pelagodroma marina*—A
Band-rumped Storm-Petrel *Oceanodroma castro*—A
White-tailed Tropicbird *Phaethon lepturus*—A
Brown Booby *Sula leucogaster*—A
Magnificent Frigatebird *Fregata magnificens*—C
Reddish Egret *Egretta rufescens*—C
White-faced Ibis *Plegadis chihi*—C
Roseate Spoonbill *Ajaia ajaja*—C
Greater Flamingo *Phoenicopterus ruber*—C
Lesser White-fronted Goose *Anser erythropus*—A
Ross's Goose *Chen rossii*—C
Barnacle Goose *Branta leucopsis*—A
Trumpeter Swan *Cygnus buccinator*—X
Cinnamon Teal *Anas cyanoptera*—C
White-cheeked Pintail *Anas bahamensis*—A
Tufted Duck *Aythya fuligula*—A
Barrow's Goldeneye *Bucephala islandica*—C
Masked Duck *Nomonyx dominicus*—A
American Swallow-tailed Kite *Elanoides forficatus*—C
Mississippi Kite *Ictinia mississippiensis*—C
White-tailed Eagle *Haliaeetus albicilla*—A

Swainson's Hawk *Buteo swainsoni*—C
Ferruginous Hawk *Buteo regalis*—C
Eurasian Kestrel *Falco tinnunculus*—A
Gyrfalcon *Falco rusticolus*—C
Greater Prairie-Chicken (Heath Hen) *Tympanuchus cupido*—X
Corn Crake *Crex crex*—A
Paint-billed Crake *Neocrex erythrops*—A
Spotted Rail *Pardirallus maculatus*—A
Limpkin *Aramus guarauna*—C
Sandhill Crane *Grus canadensis*—C
Whooping Crane *Grus americana*—X
Northern Lapwing *Vanellus vanellus*—A
Mongolian Plover *Charadrius mongolus*—A
Mountain Plover *Charadrius montanus*—C
Spotted Redshank *Tringa erythropus*—A
Eskimo Curlew *Numenius borealis*—X
Black-tailed Godwit *Limosa limosa*—A
Bar-tailed Godwit *Limosa lapponica*—A
Temminck's Stint *Calidris temminckii*—A
Sharp-tailed Sandpiper *Calidris acuminata*—A
Curlew Sandpiper *Calidris ferruginea*—C
Spoonbill Sandpiper *Eurynorhynchus pygmeus*—A
Great Snipe *Gallinago media*—A
Eurasian Woodcock *Scolopax rusticola*—A
South Polar Skua *Stercorarius maccormicki*—C
Mew Gull *Larus canus*—C
California Gull *Larus californicus*—C
Yellow-legged Gull *Larus cachinnans*—A
Thayer's Gull *Larus thayeri*—C
Sabine's Gull *Xema sabini*—C
Ivory Gull *Pagophila eburnea*—C
Elegant Tern *Sterna elegans*—C
White-winged Tern *Chlidonias leucopterus*—C
Whiskered Tern *Chlidonia hybridus*—A
Brown Noddy *Anous stolidus*—C
Common Murre *Uria aalge*—C
Thick-billed Murre *Uria lomvia*—C
Black Guillemot *Cepphus grylle*—C
Atlantic Puffin *Fratercula Arctica*—C
White-winged Dove *Zenaida asiatica*—C
Passenger Pigeon *Ectopistes migratorius*—X
Monk Parakeet *Myiopsitta monachus*—C
Carolina Parakeet *Conuropsis carolinensis*—X
Groove-billed Ani *Crotophaga sulcirostris*—C

Northern Hawk Owl *Surnia ulula*—C
Burrowing Owl *Athene cunicularia*—C
Boreal Owl *Aegolius funereus*—C
Black-chinned Hummingbird *Archilochus alexandri*—C
Calliope Hummingbird *Stellula calliope*—C
Rufous Hummingbird *Selasphorus rufus*—C
Three-toed Woodpecker *Picoides tridactylus*—C
Black-backed Woodpecker *Picoides arcticus*—C
Ivory-billed Woodpecker *Campehilus principalis*—X
Western Wood-Pewee *Contopus sordidulus*—C
Pacific-slope Flycatcher *Empidonax oberholseri*—C
Say's Phoebe *Sayornis saya*—C
Vermilion Flycatcher *Pyrocephalus rubinus*—C
Ash-throated Flycatcher *Myiarchus cinerascens*—C
Great Kiskadee *Pitangus sulphuratus*—A
Cassin's Kingbird *Tyrannus vociferans*—A
Gray Kingbird *Tyrannus dominicensis*—C
Scissor-tailed Flycatcher *Tyrannus forficatus*—C
Fork-tailed Flycatcher *Tyrannus savana*—C
Bell's Vireo *Vireo bellii*—C
Gray Jay *Perisoreus canadensis*—C
Black-billed Magpie *Pica pica*—C
Violet-green Swallow *Tachycineta thalassina*—C
Cave Swallow *Petrochelidon fulva*—C
Boreal Chickadee *Poecile hudsonica*—C
Rock Wren *Salpinctes obsoletus*—C
Northern Wheatear *Oenanthe oenanthe*—C
Mountain Bluebird *Sialia currucoides*—C
Varied Thrush *Ixoreus naevius*—C
Sage Thrasher *Oreoscoptes montanus*—C
Sprague's Pipit *Anthus spragueii*—C
Bohemian Waxwing *Bombycilla garrulus*—C
Bachman's Warbler *Vermivora bachmanii*—X
Black-throated Gray Warbler *Dendroica nigrescens*—C
Townsend's Warbler *Dendroica townsendi*—C

Sutton's Warbler *Dendroica potomac* (Hybrid? *Dendroica dominica* x *Parula americana*)
Kirtland's Warbler *Dendroica kirtlandii*—C
Western Tanager *Piranga ludoviciana*—C
Green-tailed Towhee *Pipilo chlorurus*—C
Spotted Towhee *Pipilo maculatus*—C
Canyon Towhee *Pipilo fuscus*—C
Black-throated Sparrow *Amphispiza bilineata*—C
Harris's Sparrow *Zonotrichia querula*—C
Golden-crowned Sparrow *Zonotrichia atricapilla*—C
Baird's Sparrow *Ammodramus bairdii*—C
Smith's Longspur *Calcarius pictus*—C
Chestnut-collared Longspur *Calcarius ornatus*—C
Black-headed Grosbeak *Pheucticus melanocephalus*—C
Lazuli Bunting *Passerina amoena*—C
Painted Bunting *Passerina ciris*—C
Western Meadowlark *Sturnella neglecta*—C
Bullock's Oriole *Icterus bullockii*—C
Brambling *Fringilla montifringilla*—A
Hoary Redpoll *Carduelis hornemanni*—C

MAMMALS

Seminole Bat *Lasiurus seminolus*—C
Northern Yellow Bat *Lasiurus intermedius*—C
Manatee *Trichechus manatus*—C
Gray Wolf *Canis lupus*—X
Red Wolf *Canis rufus*—X
American Marten *Martes americana*—X
Wolverine *Gulo gulo*—X
Badger *Taxidea taxus*—X?
Mountain Lion *Puma concolor*—X
Lynx *Lynx canadensis*—X
Moose *Alces alces*—X
American Bison *Bos bison*—X

GLOSSARY

Accipiter—Short-winged, long-tailed, bird-eating hawks of the eponymous genus.

Acorn—Nut-like fruit of oaks (*Quercus*).

Aestivation—Torpor or dormancy entered during warm weather periods.

Albinism—Partial or complete congenital absence of pigment in hair, feathers, or skin.

Alar membrane—Wing membrane in bats.

Alder—Deciduous shrubs and trees of the genus *Alnus*.

Altricial—Offspring born in a relatively early developmental condition, requiring parental care for warmth, food, and protection for days or weeks post-parturition.

Amphipod—Marine and freshwater invertebrates of the Order Amphipoda (Phylum Arthropoda, Class Crustacea).

Anal gland—Glands located near the anus. In some species, these glands produce noxious secretions or sprays, presumably to discourage would-be predators; in others, the secretions may be used to mark territory boundaries.

Annelid—Segmented worms of the Phylum Annelida, which includes polychaetes, earthworms, and leeches.

Anuran—Members of the Order Anura (Phylum Chordata, Class Amphibia), i.e., frogs and toads.

Aphid—Small insects that feed on the juices of plants; also called plant lice (Phylum Arthropoda, Class Insecta, Order Homoptera, Family Aphididae).

Aposematic coloration— Bright or bold coloration associated with poisonous prey, e.g., the Eastern Newt; assumed to serve as a warning to would-be predators.

Appalachian Plateau—A geologic province of the Mid-Atlantic region characterized by a rugged topography of uplands and peaks dissected by stream and river valleys.

Apple—Deciduous tree, *Malus pumila*, native to the Old World; introduced into the New World for fruit production.

Aquaculture—Human-controlled production of aquatic organisms, e.g., catfish or shrimp, usually for commercial purposes.

Arachnid—Members of the Class Arachnida (Phylum Arthropoda), which contains spiders, ticks, mites, scorpions, and pseudoscorpions.

Arctic—Region of the Earth located north of the Arctic Circle (66° 32' N Latitude); also, that area of the world found between timberline and the North Pole.

Ardeid—Members of the Family Ardeidae (Class Aves), including the herons, bitterns, and egrets.

Area effect—An hypothesis proposed to explain the apparent direct correlation between size of a piece of habitat and the number of breeding birds found there.

Arthropod—Members of Phylum Arthropoda, e.g., spiders, insects, and crustaceans.

Ascidian—Marine invertebrates of the Class Ascidiacea (Phylum Chordata); tunicates or sea squirts.

Ash—Deciduous trees and shrubs of the genus *Fraxinus*.

Autotomize—Spontaneous separation of a body part, e.g., limb or tail, usually in response to predatory attack.

Axillary (pl. axillaries)—In birds, the feathers located in the armpit area, between the body and the humerus.

Bachelor roost—In bats, a roost site in which only adult males congregate.

Barbel—Slender, fleshy appendage.

Barnacle—Marine invertebrates of the Subclass Cirripedia (Class Crustacea), many of which attach to submerged surfaces as adults and form a hard shell.

Bask—A process used to raise body temperature by lying exposed to the sun.

Basswood—Deciduous trees of the genus *Tilia*; also called linden.

Bayberry—Wax myrtles of the genus *Lyrica*, especially *Lyrica pensylvanica*.

Beard moss—Pendulous, grayish green lichen of the genus *Usnea*.

Beech—Deciduous trees of the genus *Fagus*.

Beetle—Insect of the Order Coleoptera.

Benthos—The bottom of a body of water.

Bib—In birds, the distinctly colored feathers of the chin, throat, and upper breast (see Figure 15).

Bimodal—A statistical distribution that includes two modal peaks, where a mode is considered to be the most frequent value of an observation.

Biotic Province—Biogeographic regions defined by Dice (1943) based on similarities of temperature, topography, and soil.

Birch—Deciduous trees of the genus *Betula*.

Bivalve—Member of the mollusc class Bivalvia (Pelecypoda) having a shell of two hinged parts (valves), e.g., a clam.

Black spruce—A coniferous tree, *Picea mariana*.

Blackberry—Deciduous shrubs of the genus *Rubus* that produce black fruits.

Blue Ridge—A geologic province of the Mid-Atlantic region characterized by long, low, parallel mountain ridges.

Bog—A wetland habitat characterized by a wet, spongy, acidic substrate, often including living and decaying moss (peat) of the genus *Sphagnum*.

Boreal—A portion of the north temperate region dominated by coniferous forest and bog.

Bottomland—Habitats located in or near the floodplain of a river or stream.

Box elder—A deciduous tree, *Acer negundo*.

Brine shrimp—Marine crustaceans of the genus *Artemia*.

Bulb—Swollen, underground stem, often surrounded by fleshy leaves, modified for storage of food for the plant.

Bulrush—A wetland plant of the genus *Scirpus*, characterized by long slim, grass-like leaves.

Cache—To store food in a hole or crevice for later consumption; a practice used by a number of birds and mammals, especially squirrels.

Cambium—Germinative layer in vascular plants between the xylem and the phloem.

Canid—Members of the Family Canidae, which includes dogs, wolves, and foxes.

Canopy feeding—Arching of the wings above the head to shade water for visibility; a practice used by some species of herons and egrets.

Cap—In birds, the central portion of the crown (See Figure 15).

Carapace—Fused dorsal plates (shell) of a turtle.

Carboniferous Period—According to the Geologic Time Scale, that portion of the Paleozoic Era extending from 290–354 million years ago.

Caterpillar—Wormlike larvae of members of the insect Order Lepidoptera, which includes moths and butterflies.

Catkin—Cylindrical, drooping flower clusters characteristic of many willows, birches, and oaks.

Cattail—Perennial marsh plants of the genus *Typha* having long leaves and flowers and fruits borne in a dense, brown cylinder at the end of a long stalk.

Cedar—Aromatic, coniferous evergreens; in North America, used generally in reference to trees of the genera *Thuja*, *Chamaecyparis*, or *Juniperus*.

Centipede—Members of the Class Chilopoda (Phylum Arthropoda).

Cere—In birds, naked, colored skin at the base of the upper mandible; characteristic of hawks and parrots (see Figure 15).

Cherry—Deciduous trees and shrubs of the genus *Prunus* that produce fleshy, edible fruits.

Chestnut blight—A fungus, *Cryphonectria parasitica*, native to Eurasia; introduced into North America on Asian nursery stock in about 1900, it has caused near complete elimination of mature individuals of the American Chestnut (*Castanea dentata*), once a dominant tree of the eastern deciduous forest.

Chin—In birds, the feathers at the base of the lower mandible (see Figure 15).

Chincoteague—A barrier island along the Virginia coast located in the southern Delmarva Peninsula.

Chironomid—Small, flying insects of the midge family, Chironomidae (Order Diptera).

Cicada—Various insects of the family Cicadidae (Order Homoptera) with a prolonged fossorial, larval stage (years) and a brief adult breeding phase (days).

Ciconiiforms—Members of the avian order Ciconiiformes, which includes herons, egrets, ibises, and storks.

Circadian—In reference to activity periods, those individuals active at any time during the 24-hour daily cycle.

Cladocera—An order of the arthropod Class Crustacea containing a number of small, aquatic organisms, most of which are found in fresh water.

Clearcut—A site from which all trees, regardless of size, have been cut.

Cloaca—A canal of the body into which the digestive, urinary, and reproductive canals empty and discharge from the animal; found in amphibians, reptiles, birds, and some mammals (Monotremes).

Coastal Plain—A geologic province of the Mid-Atlantic region extending from the continental coast various distances inland; characterized by low elevation and relatively flat topography.

Coleoptera—An order of Class Insecta containing the beetles.

Collembola—An order of small insects of soil and leaf litter containing the springtails.

Communal roost—A site where large numbers of individuals collect to rest or sleep, presumably for the purposes of protection from predators.

Compositae—A large family of flowering (angiosperm) plants that includes daisies and asters.

Conifer—Cone-bearing (gymnosperm), mostly evergreen trees and shrubs with needle-shaped leaves; includes pines, spruces, and firs.

Conspecific—A member of the same species.

Constrictor— Snakes in which individuals dispatch prey by coiling around and suffocating it.

Contour feather—In birds, the outer feathers of the body.

Copepod—Tiny marine and freshwater crustaceans of the Subclass Copepoda.

Coprophagous—To feed on feces; in some mammalian groups, e.g., rabbits, complete digestion of some plant foods requires more than one passage through the animal's digestive system.

Covey—A flock of quail.

Crabapple—Deciduous shrubs and trees of the genus *Malus*.

Cranial ridge— In some toads, a raised fold of skin on the head.

Creche—A group of young of different parents formed at a breeding colony.

Crepuscular—In reference to animal activity periods, those individuals active during morning and evening twilight.

Crissum—In birds, those feathers located under the base of the tail (see Figure 15).

Crop—In birds, a pouch-like enlargement of the esophagus that serves for short-term food storage, especially in seed-eating birds.

Crop milk—Epithelial cells sloughed from the crop of adult pigeons and fed to young.

Crown—In birds, feathers on the top of the head (see Figure 15).

Crustacean (pl. crustacea)—Members of the arthropod class, Crustacea, which includes crabs, crayfish, copepods, isopods, and amphipods.

Cypress bay—A south-temperate and tropical wetland habitat characterized by woods and thickets interspersed with open, shallow pools in which cypress (*Cupressus*), a deciduous conifer, are the dominant trees.

Cypress—Deciduous, coniferous shrubs and trees of the genus *Cupressus*.

Deciduous woodland—A temperate or highland habitat dominated by trees that lose all of their leaves seasonally.

Delaware Bay—A large, coastal bay located at the mouth of the Delaware River between Delaware and southern New Jersey.

Delmarva Peninsula—A peninsula bordered by Delaware Bay, the Atlantic Ocean, and Chesapeake Bay that includes Delaware, parts of eastern Maryland, and southeastern Virginia.

Detritus—Loose matter on soil surface resulting from accumulated organic and inorganic materials.

Devonian Period—According to the Geologic Time Scale, that portion of the Paleozoic Era extending from 354–417 million years ago.

Diptera—An order of small, flying insects that includes flies, mosquitoes, gnats, and midges.

Diurnal—In reference to animal activity periods, those individuals active during the daylight hours.

Dogwood—Deciduous shrubs and trees of the genus *Cornus*.

Down—In birds, soft feathers lacking a vane and with a short or vestigial rachis. Often these are the first feathers formed by young birds. In addition these feathers serve as insulation under contour feathers of adults of some species (e.g., ducks).

Dragonfly—Medium-sized to large flying insects of the order Odonata; adults forage over freshwater wetlands; larvae are aquatic.

Duckweed—Floating, stemless, freshwater plants of the genus *Lemna*.

Dutch elm fungus—A fungus (*Ophiostoma ulmi*), native to Eurasia, that was first discovered in the New World in North Dakota in 1930. The fungus is transmitted by bark beetles, and has spread throughout the continent, devastating adult populations of the American elm (*Ulmus americana*), formerly a dominant tree species of eastern deciduous forests.

Earpatch—In birds, those feathers located around the ear opening (see Figure 15).

Earthworm—Fossorial annelid worms of the Class Oligochaeta, especially those in the Family Lumbricidae.

Eastern hemlock—An evergreen conifer, *Tsuga canadensis*, characteristic of highland forest in the Appalachian region.

Echinoderm—Members of the marine invertebrate phylum Echinodermata, which includes starfish, sea cucumbers, and sea urchins.

Echolocation—Use of hearing to locate prey by reception of sounds bounced off the prey that are emitted by the hunter; the common prey location technique employed by insectivorous bats.

Eelgrass—Submerged, benthic plants of the genus *Zoster* or *Vallisneria* with long, grass-like leaves that grow in shallow marine environments, e.g., *Zoster marina*.

Eft—The terrestrial, sexually immature stage of a newt.

Eimer organ—Specialized epidermal sensory cells located on the 22 fleshy appendages that constitute the nose of the Star-nosed Mole; used for detection of prey.

Electrosensory detection—Use of electronic signals to locate prey, as in the Star-nosed Mole.

Elm—Deciduous shrubs and trees of the genus *Ulmus*.

Endemic—Native to a particular region.

Epaulet—In birds, the feathers located at the second distal joint of the wing between bones of the forearm and wrist; also called coverts; e.g., the red feathers on the wing of the Red-winged Blackbird.

Ephemeroptera—An order of small, flying insects that includes mayflies.

Epibenthos—Organisms living on or near the bottom of a waterbody.

Epiphyte—A plant that derives structural support from another plant, but takes water and nutrients from precipitation and the air, e.g., Spanish moss.

Estuary—The coastal mouth of a river where fresh and salt water intermingle subject to tidal movement.

Eyebrow—In birds, a line of distinctly colored feathers located above the eye (see Figure 15); also called the supercilium.

Eyeline—In birds, a line of distinctly colored feathers extending behind the eye (see Figure 15); also called a postorbital stripe.

Eyering—In birds, a ring of distinctly colored feathers around the eye.

Facial disk—In birds, especially owls, a distinct circlet of feathers radiating around the eye.

Fen—A bog-like, marshy wetland in which the water is alkaline rather than acidic (as in a bog). Fen plant communities differ significantly from bog plant communities.

Feral—In reference to a domestic species that has self-sustaining populations breeding in the wild without human protection or assistance.

Fir—Evergreen conifers of the genus *Abies* that are among the dominant tree species of the boreal forest.

Flank—In birds, the region below the folded wing, along the side of the belly (see Figure 15).

Fledge—In birds, the time at which young are able to fly; often coincides with when they leave the nest for altricial birds.

Flight feathers—In birds, the remiges (primary and secondary feathers of the wing) and rectrices (tail feathers) (see Figure 15).

Flycatch—In birds, a foraging movement involving capture of insects in the air.

Foot quivering—In birds, especially some thrushes and herons, the rapid shaking of a foot in litter or shallow water in an evident attempt to startle and dislodge prey.

Forecrown—In birds, distinctly colored feathers along the front edge of the crown (see Figure 15).

Forehead—In birds, distinctly colored feathers between the eyes (see Figure 15).

Forewing lining—In birds, the contour feathers along the undersurface of the forearm portion of the wing (see Figure 15).

Form—In taxonomy, a morphologically distinct group of individuals that readily interbreed with other groups of a population, e.g., the "Canebrake" form of the Timber Rattlesnake.

Fossorial—In reference to organisms that live in the soil.

Gametophyte (liverworts)—The sex cell (gamete) producing form of these bryophytes; small, low leafy plants that grow on the surface of rocks and rotting logs, often in moist woodlands or along stream banks.

Gaping—A foraging motion, used by some icterids (e.g., *Agelaius*), opening bill to force an opening in soil, rotted wood, or vegetation.

Gastropod—Members of the mollusc class Gastropoda, most of which have a single, spiral shell into which they can withdraw if threatened; includes snails, slugs (shell-less gastropods), whelks, and conchs.

Geologic Province—A region defined by specific characteristics of geological composition and history.

Gestation—In mammals, the period of development of an embryo in the uterus from fertilization until birth.

Glean—A foraging behavior involving the picking of small prey items from a surface, usually performed while standing on the surface.

Graminae—The family of monocotyledonous, angiosperm plants that includes the grasses.

Granivorous—Feeding on seeds and grains.

Graze—A foraging behavior involving feeding on grass or other low, herbaceous plants.

Gular pouch—In some birds, e.g., the Brown Pelican, a specialized expansion of skin between the lower mandibles facilitating capture and temporary holding of fish.

Hack—In reference to birds, the training of hand-raised individuals, especially falcons, to hunt under wild conditions; a process used for reintroducing individuals of extirpated species.

Hallux—In birds, the first digit of the foot; the sole backward-directed toe in most bird species (cuckoos, woodpeckers, and some other groups are exceptions).

Handstand—Position often used by skunks for delivery of spray from anal scent glands in which the animal balances on its forepaws with its anus directed toward the predator.

Hardwood—Broad-leaved trees, e.g., beech, maple, and oak; used in contrast to conifers, e.g., pines and firs.

Harem—A mating system for some ungulates, e.g., horses, and birds, e.g., grackles, in which a single adult male sequesters a group of females and offspring from other adult males; may also involve protection of the group by the male from predators.

Hawk—In birds, a foraging motion involving capture of flying prey by launching from one site and landing at a different site, post-capture.

Hawthorn—Thorny trees and shrubs of the genus *Crataegus.*

Hazel—Shrubs and understory trees of the genus *Corylus,* producing nut-like fruits called hazelnuts.

Hedgerow—A line of trees or shrubs between open areas, often separating a roadway from an agricultural field or pasture. Also called a "shelterbelt."

Helper—In birds, an individual, not a member of the breeding pair, that assists in care of the young; often a helper is an older offspring of the same pair.

Hemiptera—An insect order that includes the true bugs, e.g., bedbugs, waterbugs, and box elder bugs.

Hemlock—Evergreen coniferous trees of the genus *Tsuga.*

Hemotoxic venom—In snakes, a poisonous glandular secretion delivered by a bite that attacks the circulatory system of the victim; best adapted for use on warm-blooded prey, e.g., birds and mammals.

Herbivorous—Feeding on plants.

Hibernaculum (pl. hibernacula)—A site providing shelter for a hibernating animal.

Hibernation—Entry into torpor or dormancy by an organism during cold periods.

Hickory—Deciduous trees of the genus *Carya*; produce nut-like fruits (hickory nuts).

Holocene Epoch—According to the Geologic Time Scale, that portion of the Cenozoic Era extending from 10,000 years ago until the present.

Homoptera—An insect order containing cicadas, leaf hoppers, and scale insects, among others.

Hood—In birds, distinctly colored feathers covering all or most of the head and upper breast, as in the Hooded Warbler.

Horse chestnut—Deciduous trees and shrubs of the genus *Aesculus*; also called buckeye; dominant members of the Mixed Mesophytic Forest of the Appalachian Plateau.

Hover-glean—In avian foraging, to pick prey from a surface while in flight.

Huddle—A group of individuals clustered together in a nest or roost for warmth, e.g., in White-footed Mice.

Hymenoptera—An order of insects that includes wasps, bees, and ants.

Hypersaline environment—An aquatic habitat containing high concentrations of salts.

In-shore marine—That portion of the ocean located near the shoreline.

Incubation—In birds, the process in which one or both parents sit on their eggs from the time of laying until they hatch.

Invertebrate—All forms of animal life that lack a bony, segmented spinal column.

Irruption—In birds, irregular seasonal movement, apparently resource-related, in which large numbers of individuals appear in areas outside their normal range.

Isopod—Crustacea of the order Isopoda.

Isoptera—An order of insects that includes termites.

Juniper—Evergreen coniferous trees and shrubs of the genus *Juniperus.*

Jurassic Period—According to the Geologic Time Scale, that portion of the Mesozoic Era extending from 144 to 206 million years ago.

Juvenile—An individual that has not reached sexual maturity.

Katydid—Members of the family Tettiigoniidae (Phylum Arthropoda, Class Insecta, Order Orthoptera).

Keeled scale—A scale having a slightly raised, central ridge; possession of keeled versus unkeeled scales in different body parts are important characters separating different species of snakes.

Kelp—Benthic marine seaweed of the algal orders Laminariales and Fucales; generally characterized by long, brown fronds that reach up to 60 m (200 ft) in length from the sea floor.

Keratin—A hard, insoluble fibrous protein that constitutes the chief structural component of reptilian scales, feathers, and hair.

Kettle—In birds, the loose groupings of migrating individuals, especially hawks, associating for the purpose of riding thermal updrafts in order to gain altitude so that they can glide in the preferred migratory direction.

Kleptoparasitism—A foraging activity involving the stealing of prey items from other predators.

Lagoon—A shallow coastal bay, largely or completely separated from the sea by a sand bar or coral reef.

Lamella (pl. lamellae)—A hard, thin layer of tissue; in ducks, specialized parallel plates along the rim of the bill used for straining edible plants and invertebrates from water or mud.

Leaf hopper—Members of the family Cicadellidae (Phylum Arthropoda, Class Insecta, Order Homoptera).

Lepidoptera—An order of insects that includes the butterflies and moths.

Life Zone—A concept originated by C. Hart Merriam in 1894 to differentiate major regions of the Earth's surface using temperature as the principal determinant.

Litter—Materials accumulated on the surface of the forest floor, chiefly organic matter at various stages of decay, e.g., leaves and sticks.

Liverwort—A primitive plant of the order Bryophyta (Class Hepaticae), which forms green scaly or leafy growths over tree trunks, rocks, and other debris in humid environments.

Loblolly pine—An evergreen, coniferous tree, *Pinus taeda;* one of the dominant species of the southeastern pine forests and savannas.

Lodge—In mammals, e.g., the American Beaver, a structure built of sticks and other vegetation that is used for resting and breeding.

Longleaf pine—An evergreen, coniferous tree, *Pinus palustris* (also called southern yellow pine); the dominant tree species of southeastern pine savanna.

Lores—In birds, distinctly colored feathers between the eye and the bill (see Figure 15).

Malar stripe—In birds, distinctly colored feathers extending from the base of the bill along the side of the throat (see Figure 15).

Manateegrass—A flowering plant, *Syringodium filiforme*, with long, thin leaf blades that grows on the muddy or sandy bottoms of shallow subtropical and tropical coastal waters.

Maple—Deciduous, broad-leaved trees and shrubs of the genus *Acer*.

Mast—Nut-like fruits of forest trees, e.g., acorns, hazelnuts, and beechnuts.

Maternity roost—In bats, protected sites, e.g., caves, where adult females of one or more species congregate to bear their young.

Melanism—Abnormal amounts of dark pigment in skin, hair, or feathers; a condition common in some populations of Gray Squirrels and other species.

Mesozoic Era—According to the Geologic Time Scale, a segment of time in the Earth's history extending from 65 to 248 million years ago.

Metamorphosis—In amphibians, the process of changing from a larval to a juvenile form, which usually involves loss of gills and tail fins, development of legs and eyelids, and shift from an aquatic to a terrestrial habit; also called transformation.

Midge—Tiny flying insects of the family Chironomidae (Phylum Arthropoda, Class Insecta, Order Diptera).

Migrant—A species all or part of whose members undertake annual, seasonal movements between geographically different breeding and wintering sites.

Millipede—Arthropods of the Class Diplopoda.

Mixed woodland—A habitat in which a mixture of coniferous and deciduous hardwood trees are dominant.

Mixed-species flock—In birds, a group composed of one or more individuals of 2 or more different species for the purpose of roosting or foraging; foraging mixed-species flocks often show some level of coherence in movements and daily membership.

Mollusc (also mollusk)—Invertebrates of Phylum Mollusca, which includes the sea shells, snails, squid, and slugs.

Monogamous—A mating system in which a male and female of a species form a seasonal or lifetime bond for the purpose of bearing and rearing offspring.

Monotreme—Members of the primitive, egglaying, mammalian order Monotremata, which includes the platypus and echidna.

Moraine—Collection of stones, boulders, and other debris deposited by waters carrying glacial ice melt.

Mountain ash—Deciduous trees and shrubs of the genus *Sorbus* that bear clusters of orange or red fruits during the winter; also called rowan or Thor's helper.

Mud flat—A habitat characterized by recently exposed benthos of a shallow body of water, usually by tidal or wind action.

Multiflora rose—A perennial, deciduous shrub, *Rosa multiflora*, introduced as an ornamental into North America from Eurasia in 1866; bears small red fruits through the winter.

Musk—A foul-smelling substance produced by specialized glands of some reptiles and mammals; may serve to deter predators (snakes) or advertise territory.

Mussel—In North America, freshwater, bivalve molluscs; larger (major) mussels are typified by the genera *Anodontia* and *Unio* (Order Unionoidea); minor mussels are from the Order Veneroidea.

Mustache—In birds, a distinctly colored line of feathers extending from the angle of the jaw along the side of the chin.

Nape—In birds, the feathers located along the back of the neck (see Figure 15).

Neotropics—That portion of the New World located between the Tropic of Cancer (23° 26' 22" N) and the Tropic of Capricorn (23° 26' 22" S).

Neuorotoxic venom—In snakes, a poisonous glandular secretion delivered by a bite that attacks the central nervous system of the victim; best adapted for use on cold-blooded prey, e.g., fish, reptiles, and amphibians.

Newt—A type of salamander that has a terrestrial juvenile stage and an aquatic adult stage, e.g., the Eastern Newt, *Notophthalmus viridescens*.

Nocturnal—In reference to animal activity periods, those individuals active during the hours of darkness.

Oak—Broad-leaved tress of the genus *Quercus*; oaks are among the dominant trees of several eastern woodland habitats.

Odonota—An order of insects that includes the dragonflies.

Old field—A type of second growth habitat that develops from recent abandonment of agricultural fields; characterized by rank growth of grasses and herbs with scattered shrubs and saplings; among the most abundant habitats of the Mid-Atlantic region during the 1800s and early to mid-1900s, now much reduced in coverage as second growth areas revert to forest.

Olfaction—The sense or act of smelling; especially use of that sense for detection of prey.

Orthoptera—An order of insects that includes the grasshoppers, crickets, and cockroaches.

Paleozoic Era—According to the Geologic Time Scale, a segment of time in the Earth's history extending from 248 to 543 million years ago.

Palmetto—Low palms, especially those of the genus *Sabal*; *S. palmetto* is a dominant component of the understory of southeastern pine savanna.

Pangaea—Hypothesized super-continent that included all of the Earth's landmass prior to the Triassic Period, 248 million years ago.

Panicum—A genus of grasses (Family Gramineae).

Parkland—An open habitat characterized by short grass or herbaceous growth with scattered trees and thickets.

Parotid gland—In some toads, large bumps located behind the eye; the glands contain poisonous secretions presumably to deter would-be predators; size and shape of the glands are sometimes important identification characters.

Patagium—In bats and flying squirrels, the skin extending between the body and limbs to form the wing or gliding surface.

Pelage—Hair or fur covering the skin of mammals.

Pelagic—Pertaining to the open ocean.

Pellet—In owls, shorebirds, and some other groups, the indigestible remains of prey (e.g., shell, sand, bones, feathers, fur) regurgitated as a compact bolus.

Phytoplankton—Floating microorganisms able to manufacture food through photosynthesis; mostly algae.

Piedmont—A geologic province of the Mid-Atlantic region characterized by gently rolling topography, deeply weathered bedrock, and a lack of solid outcrops.

Pine—Evergreen, coniferous trees of the genus *Pinus*.

Pine barrens—Pine flatwoods, especially in New Jersey.

Pine flatwoods—A habitat characteristic of low, flat areas with poorly drained soils of the Coastal Plain of the southeastern United States; the overstory is dominated by evergreen coniferous trees of the genus *Pinus,* while the understory is palmetto and various low shrubs, herbs, and grasses; fire is assumed to be important in maintaining the structure and composition.

Pit organ—In pit vipers (e.g., rattlesnakes), an organ of temperature-sensitive cells used for detection of warm-blooded prey; located in indentations (pits) between the eye and the nostril.

Pitch pine—A small, evergreen coniferous tree (*Pinus rigida*) found in sandy soils.

Plankton—Floating microscopic organisms; composed of phytoplankton, mostly algae that manufacture food through photosynthesis, and zooplankton, which obtain their food by eating other microscopic organisms.

Plastron—Ventral shell of a turtle.

Pleistocene Epoch—According to the Geologic Time Scale, that portion of the Quaternary Period of the Cenozoic Era extending from 10,000 to 1.8 million years ago.

Plow—In shorebirds, a foraging behavior involving pushing bill through water to capture small aquatic invertebrates.

Pocosin—A habitat of the Coastal Plain extending from southeastern VA to FL; characterized as a shrub and forest bogland in which white cedar (*Chamaecyperis thyoides*) and loblolly pine (*Pinus taeda*) are the predominant trees.

Polyandrous—A mating system in which females mate with more than one male during a breeding season.

Polychaete—Mostly marine, annelid worms of the Class Polychaeta.

Polygamous—A mating system in which either the male or female may have more than one mate at a time during the breeding season.

Polygynous—A mating system in which a male mates with more than one female during a breeding season.

Pondweed—Floating or submerged freshwater plants of the genus *Potamogeton.*

Postorbital stripe—See "Eyeline."

Pre-Alternate molt—In birds, the second molt to take place in an annual cycle; usually preceding the breeding season, during which the bird molts from a "Basic" nonbreeding plumage to an "Alternate" breeding plumage.

Pre-Basic molt—In birds, the first or only molt to take place in an annual cycle; usually following the breeding season, during which the bird molts from an "Alternate" breeding plumage to a "Basic" non-breeding plumage.

Precocial—A term used in reference to offspring that are born at a relatively advanced state of development, and able to leave the site where they were born within a few hours.

Prehensile tail—A tail adapted for grasping or holding, especially twigs and branches during locomotion.

Primary (pl. primaries)—In birds, the flight feathers (remiges) attached to the hand (see Figure 15).

Promiscuous—A mating system in which no apparent pair bond is formed between individuals, and more than one mate is taken during a season by both the male and female.

Pseudoscorpion—Small arachnids of the Order Pseudoscorpionida.

Rattle—Dried, horny, skin segments at the end of a rattlesnake's tail.

Rectrix (pl. rectrices)—In birds, the flight feathers of the tail (see Figure 15).

Red spruce—An evergreen coniferous tree (*Picea rubens*) characteristic of highland and boreal forest.

Reeve—A female Ruff sandpiper (*Philomachus pugnax*).

Relic—In ecological terms, a formerly extensive habitat reduced in distribution to isolated patches; some highland habitats of the Mid-Atlantic region, e.g., boreal bogs, are assumed to be relics of the most recent Ice Age.

Resident—In terms of animal distribution, a species whose members appear to occupy the same range throughout the annual cycle.

Rhizome—A horizontal, underground, plant stem; often with roots and modified, scale-like leaves.

Riparian forest—A woodland found in the floodplain of a river or stream.

Rump—In birds, distinctly colored feathers on the lower back (see Figure 15).

Sally—In birds, a foraging motion involving capture of flying prey by launching from one site and returning to the same site, post-capture.

Saltmarsh grasses—Plants, e.g., *Puccinellia americana* and *Spartina* spp., found in saline shallow environments, normally along the seacoast and subject to tidal inundation.

Sap well—An indentation in the bark of a tree produced by the peck of a sapsucker's bill (e.g., Yellow-bellied Sapsucker) in which sap collects for later consumption by the bird.

Savanna—A grassland with scattered trees.

Scapular—In birds, the feathers of the humeral tract, located along the side of the upper back where the wing attaches (see Figure 15).

Scrape—In birds, a nest site made for egg deposition by digging a shallow bowl in the substrate, usually sand, shell, dirt, or litter.

Scrub pine—An evergreen coniferous tree (*Pinus virginiana*) characteristic of dry forests along slopes and ridges of the Mid-Atlantic Piedmont; can form extensive stands in second growth situations; also called Virginia pine.

Sea urchin—A benthic marine invertebrate (Phylum Echinodermata, Class Echinoidea).

Secondary—In birds, the flight feathers (remiges) attached to the ulna of the forearm (see Figure 15).

Sedge—Grass-like marsh vegetation of the family Cyperaceae.

Semi-precocial—In birds, offspring that are born at a relatively advanced state of development, and able to leave the site where they were hatched and feed themselves within a few days.

Seral stage—An intermediate stage in plant succession between bare ground and mature or "climax" stage.

Serial polyandry—A mating system in which a female mates with one male for a given clutch of eggs, but may take additional mates for subsequent clutches.

Serial polygyny—A mating system in which a male mates with one female for a given clutch of eggs, but may take additional mates for subsequent clutches.

Serviceberry—An understory shrub, *Amelanchier canadensis*.

Seta (pl. setae)—The stalk of a moss capsule.

Shoalgrass—A flowering plant, *Halodule wrightii*, with long, thin leaf blades that grows on the muddy or sandy bottoms of shallow coastal waters.

Shoulder— In birds, the upper wing coverts (see Figure 15).

Slug—Small, mostly terrestrial, gastropod molluscs that lack a shell.

Smooth cordgrass—A grass (*Spartina alterniflora*) typi-cal of saltmarshes subject to periodic tidal inundation.

Social parasitism—In birds, a breeding behavior involving the laying of one's eggs in the nest of another individual, often of another species; Brown-headed Cowbirds are obligate social parasites, i.e., they depend entirely on members of other species to raise their young.

Spanish moss—An epiphytic bromeliad (Family Bromeliaceae), *Tillandsia usneoides*.

Spat—The tiny larvae of mussels and other bivalve molluscs.

Spectacles—In birds, the plumage category in which eyerings are joined by a bar of feathers of the same color across the forehead or forecrown.

Spermatophore—A gelatinous capsule of spermatozoa produced by some male invertebrates and amphibians for transfer to the reproductive parts of the female.

Sphenodon—A genus of lizard-like reptiles found on islands off the coast of New Zealand; the only living representatives of the Order Rhynchocephalia, which flourished during the Mesozoic Era; also called Tuatara.

Spin—A foraging movement used by phalaropes where the bird makes sharp clockwise or counter-clockwise movements while swimming to capture prey.

Spruce budworm—Caterpillar of the moth species, *Choristoneura fumiferana*, which causes widespread defoliation of spruces (*Picea*) during periodic outbreaks.

Spruce—Evergreen coniferous trees of the genus *Picea*; dominant members of some highland and boreal forest habitats.

Squid—Marine molluscs of the class Cephalopoda, Order Teuthoidea.

Steppe—Semi-arid grassland.

Stitching—Rapid up-and-down head movements in foraging, as seen in the Spotted Sandpiper.

Strip mine—A shallow, open mine created by digging away the top layers of soil and rock; often a coal mine in which the seams run close to ground level.

Subnivean—Under the snow, in reference to winter activity, especially by rodents.

Subspecies—A taxonomic category recognizing systematic, recognizable morphological features characteristic of a population that differ from other populations of a species.

Subtropical—Areas with hot summers and mild winters where frost is infrequent.

Sugar maple—A broad-leaved deciduous tree, *Acer saccharum*; one of the dominant species in woodlands of the north temperate region; also called hard maple.

Sumac—Deciduous shrubs of the genus *Rhus*; common in old field, edge, and second growth habitats.

Superciliary stripe—See "Eyebrow."

Superspecies—In systematics, a group of genetically distinct, but closely related, species.

Supraorbital—In birds, a patch of distinctly colored feathers located above the eye.

Surface diving—A foraging technique involving periodic submersion of all or part of the body of a surface-swimming individual in pursuit of prey.

Swale—A shallow depression subject to periodic flooding.

Swamp—A forested wetland; a woodland inundated with shallow water for prolonged periods.

Sycamore—Broad-leaved, deciduous trees of the genus *Platanus*.

Tactile probing—In sandpipers, a foraging technique in which the bird pushes its bill into a substrate (e.g., sand or mud) to detect hidden prey by touch; dependent on possession of a long bill with sensory cells at the tip.

Tadpole—Aquatic larva of frogs and toads; possesses gills and a finned tail.

Tail Membrane—The skin between the hind limbs and rear of the body in bats.

Talus—Accumulation of rocky debris below a cliff or steep bank.

Tamarack—A deciduous, coniferous tree, *Larix laricina*; typical of bogs in highland and boreal regions; also called larch.

Tarn—A small mountain lake, especially of glacial origin.

Tarsus (pl. tarsi)—In birds, the fused bones of the lower leg to which the toes attach; more correctly called the tarsometatarsus (see Figure 15).

Taxon (pl. taxa)—Any taxonomic category, e.g., species or class.

Taxonomic Order—Organization of species based on their perceived evolutionary relationships as determined by experts.

Temperate—Areas of the Earth's surface located between the Tropic of Cancer and the Arctic Circle and the Tropic of Capricorn and the Antarctic Circle.

Tercel—Male hawk.

Terrestrial— An adjective used in reference to an organism that spends all or part of its life cycle on land.

Therapsid—Members of the extinct reptilian Order Therapsida, thought to be ancestral to mammals.

Thermoregulation—Physiological ability to control one's body temperature independent of the environmental temperature; only birds and mammals possess such an ability.

Tidal rip—Water made rough by the intersection of opposing currents during tidal movements.

Torpor—A dormant, inactive state, often involving decreased metabolism, slowed breathing rate, and lowered body temperature.

Tragus—In bats, an erect, fleshy flap of skin at the inner base of the ear; involved with echolocation; provides useful characters for identification of some species.

Transformation—See "Metamorphosis."

Transient—A migrant in transit between geographically separate breeding and wintering areas.

Transitional Zone—In Merriam's Life Zone classification system, the habitats located between the Upper Austral and Canadian zones, and literally "transitional" between these zones, possessing characteristics of each.

Triassic Period—According to the Geologic Time Scale, that portion of the Mesozoic Era extending from 206 to 248 million years ago.

Tropical—That portion of the Earth's surface located between the Tropic of Cancer and the Tropic of Capricorn.

Tuatara—See "Sphenodon."

Tuber—Fleshy, swollen, modified underground stem of a plant, e.g., a potato, with buds from which new plant shoots can grow.

Tulip poplar—A tall, deciduous, broad-leaved tree, *Liriodendron tulipifera*.

Undertail coverts—See "Crissum."

Underwing—In birds, the interior surface of the wing, hidden when the wing is folded against the body.

Ungule—The hoof of an ungulate (e.g., horses, cows).

Upper wing—In birds, the upper, outer surface of the wing, exposed when the wing is folded against the body (see Figure 15).

Upper wing covert—In birds, the feathers covering the proximal portion of the humerus near the shoulder joint.

Upper tail coverts—In birds, the feathers covering the upper, basal portion of the tail (see Figure 15).

Upwelling—Rising of cold, nutrient-rich currents to the water surface; often accompanied by swarming of invertebrates and fish.

Valley and Ridge—A geologic province of the Mid-Atlantic characterized by parallel ridges and valleys underlain by folded sedimentary and Paleozoic rock.

Venomous—In animals, the possession of the ability to manufacture and deliver a poison.

Vertebrate—Members of the Subphylum Vertebrata (Phylum Chordata), which includes fish, amphibians, reptiles, birds, and mammals; characterized by having a segmented, bony spinal column and a well-defined head.

Viviparous—Giving birth to live offspring that developed within the mother's body; characteristic of most mammals and some snakes.

Vole—Mice with blunt noses, short, hidden ears, and short tails of the genus *Microtus* and related genera of the rodent family Muridae.

Walnut—Deciduous, broad-leaved trees of the genus *Juglans*; an important mast species, producing nut-like fruits (walnuts) in fall.

Water hyacinth—Surface-floating plant, *Eichhornia crassipes*, typical of quiet freshwater streams, rivers, pools, and canals; introduced from South America to Florida in the 1880s, the plant has become a pest, clogging many waterways.

Waterweed—Benthic flowering plants of the genus *Elodea* (Family Hydrocharitaceae) of shallow freshwater wetlands.

Wax myrtle—An evergreen shrub, *Myrica cerifera*.

Wean—In mammals, gradual replacement of mother's milk in the diet by other foods; generally associated with achieving independence from the parent(s).

White pine—A tall, evergreen coniferous tree (*Pinus strobus*).

Widgeongrass—A benthic seagrass, *Ruppia maritima*, characteristic of shallow, saline environments.

Wild celery—Submerged, benthic plants of the genus *Vallisneria* with long, grass-like leaves that grow in shallow marine environments.

Wing membrane—In bats, the skin between the forelimb and body, which forms the wing (patagium).

Wing lining—In birds, the contour feathers on the inner surface of the wing (see Figure 15).

Wing bar—In birds, a patch of distinct feather coloration on the outer surface (greater coverts) of the folded wing extending (see Figure 15).

Wisconsin ice sheet—Ice which covered large portions of north temperate North America during the most recent Ice Age (the Wisconsin), which lasted from 80,000 to 10,000 years before present.

Wooly adelgid—A homopteran insect, *Adelges tsugae*, introduced from Asia into the Pacific Northwest in 1924; the insect causes extensive damage and death to eastern hemlocks.

Zooplankton—Floating microscopic organisms, which obtain their food by eating other microscopic organisms.

REFERENCES

Adkisson, C. S. 1996. Red Crossbill. *In* The Birds of North America, # 256 (A. Poole and F. Gill, Eds.), National Academy of Sciences, Philadelphia, Pennsylvania.

Adkisson, C. S. 1999. Pine Grosbeak. *In* The Birds of North America, # 456 (A. Poole and F. Gill, Eds.), National Academy of Sciences, Philadelphia, Pennsylvania.

Altman, R., and R. Sallabanks. 2000. Olive-sided Flycatcher. *In* The Birds of North America, # 502 (A. Poole and F. Gill, Eds.), National Academy of Sciences, Philadelphia, Pennsylvania.

American Ornithologists' Union. 1998. Check-list of North American birds. Seventh Edition. American Ornithologists' Union, Lawrence, Kansas.

American Ornithologists' Union. 2000. Forty-second supplement to the American Ornithologists' Union Checklist of North American birds. Auk 117:847–858.

American Ornithologists' Union. 2002. Forty-third supplement to the American Ornithologists' Union Check-list of North American birds. Auk 119:897–906.

American Ornithologists' Union. 2003. Forty-fourth supplement to the American Ornithologists' Union Checklist of North American birds. Auk 120:923–931.

American Ornithologists' Union. 2004. Forty-fifth supplement to the American Ornithologists' Union Check-list of North American birds. Auk 121:985–995.

American Ornithologists' Union. 2005. Forty-sixth supplement to the American Ornithologists' Union Check-list of North American birds. Auk 122:1026–1031.

American Society of Mammalogists. 1969–2004. Mammalian Species. Available online at http://www.science.smith.edu/departments/Biology/VHAYSSEN/msi/default.html.

Ammon, E. M. 1995. Lincoln's Sparrow. *In* The Birds of North America, # 190 (A. Poole, and F. Gill, Eds.), National Academy of Sciences, Philadelphia, Pennsylvania.

Ammon, E. M., and W. M. Gilbert. 1999. Wilson's Warbler. *In* The Birds of North America, # 478 (A. Poole and F. Gill, Eds.), National Academy of Sciences, Philadelphia, Pennsylvania.

Arcese, P., M. K. Sogge, A. B. Marr, and M. A. Patten. 2002. Song Sparrow. *In* The Birds of North America, # 704 (A. Poole and F. Gill, Eds.), National Academy of Sciences, Philadelphia, Pennsylvania.

Armitage, D. 2004. *Rattus norvegicus*. Animal Diversity Web, University of Michigan Museum of Zoology, http://animaldiversity.ummz.umich.edu/site/accounts/information/ Rattus_norvegicus.html.

Askins, R. A. 2000. Restoring North America's birds: Lessons from landscape ecology. Yale University Press, New Haven, Connecticut.

Askins, R. A., J. F. Lynch, and R. Greenberg. 1990. Population declines in migratory birds in eastern North America. Current Ornithology 7:1–57.

Audubon, J. J. 1840–1844. Birds of America. J. B. Chevalier, Philadelphia, Pennsylvania [e-text available online from Richard R. Buonanno at http://www.abirdshome.com/Audubon/].

Audubon, J. J., and J. Bachman. 1846–1854. The viviparous quadrupeds of North America. Vols. 1 and 2 (H. Ludwig), Vol. 3 (Craighead), New York (reprinted in 1977 by Volair Ltd., Kent, Ohio).

Austin, J. E., C. M. Custer, and A. D. Afton. 1998. Lesser Scaup. *In* The Birds of North America, # 338 (A. Poole and F. Gill, Eds.), National Academy of Sciences, Philadelphia, Pennsylvania.

Austin, J. E., and M. R. Miller. 1995. Northern Pintail. *In* The Birds of North America, # 163 (A. Poole and F. Gill, Eds.), National Academy of Sciences, Philadelphia, Pennsylvania.

Avery, M. L. 1995. Rusty Blackbird. *In* The Birds of North America, # 200 (A. Poole and F. Gill, Eds.), National Academy of Sciences, Philadelphia, Pennsylvania.

Baird, P. H. 1994. Black-legged Kittiwake. *In* The Birds of North America, # 92 (A. Poole and F. Gill, Eds.), National Academy of Sciences, Philadelphia, Pennsylvania.

Baker, R. J., L. C. Bradley, R. D. Bradley, J. W. Dragoo, M. D. Engstrom, R. S. Hoffmann, C. A. Jones, F. Reid, D. W. Rice, and C. Jones. 2003. Revised checklist of North American mammals north of Mexico, 2003. Occasional Papers, Museum of Texas Tech University, No. 229.

Ballenger, L. 1999. *Mus musculus*. Animal Diversity Web, University of Michigan Museum of Zoology, http://animaldiversity.ummz.umich.edu/site/accounts/information/ Mus_musculus.html.

Baltz, M. E., and S. C. Latta. 1998. Cape May Warbler. *In* The Birds of North America, # 332 (A. Poole and F. Gill, Eds.), National Academy of Sciences, Philadelphia, Pennsylvania.

Bannor, B. K., and E. Kiviat. 2002. Common Moorhen. *In* The Birds of North America, # 685 (A. Poole and F. Gill, Eds.), National Academy of Sciences, Philadelphia, Pennsylvania.

Barr, J. F., C. Eberl, and J. W. McIntyre. 2000. Red-throated Loon. *In* The Birds of North America, # 513 (A. Poole and F. Gill, Eds.), National Academy of Sciences, Philadelphia, Pennsylvania.

Bartram, W. 1791. Bartram's travels. James and Johnson, Philadelphia [e-text available through the University of North Carolina, Chapel Hill, Museum and Library Services, http://docsouth.unc.edu/nc/bartram/bartram.html].

Baskett, T. S., M. W. Sayre, R. E. Tomlinson, and R. E. Mirachi, Eds. 1993. Ecology and management of the Mourning Dove. Stackpole Books, Harrisburg, Pennsylvania.

Beason, R. C. 1995. Horned Lark. *In* The Birds of North America, # 195 (A. Poole and F. Gill, Eds.), National Academy of Sciences, Philadelphia, Pennsylvania.

Bechard, M. J., and T. R. Swem. 2002. Rough-legged Hawk. *In* The Birds of North America, # 641 (A. Poole and F. Gill, Eds.), National Academy of Sciences, Philadelphia, Pennsylvania.

Bekoff, M. 1977. *Canis latrans*. Mammalian Species # 79. American Society of Mammalogists, http://www.science.smith.edu/departments/Biology/VHAYSSEN/msi/.

Beneski, J. T., and D. W. Stinson. 1987. *Sorex palustris*. Mammalian Species # 296. American Society of Mammalogists, http://www.science.smith.edu/departments/ Biology/VHAYSSEN/msi/.

Benkman, C. W. 1992. White-winged Crossbill. *In* The Birds of North America, # 27 (A. Poole, P. Stettenheim, and F. Gill, Eds.), National Academy of Sciences, Philadelphia, Pennsylvania.

Bent, A. C. 1953. Life histories of North American wood warblers. U. S. National Museum Bulletin 203. Smithsonian Institution, Washington, D. C.

Best, T. L. 1996. *Lepus californicus*. Mammalian Species # 530. American Society of Mammalogists, http://www.science.smith.edu/departments/Biology/VHAYSSEN/msi/.

Best, T. L., and J. B. Jennings. 1997. *Myotis leibii*. Mammalian Species # 547. American Society of Mammalogists, http://www.science.smith.edu/departments/Biology/VHAYSSEN/msi/.

Bhagat, S., and T. Dewey. 2002. *Canis lupus familiaris*. Animal Diversity Web, University of Michigan Museum of Zoology, http://animaldiversity.ummz.umich.edu/site/accounts/information/Canis_lupus_familiaris.html.

Bildstein, K. L., and K. Meyer. 2000. Sharp-shinned Hawk. *In* The Birds of North America, # 482 (A. Poole and F. Gill, Eds.), National Academy of Sciences, Philadelphia, Pennsylvania.

Blockstein, D. E. 1989. Crop milk and clutch size in Mourning Doves. Wilson Bulletin 101:11– 25.

Boarman, W. I., and B. Heinrich. 1999. Common Raven. *In* The Birds of North America, # 476 (A. Poole and F. Gill, Eds.), National Academy of Sciences, Philadelphia, Pennsylvania.

Bollinger, E. K., and T. A. Gavin. 1992. Eastern Bobolink populations: ecology and conservation in an agricultural landscape. Pp. 497–506 *In* Ecology and conservation of neotropical migrant landbirds (J. M. Hagan, III, and D. W. Johnston, Eds.), Smithsonian Institution Press, Washington, D. C.

Bookhout, T. A. 1994. Yellow Rail. *In* The Birds of North America, # 139 (A. Poole and F. Gill, Eds.), National Academy of Sciences, Philadelphia, Pennsylvania.

Bordage, D., and J.-P. Savard. 1995. Black Scoter. *In* The Birds of North America, # 177 (A. Poole and F. Gill, Eds.), National Academy of Sciences, Philadelphia, Pennsylvania.

Bowman, R. 2002. Common Ground-Dove. *In* The Birds of North America, # 645 (A. Poole, and F. Gill, Eds.), National Academy of Sciences, Philadelphia, Pennsylvania.

Brady, J., T. Kunz, M. D. Tuttle, and D. Wilson. 1982. Gray Bat recovery plan. U.S. Fish and Wildlife Service, Washington, D. C.

Brennan, L. A. 1999. Northern Bobwhite. *In* The Birds of North America, # 397 (A. Poole and F. Gill, Eds.), National Academy of Sciences, Philadelphia, Pennsylvania.

Brisbin, I. L., Jr., and T. B. Mowbray. 2002. American Coot. *In* The Birds of North America, # 697 (A. Poole and F. Gill, Eds.), National Academy of Sciences, Philadelphia, Pennsylvania.

Briskie, J. V. 1994. Least Flycatcher. *In* The Birds of North America, # 99 (A. Poole and F. Gill, Eds.), National Academy of Sciences, Philadelphia, Pennsylvania.

Brown, C. R. 1997. Purple Martin. *In* The Birds of North America, # 287 (A. Poole and F. Gill, Eds.), National Academy of Sciences, Philadelphia, Pennsylvania.

Brown, C. R., and M. B. Brown. 1995. Cliff Swallow. *In* The Birds of North America, # 149 (A. Poole and F. Gill, Eds.), National Academy of Sciences, Philadelphia, Pennsylvania.

Brown, C. R., and M. B. Brown. 1999. Barn Swallow. *In* The Birds of North America, # 452 (A. Poole and F. Gill, Eds.), National Academy of Sciences, Philadelphia, Pennsylvania.

Brown, P. W., and L. H. Fredrickson. 1997. White-winged Scoter. *In* The Birds of North America, # 274 (A. Poole and F. Gill, Eds.), National Academy of Sciences, Philadelphia, Pennsylvania.

Brown, R. E., and J. G. Dickson. 1994. Swainson's Warbler. *In* The Birds of North America, # 126 (A. Poole and F. Gill, Eds.), National Academy of Sciences, Philadelphia, Pennsylvania.

Brua, R. B. 2002. Ruddy Duck. *In* The Birds of North America, # 696 (A. Poole and F. Gill, Eds.), National Academy of Sciences, Philadelphia, Pennsylvania.

Buckley, N. J. 1999. Black Vulture. *In* The Birds of North America, # 411 (A. Poole and F. Gill, Eds.), National Academy of Sciences, Philadelphia, Pennsylvania.

Buckley, P. A., and F. G. Buckley. 2002. Royal Tern. *In* The Birds of North America, # 700 (A. Poole and F. Gill, Eds.), National Academy of Sciences, Philadelphia, Pennsylvania.

Buehler, D. A. 2000. Bald Eagle. *In* The Birds of North America, # 506 (A. Poole and F. Gill, Eds.), National Academy of Sciences, Philadelphia, Pennsylvania.

Buhlmann, K. A. 1995. Habitat use, terrestrial movements, and conservation of the turtle, *Deirochelys reticularia* in Virginia. Journal ofHerpetology 29:173–181.

Bull, E. L., and J. A. Jackson. 1995. Pileated Woodpecker. *In* The Birds of North America, # 148 (A. Poole and F. Gill, Eds.), National Academy of Sciences, Philadelphia, Pennsylvania.

Burger, J. 1996. Laughing Gull. *In* The Birds of North America, # 225 (A. Poole and F. Gill, Eds.), National Academy of Sciences, Philadelphia, Pennsylvania.

Burger, J., and M. Gochfeld. 1994. Franklin's Gull. *In* The Birds of North America, # 116 (A. Poole and F. Gill,

Eds.), National Academy of Sciences, Philadelphia, Pennsylvania.

Burger, J., and M. Gochfeld. 2002. Bonaparte's Gull. *In* The Birds of North America, # 634 (A. Poole and F. Gill, Eds.), National Academy of Sciences, Philadelphia, Pennsylvania.

Buskirk, W. H., G. V. N. Powell, , J. F. Wittenberger, R. E. Buskirk, and T. U. Powell. 1972. Interspecific bird flocks in tropical highland Panama. Auk 89:612–624.

Butler, R. W. 1992. Great Blue Heron. *In* The Birds of North America, # 25 (A. Poole, P. Stettenheim, and F. Gill, Eds.), National Academy of Sciences, Philadelphia, Pennsylvania.

Cabe, P. R. 1993. European Starling. *In* The Birds of North America, # 48 (A. Poole, P. Stettenheim, and F. Gill, Eds.), National Academy of Sciences, Philadelphia, Pennsylvania.

Caceres, M. C., and R. M. R. Barclay. 2000. *Myotis septentrionalis*. Mammalian Species # 634. American Society of Mammalogists, http://www.science.smith.edu/ departments/Biology/VHAYSSEN/msi/.

Cade, T. J., and E. C. Atkinson. 2002. Northern Shrike. *In* The Birds of North America, # 671 (A. Poole and F. Gill, Eds.), National Academy of Sciences, Philadelphia, Pennsylvania.

Cameron, G. N., and S. R. Spencer. 1981. *Sigmodon hispidus*. Mammalian Species # 158. American Society of Mammalogists, http://www.science.smith.edu/ departments/ Biology/VHAYSSEN/msi/.

Camprich, D. A., and F. R. Moore. 1995. Gray Catbird. *In* The Birds of North America, # 167 (A. Poole and F. Gill, Eds.), National Academy of Sciences, Philadelphia, Pennsylvania.

Cannings, R. J. 1993. Northern Saw-whet Owl. *In* The Birds of North America, # 42 (A. Poole, P. Stettenheim, and F. Gill, Eds.), National Academy of Sciences, Philadelphia, Pennsylvania.

Cannings, S. 2004. Eurasian Wigeon. NatureServe Explorer Species Database. http://www.natureserve.org/explorer/index.htm.

Cannings, S., and G. Hammerson. 2004. Cinereus Shrew. NatureServe Explorer Species Database. http://www.natureserve.org/explorer/index.htm.

Carey, M., D. E. Burhans, and D. A. Nelson. 1994. Field Sparrow. *In* The Birds of North America, # 103 (A. Poole and F. Gill, Eds.), National Academy of Sciences, Philadelphia, Pennsylvania.

Carter, R. W. G. 1988. Coastal environments: An introduction to the physical, ecological, and cultural systems of coastlines. Academic Press, New York.

Catesby, M. 1731–1748. The natural history of Carolina, Florida, and the Bahama Islands. Privately printed, London. [e-text version edited and with an introduction by K. Amacker available on line at http://xroads. virginia.edu/~ma02/amacker/etext/home.htm].

Cavitt, J. F., and C. A. Haas. 2000. Brown Thrasher. *In* The Birds of North America, # 557 (A. Poole and F. Gill, Eds.), National Academy of Sciences, Philadelphia, Pennsylvania.

Chapman, J. A. 1975. *Sylvilagus transitionalis*. Mammalian Species # 55. American Society of Mammalogists, http://www.science.smith.edu/departments/Biology/ VHAYSSEN/msi/.

Chapman, J. A., K. L. Cramer, N. J. Dippenaar and T. J. Robinson. 1992. Systematics and biogeography of the New England cottontail, *Sylvilagus transitionalis* (Bangs, 1895), with the description of a new species from the Appalachian Mountains. Proceedings of the Biological Society of Washington 105 (4): 841–866.

Chapman, J. A., J. G. Hockman, and M. M.. Ojeda C. 1980. *Sylvilagus floridanus*. Mammalian Species # 136. American Society of Mammalogists, http://www. science.smith.edu/departments/Biology/ VHAYSSEN/msi/.

Chilton, G., M. C. Baker, C. D. Barrentine, and M. A. Cunningham. 1995. White-crowned Sparrow. *In* The Birds of North America, # 183 (A. Poole and F. Gill, Eds.), National Academy of Sciences, Philadelphia, Pennsylvania.

Ciaranca, M. A., C. C. Allin, and G. S. Jones. 1997. Mute Swan. *In* The Birds of North America, # 273 (A. Poole and F. Gill, Eds.), National Academy of Sciences, Philadelphia, Pennsylvania.

Cimprich, D. A., F. R. Moore, and M. P. Guilfoyle. 2000. Red-eyed Vireo. *In* The Birds of North America, # 527 (A. Poole and F. Gill, Eds.), National Academy of Sciences, Philadelphia, Pennsylvania.

Cink, C. L. 2002. Whip-poor-will. *In* The Birds of North America, # 619 (A. Poole and F. Gill, Eds.), National Academy of Sciences, Philadelphia, Pennsylvania.

Cink, C. L., and C. T. Collins. 2002. Chimney Swift. *In* The Birds of North America, # 646 (A. Poole and F. Gill, Eds.), National Academy of Sciences, Philadelphia, Pennsylvania.

Clausen, M. K., and G. Hammerson. 2005. Weller's Salamander. NatureServe Explorer Species Database. http://www.natureserve.org/explorer/index.htm.

Collins, J. T. 1990. Standard common and current scientific names for North American amphibians and reptiles. Third Ed. Society for the Study of Amphibians and Reptiles. Herpetological Circular No. 19.

Collins, J. T., and T. W. Taggart. 2002. Standard common and current scientific names for North American amphibians, turtles, reptiles, and crocodilians. 5th Ed. The Center for North American Herpetology, Lawrence, Kansas.

Colwell, M. A., and J. R. Jehl, Jr. 1994. Wilson's Phalarope. *In* The Birds of North America, # 83 (A. Poole and F. Gill, Eds.), National Academy of Sciences, Philadelphia, Pennsylvania.

Conant, R. 1984. Reptiles and amphibians of eastern and central North America. Houghton Mifflin Co., Boston, Massachusetts.

Confer, J. L. 1992. Golden-winged Warbler. *In* The Birds of North America, # 20 (A. Poole, P. Stettenheim, and F. Gill, Eds.), National Academy of Sciences, Philadelphia, Pennsylvania.

Conway, C. J. 1995. Virginia Rail. *In* The Birds of North

America, # 173 (A. Poole and F. Gill, Eds.), National Academy of Sciences, Philadelphia, Pennsylvania.

Conway, C. J. 1999. Canada Warbler. *In* The Birds of North America, # 421 (A. Poole and F. Gill, Eds.), National Academy of Sciences, Philadelphia, Pennsylvania.

Cooper, J. M. 1994. Least Sandpiper. *In* The Birds of North America, # 115 (A. Poole and F. Gill, Eds.), National Academy of Sciences, Philadelphia, Pennsylvania.

Corbat, C. A., and P. W. Bergstrom. 2000. Wilson's Plover. *In* The Birds of North America, # 516 (A. Poole and F. Gill, Eds.), National Academy of Sciences, Philadelphia, Pennsylvania.

Coulter, M. C., J. A. Rodgers, J. C. Ogden, and F. C. Depkin. 1999. Wood Stork. *In* The Birds of North America, # 408 (A. Poole and F. Gill, Eds.), National Academy of Sciences, Philadelphia, Pennsylvania.

Crocoll, S. T. 1994. Red-shouldered Hawk. *In* The Birds of North America, # 107 (A. Poole and F. Gill, Eds.), National Academy of Sciences, Philadelphia, Pennsylvania.

Crother, B. I. 2000. Scientific and standard English names of amphibians and reptiles of North American north of Mexico, with comments regarding confidence in our understanding. Society for the Study of Amphibians and Reptiles, Circular No. 29.

Cullen, S. A., J. R. Jehl, Jr., and G. L. Nuechterlein. 1999. Eared Grebe. *In* The Birds of North America, # 432 (A. Poole and F. Gill, Eds.), National Academy of Sciences, Philadelphia, Pennsylvania.

Cuthbert, F. J., and L. R. Wires. 1999. Caspian Tern. *In* The Birds of North America, # 403 (A. Poole and F. Gill, Eds.), National Academy of Sciences, Philadelphia, Pennsylvania.

Dawson, W. R. 1997. Pine Siskin. *In* The Birds of North America, # 280 (A. Poole and F. Gill, Eds.), National Academy of Sciences, Philadelphia, Pennsylvania.

Davis, W. E., Jr. 1993. Black-crowned Night-Heron. *In* The Birds of North America, # 74 (A. Poole, P. Stettenheim, and F. Gill, Eds.), National Academy of Sciences, Philadelphia, Pennsylvania.

Davis, W. E., Jr., and J. Kricher. 2000. Glossy Ibis. *In* The Birds of North America, # 545 (A. Poole and F. Gill, Eds.), National Academy of Sciences, Philadelphia, Pennsylvania.

Davis, W. E., Jr., and J. A. Kushlan. 1994. Green Heron. *In* The Birds of North America, # 129 (A. Poole and F. Gill, Eds.), National Academy of Sciences, Philadelphia, Pennsylvania.

Decher, J., and J. R. Choate. 1995. *Myotis grisescens*. Mammalian Species # 510. American Society of Mammalogists, http://www.science.smith.edu/departments/Biology/VHAYSSEN/msi/.

DeGraaf, R. M., and D. D. Rudis. 1983. Amphibians and reptiles of New England. University of Massachusetts Press, Amherst, Massachusetts.

DeGraaf, R. M., and J. H. Rappole. 1995. Neotropical migratory birds. Cornell University Press, Ithaca, New York.

DeGraaf, R. M., and M. Yamasaki. 2001. New England wildlife: Habitat, natural history, and distribution.

University Press of New England, Hanover, New Hampshire.

DeJong, M. J. 1996. Northern Rough-winged Swallow. *In* The Birds of North America, # 234 (A. Poole and F. Gill, Eds.), National Academy of Sciences, Philadelphia, Pennsylvania.

Derrickson, K. C., and R. Breitwisch. 1992. Northern Mockingbird. *In* The Birds of North America, # 7 (A. Poole, P. Stettenheim, and F. Gill, Eds.), National Academy of Sciences, Philadelphia, Pennsylvania.

Di Giacomo, A. S., A. G. Di Giacomo, and J. R. Contreras. 2003. Status and conservation of the Bobolink (*Dolichonyx oryzivorus*) in Argentina. U.S. Department of Agriculture Forest Service General Technical Report PSW-GTR 191.

Dice, L. R. 1943. The Biotic Provinces of North America. University of Michigan Press, Ann Arbor, Michigan.

Dolan, P. G., and D. C. Carter. 1977. *Glaucomys volans*. Mammalian Species # 78. American Society of Mammalogists, http://www.science.smith.edu/departments/Biology/VHAYSSEN/msi/.

Drilling, N., R. Titman, and F. McKinney. 2002. Mallard. *In* The Birds of North America, # 658 (A. Poole and F. Gill, Eds.), National Academy of Sciences, Philadelphia, Pennsylvania.

Drost, C. A., and G. M. Fellers. 2005. Non-native animals on public lands. National Biological Service. http://biology.usgs.gov/s+t/nonframe/x180.htm.

Dubowy, P. J. 1996. Northern Shoveler. *In* The Birds of North America, # 217 (A. Poole and F. Gill, Eds.), National Academy of Sciences, Philadelphia, Pennsylvania.

Dugger, B. D., and K. M. Dugger. 2002. Long-billed Curlew. *In* The Birds of North America, # 628 (A. Poole and F. Gill, Eds.), National Academy of Sciences, Philadelphia, Pennsylvania.

Dugger, B. D., K. M. Dugger, and L. H. Fredrickson. 1994. Hooded Merganser. *In* The Birds of North America, # 98 (A. Poole and F. Gill, Eds.), National Academy of Sciences, Philadelphia, Pennsylvania.

Dunn, E. H., and D. J. Agro. 1995. Black Tern. *In* The Birds of North America, # 147 (A. Poole and F. Gill, Eds.), National Academy of Sciences, Philadelphia, Pennsylvania.

Dunning, J. B. 1993. Bachman's Sparrow. *In* The Birds of North America, # 38 (A. Poole, P. Stettenheim, and F. Gill, Eds.), National Academy of Sciences, Philadelphia, Pennsylvania.

Eadie, J. M., M. L. Mallory, and H. G. Lumsden. 1995. Common Goldeneye. *In* The Birds of North America, # 170 (A. Poole and F. Gill, Eds.), National Academy of Sciences, Philadelphia, Pennsylvania.

Eaton, S. W. 1958. A life history study of the Louisiana Waterthrush. Wilson Bulletin 70:210–235.

Eaton, S. W. 1992. Wild Turkey. *In* The Birds of North America, # 22 (A. Poole, P. Stettenheim, and F. Gill, Eds.), National Academy of Sciences, Philadelphia, Pennsylvania.

Eaton, S. W. 1995. Northern Waterthrush. *In* The Birds of

North America, # 182 (A. Poole and F. Gill, Eds.), National Academy of Sciences, Philadelphia, Pennsylvania.

Eckerle, K. P., and C. F. Thompson. 2001. Yellow-breasted Chat. *In* The Birds of North America, # 575 (A. Poole and F. Gill, Eds.), National Academy of Sciences, Philadelphia, Pennsylvania.

Eddleman, W. R., and C. J. Conway. 1998. Clapper Rail. *In* The Birds of North America, # 340 (A. Poole and F. Gill, Eds.), National Academy of Sciences, Philadelphia, Pennsylvania.

Eddleman, W. R., R. E. Flores, and M. L. Legare. 1994. Black Rail. *In* The Birds of North America, # 123 (A. Poole and F. Gill, Eds.), National Academy of Sciences, Philadelphia, Pennsylvania.

Ellison, W. G. 1992. Blue-gray Gnatcatcher. *In* The Birds of North America, # 23 (A. Poole, P. Stettenheim, and F. Gill, Eds.), National Academy of Sciences, Philadelphia, Pennsylvania.

Elphick, C. S., and J. Klima. 2002. Hudsonian Godwit. *In* The Birds of North America, # 629 (A. Poole and F. Gill, Eds.), National Academy of Sciences, Philadelphia, Pennsylvania.

Elphick, C. S., and T. Lee Tibbitts. 1998. Greater Yellowlegs. *In* The Birds of North America, # 355 (A. Poole and F. Gill, Eds.), National Academy of Sciences, Philadelphia, Pennsylvania.

Ely, C. R., and A. X. Dzubin. 1994. Greater White-fronted Goose. *In* The Birds of North America, # 131 (A. Poole and F. Gill, Eds.), National Academy of Sciences, Philadelphia, Pennsylvania.

Evans, R. M., and F. L. Knopf. 1993. American White Pelican. *In* The Birds of North America, # 57 (A. Poole, P. Stettenheim, and F. Gill, Eds.), National Academy of Sciences, Philadelphia, Pennsylvania.

Ewins, P. J., and D. V. C. Weseloh. 1999. Little Gull. *In* The Birds of North America, # 428 (A. Poole and F. Gill, Eds.), National Academy of Sciences, Philadelphia, Pennsylvania.

Falls, J. B., and J. G. Kopachena. 1994. White-throated Sparrow. *In* The Birds of North America, # 128 (A. Poole and F. Gill, Eds.), National Academy of Sciences, Philadelphia, Pennsylvania.

Feldhamer, G. A. 1980. *Cervus nippon*. Mammalian Species # 128. American Society of Mammalogists, http://www.science.smith.edu/departments/Biology/VHAYSSEN/msi/.

Fenton, M. B., and R. M. R. Barclay. 1980. *Myotis lucifugus*. Mammalian Species # 142. American Society of Mammalogists, http://www.science.smith.edu/departments/Biology/VHAYSSEN/msi/.

Ficken, M. S., and R. W. Ficken. 1967. Age-specific differences in the breeding behavior and ecology of the American Redstart. Wilson Bulletin 79:188–199.

Frazier, N. B., J. W. Gibbons, and J. L. Greene. 1991. Life history and demography of the Common Mud Turtle *Kinosternon subrubrum* in South Carolina, USA. Ecology 72: 2218– 2231.

Frederick, P. C. 1997. Tricolored Heron. *In* The Birds of North America, # 306 (A. Poole and F. Gill, Eds.), National Academy of Sciences, Philadelphia, Pennsylvania.

Frederick, P. C., and D. Siegel-Causey. 2000. Anhinga. *In* The Birds of North America, # 522 (A. Poole and F. Gill, Eds.), National Academy of Sciences, Philadelphia, Pennsylvania.

French, T. W. 1980. *Sorex longirostris*. Mammalian Species # 143. American Society of Mammalogists, http://www.science.smith.edu/departments/Biology/VHAYSSEN/msi/.

Fritzell, E. K., and K. J. Haroldson. 1982. *Urocyon cinereoargenteus*. Mammalian Species # 189. American Society of Mammalogists, http://www.science.smith.edu/departments/Biology/VHAYSSEN/msi/.

Fujita, M. S., and T. H. Kunz. 1984. *Pipistrellus subflavus*. Mammalian Species # 228. American Society of Mammalogists, http://www.science.smith.edu/departments/Biology/VHAYSSEN/msi/.

Gamble, L. R., and T. M. Bergin. 1996. Western Kingbird. *In* The Birds of North America, # 227 (A. Poole and F. Gill, Eds.), National Academy of Sciences, Philadelphia, Pennsylvania.

Gardali, T., and G. Ballard. 2000. Warbling Vireo. *In* The Birds of North America, # 551 (A. Poole and F. Gill, Eds.), National Academy of Sciences, Philadelphia, Pennsylvania.

Garrison, B. A. 1999. Bank Swallow. *In* The Birds of North America, # 414 (A. Poole and F. Gill, Eds.), National Academy of Sciences, Philadelphia, Pennsylvania.

Gauthier, G. 1993. Bufflehead. *In* The Birds of North America, # 67 (A. Poole, P. Stettenheim, and F. Gill, Eds.), National Academy of Sciences, Philadelphia, Pennsylvania.

Gehlbach, F. R. 1995. Eastern Screech-Owl. *In* The Birds of North America, # 165 (A. Poole and F. Gill, Eds.), National Academy of Sciences, Philadelphia, Pennsylvania.

George, S. B., J. R. Choate, and H. H. Genoways. 1986. *Blarina brevicauda*. Mammalian Species # 261. American Society of Mammalogists, http://www.-science.smith.edu/departments/Biology/VHAYSSEN/msi/.

Ghalambor, C. K., and T. E. Martin. 1999. Red-breasted Nuthatch. *In* The Birds of North America, # 459 (A. Poole and F. Gill, Eds.), National Academy of Sciences, Philadelphia, Pennsylvania.

Gibbs, J. P., S. M. Melvin, F. A. Reid. 1992a. American Bittern. *In* The Birds of North America, # 18 (A. Poole, P. Stettenheim, and F. Gill, Eds.), National Academy of Sciences, Philadelphia, Pennsylvania.

Gibbs, J. P., F. A. Reid, and S. M. Melvin. 1992b. Least Bittern. *In* The Birds of North America, # 17 (A. Poole, P. Stettenheim, and F. Gill, Eds.), National Academy of Sciences, Philadelphia, Pennsylvania.

Gilchrist, H. G. 2001. Glaucous Gull. *In* The Birds of North America, # 573 (A. Poole and F. Gill, Eds.), National Academy of Sciences, Philadelphia, Pennsylvania.

Gill, F. B. 1997. Local cytonuclear extinction of the Golden-winged Warbler. Evolution 51:519–525.

Gill, F. B., R. A. Canterbury, and J. L. Confer. 2001. Blue-winged Warbler. *In* The Birds of North America, # 584 (A. Poole and F. Gill, Eds.), National Academy of Sciences, Philadelphia, Pennsylvania.

Gillespie, H. 2004. *Rattus rattus*. Animal Diversity Web, University of Michigan Museum of Zoology, http://animaldiversity.ummz.umich.edu/site/accounts/information/ Rattus_rattus.html.

Gillihan, S. W., and B. Byers. 2001. Evening Grosbeak. *In* The Birds of North America, # 599 (A. Poole and F. Gill, Eds.), National Academy of Sciences, Philadelphia, Pennsylvania.

Giudice, J. H., and J. T. Ratti. 2001. Ring-necked Pheasant. *In* The Birds of North America, # 572 (A. Poole and F. Gill, Eds.), National Academy of Sciences, Philadelphia, Pennsylvania.

Gochfeld, M., and J. Burger. 1994. Black Skimmer. *In* The Birds of North America, # 108 (A. Poole and F. Gill, Eds.), National Academy of Sciences, Philadelphia, Pennsylvania.

Gochfeld, M., J. Burger, and I. C. T. Nisbet. 1998. Roseate Tern. *In* The Birds of North America, # 370 (A. Poole and F. Gill, Eds.), National Academy of Sciences, Philadelphia, Pennsylvania.

Good, T. P. 1998. Great Black-backed Gull. *In* The Birds of North America, # 330 (A. Poole, and F. Gill, Eds.), National Academy of Sciences, Philadelphia, Pennsylvania.

Goodrich, L. J., S. C. Crocoll, and S. E. Senner. 1996. Broad-winged Hawk. *In* The Birds of North America, # 218 (A. Poole and F. Gill, Eds.), National Academy of Sciences, Philadelphia, Pennsylvania.

Goudie, R. I., G. J. Robertson, and A. Reed. 2000. Common Eider. *In* The Birds of North America, # 546 (A. Poole and F. Gill, Eds.), National Academy of Sciences, Philadelphia, Pennsylvania.

Gould, E., W. McShea, and T. Grand. 1993. Function of the star in the star-nosed mole. Journal of Mammalogy 74:108–116.

Gowaty, P. A., and J. H. Plissner. 1998. Eastern Bluebird. *In* The Birds of North America, # 381 (A. Poole and F. Gill, Eds.), National Academy of Sciences, Philadelphia, Pennsylvania.

Gratto-Trevor, C. L. 1992. Semipalmated Sandpiper. *In* The Birds of North America, # 6 (A. Poole, P. Stettenheim, and F. Gill, Eds.), National Academy of Sciences, Philadelphia, Pennsylvania.

Gratto-Trevor, C. L. 2000. Marbled Godwit. *In* The Birds of North America, # 492 (A. Poole, and F. Gill, Eds.), National Academy of Sciences, Philadelphia, Pennsylvania.

Graves, G. R. 1997. Geographic clines of age ratios of black-throated blue warblers (*Dendroica caerulescens*). Ecology 78:2524–2531.

Graves, G. R., C. S. Romanek, and A. Rodriguez Navarro. 1997. Stable isotope signature of philopatry and dispersal in a migratory songbird. Proceedings of the National Academy of Sciences 99: 8096–8100.

Green, N. B., and T. K. Pauley. 1987. University of Pittsburgh Press, Pittsburgh, Pennsylvania.

Greenberg, R. 1984. The winter exploitation systems of Bay-breasted and Chestnut-sided warblers in Panama. University of California Publications in Zoology 116.

Greenlaw, J. S. 1996. Eastern Towhee. *In* The Birds of North America, # 262 (A. Poole, and F. Gill, Eds.), National Academy of Sciences, Philadelphia, Pennsylvania.

Greenlaw, J. S., and J. D. Rising. 1994. Sharp-tailed Sparrow. *In* The Birds of North America, # 112 (A. Poole and F. Gill, Eds.), National Academy of Sciences, Philadelphia, Pennsylvania.

Gross, D. A., and P. E. Lowther. 2001. Yellow-bellied Flycatcher. *In* The Birds of North America, # 566 (A. Poole and F. Gill, Eds.), National Academy of Sciences, Philadelphia, Pennsylvania.

Grubb, T. C., Jr., and V. V. Pravosudov. 1994. Tufted Titmouse. *In* The Birds of North America, # 86 (A. Poole and F. Gill, Eds.), National Academy of Sciences, Philadelphia, Pennsylvania.

Guthery, F. S. 1986. Beef, brush, and bobwhite. Caesar Kleberg Wildlife Research Institute, Kingsville, Texas.

Guzy, M. J., and G. Ritchison. 1999. Common Yellowthroat. *In* The Birds of North America, # 448 (A. Poole and F. Gill, Eds.), National Academy of Sciences, Philadelphia, Pennsylvania.

Haggerty, T. M., and E. S. Morton. 1995. Carolina Wren. *In* The Birds of North America, # 188 (A. Poole and F. Gill, Eds.), National Academy of Sciences, Philadelphia, Pennsylvania.

Haig, S. M. 1992. Piping Plover. *In* The Birds of North America, # 2 (A. Poole, P. Stettenheim, and F. Gill, Eds.), National Academy of Sciences, Philadelphia, Pennsylvania.

Hajna, L. 2002. Breeding areas disappear for many species. New Jersey CourierPost Online. http://www.courierpostonline.com/pinelands/pinearc2.htm.

Halkin, S. L., and S. U. Linville. 1999. Northern Cardinal. *In* The Birds of North America, # 440 (A. Poole and F. Gill, Eds.), National Academy of Sciences, Philadelphia, Pennsylvania.

Hall, G. A. 1994. Magnolia Warbler. *In* The Birds of North America, # 136 (A. Poole and F. Gill, Eds.), National Academy of Sciences, Philadelphia, Pennsylvania.

Hall, G. A. 1996. Yellow-throated Warbler. *In* The Birds of North America, # 223 (A. Poole, and F. Gill, Eds.), National Academy of Sciences, Philadelphia, Pennsylvania.

Hallett, J. G. 1978. *Parascalops breweri*. Mammalian Species # 98. American Society of Mammalogists, http://www.science.smith.edu/departments/Biology/VHAYSSEN/msi/.

Hamas, M. J. 1994. Belted Kingfisher. *In* The Birds of North America, # 84 (A. Poole, P. Stettenheim, and F. Gill, Eds.), National Academy of Sciences, Philadelphia, Pennsylvania.

Hamel, P. 2000. Cerulean Warbler. *In* The Birds of North America, # 511 (A. Poole and F. Gill, Eds.), National Academy of Sciences, Philadelphia, Pennsylvania.

Hammerson, G. 2003. Wood Turtle. NatureServe Explorer Species Database. http://www.natureserve.org/explorer/index.htm.

Hammerson, G. 2005a. Reptiles and Amphibians. NatureServe Explorer Species Database. http://www.natureserve.org/explorer/index.htm.

Hammerson, G. 2005b. Ruff. NatureServe Explorer Species Database. http://www.natureserve.org/explorer/index.htm.

Hammerson, G., and S. Cannings. 2004. Curlew Sandpiper. NatureServe Explorer Species Database. http://www.natureserve.org/explorer/index.htm.

Hammerson, G., and M. K. Clausen. 2005. Diamondback Terrapin. NatureServe Explorer Species Database. http://www.natureserve.org/explorer/index.htm.

Hammerson, G., and F. Dirrigl, Jr. 2005. Eastern Spadefoot Toad. NatureServe Explorer Species Database. http://www.natureserve.org/explorer/index.htm.

Hammerson, G., and B. Qureshi. 2005. Cheat Mountain Salamander. NatureServe Explorer Species Database. http://www.natureserve.org/explorer/index.htm.

Handley, C. O., Jr. 1991. Mammals. Pp. 539–629 *In* Virginia's endangered species (K. Terwilliger, Ed.), McDonald and Woodward Publishing Company, Blacksburg, Virginia.

Hanners, L. A., and S. R. Patton. 1998. Worm-eating Warbler. *In* The Birds of North America, # 367 (A. Poole and F. Gill, Eds.), National Academy of Sciences, Philadelphia, Pennsylvania.

Harrington, B. A. 2001. Red Knot. *In* The Birds of North America, # 563 (A. Poole and F. Gill, Eds.), National Academy of Sciences, Philadelphia, Pennsylvania.

Harris, H. S. 1975. Distributional Survey (Amphibia/Reptilia): Maryland and the District of Columbia. Bulletin Maryland Herpetological Society 11(3): 73–167.

Hatch, J. J. 2002. Arctic Tern. *In* The Birds of North America, # 707 (A. Poole and F. Gill, Eds.), National Academy of Sciences, Philadelphia, Pennsylvania.

Hatch, J. J., K. M. Brown, G. G. Hogan, and R. D. Morris. 2000. Great Cormorant. *In* The Birds of North America, # 553 (A. Poole and F. Gill, Eds.), National Academy of Sciences, Philadelphia, Pennsylvania.

Hatch, J. J., and D. V. Weseloh. 1999. Double-crested Cormorant. *In* The Birds of North America, # 441 (A. Poole and F. Gill, Eds.), National Academy of Sciences, Philadelphia, Pennsylvania.

Hejl, S. J., J. A. Holmes, and D. E. Kroodsma. 2002. Winter Wren. *In* The Birds of North America, # 623 (A. Poole and F. Gill, Eds.), National Academy of Sciences, Philadelphia, Pennsylvania.

Hejl, S. J., K. R. Newlon, M. E. McFadzen, J. S. Young, and C. K. Ghalambor. 2002. Brown Creeper. *In* The Birds of North America, # 669 (A. Poole and F. Gill, Eds.), National Academy of Sciences, Philadelphia, Pennsylvania.

Hepp, G. R., and F. C. Bellrose. 1995. Wood Duck. *In* The Birds of North America, # 169 (A. Poole and F. Gill, Eds.), National Academy of Sciences, Philadelphia, Pennsylvania.

Herkert, J. R., D. E. Kroodsma, and J. P. Gibbs. 2001. Sedge Wren. *In* The Birds of North America, # 582 (A. Poole and F. Gill, Eds.), National Academy of Sciences, Philadelphia, Pennsylvania.

Herkert, J. R., P. D. Vickery, and D. E. Kroodsma. 2002. Henslow's Sparrow. *In* The Birds of North America, # 672 (A. Poole and F. Gill, Eds.), National Academy of Sciences, Philadelphia, Pennsylvania.

Highton, R., G. C. Maha, and L. R. Maxson. 1989. Biochemical evolution in the slimy salamanders of the *Plethodon glutinosus* complex in the eastern United States. Illinois Biological Monographs 57:1–153.

Hill, G. E. 1993. House Finch. *In* The Birds of North America, # 46 (A. Poole, P. Stettenheim, and F. Gill, Eds.), National Academy of Sciences, Philadelphia, Pennsylvania.

Hohman, W. L., and R. T. Eberhardt. 1998. Ring-necked Duck. *In* The Birds of North America, # 329 (A. Poole, and F. Gill, Eds.), National Academy of Sciences, Philadelphia, Pennsylvania.

Hohman, W. L., and S. A. Lee. 2001. Fulvous Whistling-Duck. *In* The Birds of North America, # 562 (A. Poole and F. Gill, Eds.), National Academy of Sciences, Philadelphia, Pennsylvania.

Holmes, R. T. 1994. Black-throated Blue Warbler. *In* The Birds of North America, # 87 (A. Poole and F. Gill, Eds.), National Academy of Sciences, Philadelphia, Pennsylvania.

Holmes, R. T., and F. A. Pitelka. 1998. Pectoral Sandpiper. *In* The Birds of North America, # 348 (A. Poole and F. Gill, Eds.), National Academy of Sciences, Philadelphia, Pennsylvania.

Holt, D. W., and S. M. Leasure. 1993. Short-eared Owl. *In* The Birds of North America, # 62 (A. Poole, P. Stettenheim, and F. Gill, Eds.), National Academy of Sciences, Philadelphia, Pennsylvania.

Hopp, S. L., A. Kirby, and C. A. Boone. 1995. White-eyed Vireo. *In* The Birds of North America, # 168 (A. Poole and F. Gill, Eds.), National Academy of Sciences, Philadelphia, Pennsylvania.

Houston, C. S., and D. E. Bowen. 2001. Upland Sandpiper. *In* The Birds of North America, # 580 (A. Poole and F. Gill, Eds.), National Academy of Sciences, Philadelphia, Pennsylvania.

Houston, C. S., D. G. Smith, and C. Rohner. 1998. Great Horned Owl. *In* The Birds of North America, # 372 (A. Poole and F. Gill, Eds.), National Academy of Sciences, Philadelphia, Pennsylvania.

Howe, M. A., P. H. Geissler, and B. Harrington. 1989. Population trends of North American shorebirds based on the International Shorebird Survey. Biological Conservation 49:185– 199.

Hruby, J., and T. Dewey. 2002. *Sus scrofa*. Animal Diversity Web, University of Michigan Museum of Zoology, http://animaldiversity.ummz.umich.edu/site/accounts/information/ Sus_scrufa.html.

Hughes, J. M. 1999. Yellow-billed Cuckoo. *In* The Birds of North America, # 418 (A. Poole and F. Gill, Eds.), National Academy of Sciences, Philadelphia, Pennsylvania.

Hughes, J. M. 2001. Black-billed Cuckoo. *In* The Birds of

North America, # 586 (A. Poole and F. Gill, Eds.), National Academy of Sciences, Philadelphia, Pennsylvania.

Hulse, A. C., C. J. McCoy, and E. Censky. 2001. Amphibians and reptiles of Pennsylvania and the Northeast. Cornell University Press, Ithaca, New York.

Hunt, C. B. 1974. Natural regions of the United States and Canada. W. H. Freeman and Co., San Francisco, California.

Hunt, P. D., and B. C. Eliason. 1999. Blackpoll warbler. *In* The Birds of North America, # 431 (A. Poole and F. Gill, Eds.), National Academy of Sciences, Philadelphia, Pennsylvania.

Hunt, P. D., and D. J. Flaspohler. 1998. Yellow-rumped Warbler. *In* The Birds of North America, # 376 (A. Poole and F. Gill, Eds.), National Academy of Sciences, Philadelphia, Pennsylvania.

Hussell, D. J. T., and R. Montgomerie. 2002. Lapland Longspur. *In* The Birds of North America, # 656 (A. Poole and F. Gill, Eds.), National Academy of Sciences, Philadelphia, Pennsylvania.

Ingold, J. I. 1993. Blue Grosbeak. *In* The Birds of North America, # 79 (A. Poole and F. Gill, Eds.), National Academy of Sciences, Philadelphia, Pennsylvania.

Ingold, J. I., and R. Galati. 1997. Golden-crowned Kinglet. *In* The Birds of North America, # 301 (A. Poole and F. Gill, Eds.), National Academy of Sciences, Philadelphia, Pennsylvania.

Ingold, J. I., and G. E. Wallace. 1994. Ruby-crowned Kinglet. *In* The Birds of North America, # 119 (A. Poole and F. Gill, Eds.), National Academy of Sciences, Philadelphia, Pennsylvania.

Jackson, B. J. S., and J. A. Jackson. 2000. Killdeer. *In* The Birds of North America, # 517 (A. Poole and F. Gill, Eds.), National Academy of Sciences, Philadelphia, Pennsylvania.

Jackson, J. A. 1994. Red-cockaded Woodpecker. *In* The Birds of North America, # 85 (A. Poole, P. Stettenheim, and F. Gill, Eds.), National Academy of Sciences, Philadelphia, Pennsylvania.

Jackson, J. A., and H. R. Ouellet. 2002. Downy Woodpecker. *In* The Birds of North America, # 613 (A. Poole and F. Gill, Eds.), National Academy of Sciences, Philadelphia, Pennsylvania.

Jackson, J. A., H. R. Ouellet, and B. J. S. Jackson. 2002. Hairy Woodpecker. *In* The Birds of North America, # 702 (A. Poole and F. Gill, Eds.), National Academy of Sciences, Philadelphia, Pennsylvania.

James, R. D. 1998. Blue-headed Vireo. *In* The Birds of North America, # 379 (A. Poole and F. Gill, Eds.), National Academy of Sciences, Philadelphia, Pennsylvania.

Jefferson, T. 1784. Notes on the State of Virginia. Privately printed by Phillipe-Denis Pierre, Paris. [e-text version available online through the Electronic Text Center, University of Virginia at http://etext.lib.virginia.edu/toc/modeng/public/JefVirg.html].

Jehl, Jr., J. R., J. Klima, and R. E. Harris. 2001. Short-billed Dowitcher. *In* The Birds of North America, # 564 (A.

Poole and F. Gill, Eds.), National Academy of Sciences, Philadelphia, Pennsylvania.

Jenkins, S. H., and P. E. Busher. 1979. *Castor canadensis*. Mammalian Species # 120. American Society of Mammalogists, http://www.science.smith.edu/departments/Biology/VHAYSSEN/msi/.

Johnson, K. 1995. Green-winged Teal. *In* The Birds of North America, # 193 (A. Poole and F. Gill, Eds.), National Academy of Sciences, Philadelphia, Pennsylvania.

Johnson, L. S. 1998. House Wren. *In* The Birds of North America, # 380 (A. Poole and F. Gill, Eds.), National Academy of Sciences, Philadelphia, Pennsylvania.

Johnson, O. W., and P. G. Connors. 1996. American Golden-Plover. *In* The Birds of North America, # 201 (A. Poole and F. Gill, Eds.), National Academy of Sciences, Philadelphia, Pennsylvania.

Johnston, D. W. 2003. The history of ornithology in Virginia. University of Virginia Press, Charlottesville, Virginia.

Johnston, R. F. 1992. Rock Dove. *In* The Birds of North America, # 13 (A. Poole, P. Stettenheim, and F. Gill, Eds.), National Academy of Sciences, Philadelphia, Pennsylvania.

Jones, C. R. 1977. *Plecotus rafinesquii*. Mammalian Species # 69. American Society of Mammalogists, http://www.science.smith.edu/departments/Biology/VHAYSSEN/msi/.

Jones, C, R. S. Hoffmann, D. W. Rice, m. D. Engstrom, R. D. Bradley, D. J. Schmidly, C. A. Jones, and R. J. Baker. 1997. Revised checklist of North American mammals north of Mexico. Texas Tech University Museum, Occasional Paper # 173.

Jones, C. R., and R. W. Manning. 1989. *Myotis austroriparius*. Mammalian Species # 332. American Society of Mammalogists, http://www.science.smith.edu/departments/Biology/VHAYSSEN/msi/.

Jones, P. W., and T. M. Donovan. 1996. Hermit Thrush. *In* The Birds of North America, # 261 (A. Poole and F. Gill, Eds.), National Academy of Sciences, Philadelphia, Pennsylvania.

Jones, S. L., and J. E. Cornely. 2002. Vesper Sparrow. *In* The Birds of North America, # 624 (A. Poole and F. Gill, Eds.), National Academy of Sciences, Philadelphia, Pennsylvania.

Kennedy, E. D., and D. W. White. 1997. Bewick's Wren. *In* The Birds of North America, # 315 (A. Poole and F. Gill, Eds.), National Academy of Sciences, Philadelphia, Pennsylvania.

Keppie, D. M., and R. M. Whiting, Jr. 1994. American Woodcock. *In* The Birds of North America, # 100 (A. Poole and F. Gill, Eds.), National Academy of Sciences, Philadelphia, Pennsylvania.

Kercheval, S. 1925. A history of the Valley of Virginia. 4th Ed. Shenandoah Publishing House, Strasburg, Virginia.

Kessel, B., D. A. Rocque, and J. S. Barclay. 2002. Greater Scaup. *In* The Birds of North America, # 650 (A. Poole and F. Gill, Eds.), National Academy of Sciences, Philadelphia, Pennsylvania.

King, C. M. 1983. *Mustela erminea*. Mammalian Species #

195. American Society of Mammalogists, http://www.science.smith.edu/departments/Biology/VHAYSSEN/msi/.

Kinlaw, A. 1995. *Spilogale putorius*. Mammalian Species # 511. American Society of Mammalogists, http://www.science.smith.edu/departments/Biology/VHAYSSEN/msi/.

Kirk, D. A., and M. J. Mossman. 1998. Turkey Vulture. *In* The Birds of North America, # 339 (A. Poole and F. Gill, Eds.), National Academy of Sciences, Philadelphia, Pennsylvania.

Kirkland, G. L., Jr. 1981. *Sorex dispar*. Mammalian Species # 155. American Society of Mammalogists, http://www.science.smith.edu/departments/Biology/VHAYSSEN/msi/.

Kirkland, G. L., Jr., and F. J. Jannett, Jr. 1982. *Microtus chrotorrhinus*. Mammalian Species # 180. American Society of Mammalogists, http://www.science.smith.edu/departments/Biology/VHAYSSEN/msi/.

Klicka, J. 1994. The biological and taxonomic status of the Brownsville Yellowthroat (*Geothlypis trichas insperata*). M. S. Thesis, Texas A and I University, Kingsville, Texas.

Klima, J., and J. R. Jehl, Jr. 1998. Stilt Sandpiper. *In* The Birds of North America, # 341 (A. Poole and F. Gill, Eds.), National Academy of Sciences, Philadelphia, Pennsylvania.

Knapton, R. W. 1994. Clay-colored Sparrow. *In* The Birds of North America, # 120 (A. Poole, and F. Gill, Eds.), National Academy of Sciences, Philadelphia, Pennsylvania.

Knox, A. G., and P. E. Lowther. 2000. Common Redpoll. *In* The Birds of North America, # 543 (A. Poole and F. Gill, Eds.), National Academy of Sciences, Philadelphia, Pennsylvania.

Kochert, M. N., K. Steenhof, C. L. McIntyre, and E. H. Craig. 2002. Golden Eagle. *In* The Birds of North America, # 684 (A. Poole and F. Gill, Eds.), National Academy of Sciences, Philadelphia, Pennsylvania.

Koprowski, J. L. 1994b. *Sciurus niger*. Mammalian Species # 479. American Society of Mammalogists, http://www.science.smith.edu/departments/Biology/VHAYSSEN/msi/.

Koprowski, J. L. 1994a. *Sciurus carolinensis*. Mammalian Species # 480. American Society of Mammalogists, http://www.science.smith.edu/departments/Biology/VHAYSSEN/msi/.

Kricher, J. C. 1995. Black-and-white Warbler. *In* The Birds of North America, # 158 (A. Poole and F. Gill, Eds.), National Academy of Sciences, Philadelphia, Pennsylvania.

Kroodsma, D. E., and J. Verner. 1997. Marsh Wren. *In* The Birds of North America, # 308 (A. Poole and F. Gill, Eds.), National Academy of Sciences, Philadelphia, Pennsylvania.

Küchler, A. W. 1975. Potential natural vegetation of the coterminous United States. American Geographical Society, New York.

Kunz, T. H. 1982. *Lasionycteris noctivagans*. Mammalian Species # 172. American Society of Mammalogists, http://www.science.smith.edu/departments/Biology/VHAYSSEN/msi/.

Kunz, T. H., and R. A. Martin. 1982. *Plecotus townsendii*. Mammalian Species # 175. American Society of Mammalogists, http://www.science.smith.edu/departments/Biology/VHAYSSEN/msi/.

Kurta, A., and R. H. Baker. 1990. *Eptesicus fuscus*. Mammalian Species # 356. American Society of Mammalogists, http://www.science.smith.edu/departments/Biology/VHAYSSEN/msi/.

Kushlan, J. A., and K. L. Bildstein. 1992. White Ibis. *In* The Birds of North America, # 9 (A. Poole, P. Stettenheim, and F. Gill, Eds.), National Academy of Sciences, Philadelphia, Pennsylvania.

Kwiecinski, G. G. 1998. *Marmota monax*. Mammalian Species # 591. American Society of Mammalogists, http://www.science.smith.edu/departments/Biology/VHAYSSEN/msi/.

Lackey, J. A., D. G. Huckaby, and B. G. Ormiston. 1985. *Peromyscus leucopus*. Mammalian Species # 247.

Lamb, T., and J. Lovich. 1990. Morphometric validation of the striped mud turtle (*Kinosternon baurii*) in the Carolinas and Virginia. Copeia 1990:613–618.

Lanctot, R. B., and C. D. Laredo. 1994. Buff-breasted Sandpiper. *In* The Birds of North America, # 91 (A. Poole and F. Gill, Eds.), National Academy of Sciences, Philadelphia, Pennsylvania.

Lanyon, W. E. 1995. Eastern Meadowlark. *In* The Birds of North America, # 160 (A. Poole and F. Gill, Eds.), National Academy of Sciences, Philadelphia, Pennsylvania.

Lanyon, W. E. 1997. Great Crested Flycatcher. *In* The Birds of North America, # 300 (A. Poole and F. Gill, Eds.), National Academy of Sciences, Philadelphia, Pennsylvania.

Larivière, S. 1999. *Mustela vison*. Mammalian Species # 608. American Society of Mammalogists, http://www.science.smith.edu/departments/Biology/VHAYSSEN/msi/.

Larivière, S. 2001. *Ursus americanus*. Mammalian Species # 647. American Society of Mammalogists, http://www.science.smith.edu/departments/Biology/VHAYSSEN/msi/.

Larivière, S., and M. Pasitschniak-Arts. 1996. *Vulpes vulpes*. Mammalian Species # 537. American Society of Mammalogists, http://www.science.smith.edu/departments/Biology/VHAYSSEN/msi/.

Larivière, S., and L. R. Walton. 1985. *Lynx rufus*. Mammalian Species # 563. American Society of Mammalogists, http://www.science.smith.edu/departments/Biology/VHAYSSEN/msi/.

Larivière, S., and L. R. Walton. 1998. *Lontra canadensis*. Mammalian Species # 587. American Society of Mammalogists, http://www.science.smith.edu/departments/Biology/VHAYSSEN/msi/.

Lehmann, V. W. 1984. Bobwhites in the Rio Grande Plain of Texas. Texas A and M University Press, College Station, Texas.

Leschack, C. R., S. K. McKnight, and G. R. Hepp. 1997. Gadwall. *In* The Birds of North America, # 283 (A. Poole and F. Gill, Eds.), National Academy of Sciences, Philadelphia, Pennsylvania.

Lewis, T. 1746. The Fairfax line: Thomas Lewis's journal of 1746: with footnotes and an index by John W. Wayland, Ph.D. Printed by Henkel Press in 1925, New Market, Virginia.

Little, C. E. 1995. The dying of the trees: The pandemic in America's forests. Viking, New York.

Linzey, A. V. 1983. *Synaptomys cooperi*. Mammalian Species # 210. American Society of Mammalogists, http://www.science.smith.edu/departments/Biology/VHAYSSEN/msi/.

Linzey, D. W. 1998. The mammals of Virginia. McDonald and Woodward Publishing Company, Blacksburg, Virginia.

Linzey, D. W., and R. L. Packard. 1977. *Ochrotomys nuttalli*. Mammalian Species # 75.

Long, C. A. 1974. *Microsorex hoyi*. Mammalian Species # 33. American Society of Mammalogists, http://www.science.smith.edu/departments/Biology/VHAYSSEN/msi/.

Longcore, J. R., D. G. McAuley, G. R. Hepp, and J. M. Rhymer. 2000. American Black Duck. *In* The Birds of North America, # 481 (A. Poole and F. Gill, Eds.), National Academy of Sciences, Philadelphia, Pennsylvania.

Lotze, J.-H., and S. Anderson. 1979. *Procyon lotor*. Mammalian Species # 119. American Society of Mammalogists, http://www.science.smith.edu/departments/Biology/VHAYSSEN/msi/.

Lowther, P. E. 1993. Brown-headed Cowbird. *In* The Birds of North America, # 47 (A. Poole, P. Stettenheim, and F. Gill, Eds.), National Academy of Sciences, Philadelphia, Pennsylvania.

Lowther, P. E. 1996. Le Conte's Sparrow. *In* The Birds of North America, # 224 (A. Poole, and F. Gill, Eds.), National Academy of Sciences, Philadelphia, Pennsylvania.

Lowther, P. E. 1999. Alder Flycatcher. *In* The Birds of North America, # 446 (A. Poole and F. Gill, Eds.), National Academy of Sciences, Philadelphia, Pennsylvania.

Lowther, P. E., C. Celada, N. K. Klein, C. C. Rimmer, and D. A. Spector. 1999. Yellow Warbler. *In* The Birds of North America, # 454 (A. Poole and F. Gill, Eds.), National Academy of Sciences, Philadelphia, Pennsylvania.

Lowther, P. E., and C. L. Cink. 1992. House Sparrow. *In* The Birds of North America, # 12 (A. Poole, P. Stettenheim, and F. Gill, Eds.), National Academy of Sciences, Philadelphia, Pennsylvania.

Lowther, P. E., H. D. Douglas, III, and C. L. Gratto-Trevor. 2001. Willet. *In* The Birds of North America, # 579 (A. Poole and F. Gill, Eds.), National Academy of Sciences, Philadelphia, Pennsylvania.

Lowther, P. E., C. C. Rimmer, B. Kessel, S. L. Johnson, and W. G. Ellison. 2001. Gray-cheeked Thrush. *In* The Birds of North America, # 591 (A. Poole and F. Gill, Eds.), National Academy of Sciences, Philadelphia, Pennsylvania.

Lyon, B., and R. Montgomerie. 1995. Snow Bunting. *In* The Birds of North America, # 198 (A. Poole and F. Gill, Eds.), National Academy of Sciences, Philadelphia, Pennsylvania.

MacArthur, R. H. 1972. Geographical ecology. Harper and Row, New York.

Mack, D. E., and W. Yong. 2000. Swainson's Thrush. *In* The Birds of North America, # 540 (A. Poole and F. Gill, Eds.), National Academy of Sciences, Philadelphia, Pennsylvania.

MacPhee, R. D. E. (Ed.). 1999. Extinctions in near time: Causes, contexts, and consequences. Plenum Press, New York.

MacWhirter, B., P. Austin-Smith, Jr., and D. Kroodsma. 2002. Sanderling. *In* The Birds of North America, # 653 (A. Poole and F. Gill, Eds.), National Academy of Sciences, Philadelphia, Pennsylvania.

MacWhirter, R. B., and K. L. Bildstein. 1996. Northern Harrier. *In* The Birds of North America, # 210 (A. Poole and F. Gill, Eds.), National Academy of Sciences, Philadelphia, Pennsylvania.

Mallory, M., and K. Metz. 1999. Common Merganser. *In* The Birds of North America, # 442 (A. Poole and F. Gill, Eds.), National Academy of Sciences, Philadelphia, Pennsylvania.

Marks, J. S., D. L. Evans, and D. W. Holt. 1994. Long-eared Owl. *In* The Birds of North America, # 133 (A. Poole and F. Gill, Eds.), National Academy of Sciences, Philadelphia, Pennsylvania.

Marra, P., S. Griffing, C. Caffrey, A. M. Kilpatrick, R. McLean, C. Brand, E. Saito, A. P. Dupuis, L. Kramer, and R. Novak. 2004. West Nilve virus and wildlife. BioScience 54:393–402.

Marti, C. D. 1992. Barn Owl. *In* The Birds of North America, # 1 (A. Poole, P. Stettenheim, and F. Gill, Eds.), National Academy of Sciences, Philadelphia, Pennsylvania.

Martin, J. W., and J. R. Parrish. 2000. Lark Sparrow. *In* The Birds of North America, # 488 (A. Poole and F. Gill, Eds.), National Academy of Sciences, Philadelphia, Pennsylvania.

Martin, P. S., and R. G. Klein (Eds.) 1984. Quaternary extinctions: A prehistoric revolution. University of Arizona Press, Tucson, Arizona.

Martin, S. G. 2002. Brewer's Blackbird. *In* The Birds of North America, # 616 (A. Poole and F. Gill, Eds.), National Academy of Sciences, Philadelphia, Pennsylvania.

Martin, S. G., and T. A. Gavin. 1995. Bobolink. *In* The Birds of North America, # 176 (A. Poole and F. Gill, Eds.), National Academy of Sciences, Philadelphia, Pennsylvania.

Martof, B. S., W. M. Palmer, J. R. Bailey, and J. R. Harrison, III. 1980. Amphibians and reptiles of the Carolinas and Virginia. University of North Carolina Press, Chapel Hill, North Carolina.

Matthews, S. N., R. J. O'Connor, L. R. Iverson, and A. M. Prasad. 2004. Atlas of climate change effects in 150 bird species of the eastern United States. U.S.

Department of Agriculture Forest Service, General Technical Report NE-318.

Mazur, K. M., and P. C. James. 2000. Barred Owl. *In* The Birds of North America, # 508 (A. Poole and F. Gill, Eds.), National Academy of Sciences, Philadelphia, Pennsylvania.

McCarty, J. P. 1996. Eastern Wood-Pewee. *In* The Birds of North America, # 245 (A. Poole, and F. Gill, Eds.), National Academy of Sciences, Philadelphia, Pennsylvania.

McCay, T. S. 2001. *Blarina carolinensis*. Mammalian Species # 673. American Society of Mammalogists, http://www.science.smith.edu/departments/Biology/VHAYSSEN/msi/.

McCrimmon, Jr., D. A., J. C. Ogden, and G. T. Bancroft. 2001. Great Egret.. *In* The Birds of North America, # 570 (A. Poole and F. Gill, Eds.), National Academy of Sciences, Philadelphia, Pennsylvania.

McDonald, M. V. 1998. Kentucky Warbler. *In* The Birds of North America, # 324 (A. Poole, and F. Gill, Eds.), National Academy of Sciences, Philadelphia, Pennsylvania.

McGowan, K. J. 2001. Fish Crow. *In* The Birds of North America, # 589 (A. Poole and F. Gill, Eds.), National Academy of Sciences, Philadelphia, Pennsylvania.

McIntyre, J. W., and J. F. Barr. 1997. Common Loon. *In* The Birds of North America, # 313 (A. Poole and F. Gill, Eds.), National Academy of Sciences, Philadelphia, Pennsylvania.

McManus, J. J. 1974. *Didelphis virginiana*. Mammalian Species # 40. American Society of Mammalogists, http://www.science.smith.edu/departments/Biology/VHAYSSEN/msi/.

McNicholl, M. K., ad P. E. Lowther, and J. A. Hall. 2001. Forster's Tern. *In* The Birds of North America, # 595 (A. Poole and F. Gill, Eds.), National Academy of Sciences, Philadelphia, Pennsylvania.

McShea, W. J., H. B. Underwood, and J. H. Rappole. 1997. The science of overabundance: deer ecology and population management. Smithsonian Institution Press, Washington, D. C.

Meanley, B. 1971. Natural history of the Swainson's Warbler. North American Fauna, No. 69, U. S. Dept. Interior, Washington, D. C.

Meanley, B. 1992. King Rail. *In* The Birds of North America, # 3 (A. Poole, P. Stettenheim, and F. Gill, Eds.), National Academy of Sciences, Philadelphia, Pennsylvania.

Melvin, S. M., and J. P. Gibbs. 1996. Sora. *In* The Birds of North America, # 250 (A. Poole, and F. Gill, Eds.), National Academy of Sciences, Philadelphia, Pennsylvania.

Merriam, C. H. 1894. Laws of temperature control of the geographic distribution of terrestrial animals and plants. National Geographic Magazine 6:229–238.

Merritt, J. F. 1981. *Clethrionomys gapperi*. Mammalian Species # 146. American Society of Mammalogists, http://www.science.smith.edu/departments/Biology/VHAYSSEN/msi/.

Merritt, J. F. 1987. Guide to the mammals of Pennsylvania. University of Pittsburgh Press, Pittsburgh, Pennsylvania.

Middleton, A. L. A. 1993. American Goldfinch. *In* The Birds of North America, # 80 (A. Poole, P. Stettenheim, and F. Gill, Eds.), National Academy of Sciences, Philadelphia, Pennsylvania.

Middleton, A. L. A. 1998. Chipping Sparrow. *In* The Birds of North America, # 334 (A. Poole and F. Gill, Eds.), National Academy of Sciences, Philadelphia, Pennsylvania.

Mirarchi, R. E., and T. S. Baskett. 1994. Mourning Dove. *In* The Birds of North America, # 117 (A. Poole and F. Gill, Eds.), National Academy of Sciences, Philadelphia, Pennsylvania.

Mitchell, J. C. 1991. Amphibians and reptiles. Pp. 411–476 *In* Virginia's endangered species (K. Terwilliger, Ed.), McDonald and Woodward Publishing Company, Blacksburg, Virginia.

Mitchell, J. C. 1994. The reptiles of Virginia. Smithsonian Institution Press, Washington, D. C.

Moldenhauer, R. R., and D. J. Regelski. 1996. Northern Parula. *In* The Birds of North America, # 215 (A. Poole and F. Gill, Eds.), National Academy of Sciences, Philadelphia, Pennsylvania.

Moore, W. S. 1995. Northern Flicker. *In* The Birds of North America, # 166 (A. Poole and F. Gill, Eds.), National Academy of Sciences, Philadelphia, Pennsylvania.

Morrison, R. I. G., C. Downes, and B. Collins. 1994. Population trends of shorebirds on fall migration in eastern Canada, 1974–1991. Wilson Bulletin 106: 431–447.

Morse, D. H. 1993. Black-throated Green Warbler. *In* The Birds of North America, # 55 (A. Poole, P. Stettenheim, and F. Gill, Eds.), National Academy of Sciences, Philadelphia, Pennsylvania.

Morse, D. H. 1994. Blackburnian Warbler. *In* The Birds of North America, # 102 (A. Poole and F. Gill, Eds.), National Academy of Sciences, Philadelphia, Pennsylvania.

Morton, E. S. 1971. Food and migration habits of the Eastern Kingbird in Panama. Auk 88:44–54.

Morton, E. S. 1980. Adaptations to seasonal changes by migrant landbirds in the Panama Canal Zone. Pp. 437–453 *In* Migrant birds in the Neotropics (A. Keast, and E. S. Morton, eds.), Smithsonian Institution Press, Washington, D. C.

Morton, E. S., L. Forman, and M. Braun. 1990. Extrapair fertilizations and the evolution of colonial breeding in Purple Martins. Auk 107:275–283.

Morton, E. S., B. J. M. Stutchbury, J. S. Howlett, and H. W. Piper. 1998. Genetics monogamy in Blue-headed Vireos and a comparison with a sympatric vireo with extrapair paternity. Behavioral Ecology 9:515–524.

Moskoff, W. 1995. Solitary Sandpiper. *In* The Birds of North America, # 156 (A. Poole and F. Gill, Eds.), National Academy of Sciences, Philadelphia, Pennsylvania.

Moskoff, W. 1995. Veery. *In* The Birds of North America, # 142 (A. Poole and F. Gill, Eds.), National Academy of Sciences, Philadelphia, Pennsylvania.

Moskoff, W., and R. Montgomerie. 2002. Baird's Sandpiper. *In* The Birds of North America, # 661 (A. Poole and F. Gill, Eds.), National Academy of Sciences, Philadelphia, Pennsylvania.

Moskoff, W., and S. K. Robinson. 1996. Philadelphia Vireo. *In* The Birds of North America, # 214 (A. Poole and F. Gill, Eds.), National Academy of Sciences, Philadelphia, Pennsylvania.

Mostrom, A. M., R. L. Curry, and B. Lohr. 2002. Carolina Chickadee. *In* The Birds of North America, # 636 (A. Poole and F. Gill, Eds.), National Academy of Sciences, Philadelphia, Pennsylvania.

Mowbray, T. B. 1997. Swamp Sparrow. *In* The Birds of North America, # 279 (A. Poole and F. Gill, Eds.), National Academy of Sciences, Philadelphia, Pennsylvania.

Mowbray, T. B. 1999. American Wigeon. *In* The Birds of North America, # 401 (A. Poole and F. Gill, Eds.), National Academy of Sciences, Philadelphia, Pennsylvania.

Mowbray, T. B. 1999. Scarlet Tanager. *In* The Birds of North America, # 479 (A. Poole and F. Gill, Eds.), National Academy of Sciences, Philadelphia, Pennsylvania.

Mowbray, T. B. 2002. Canvasback. *In* The Birds of North America, # 659 (A. Poole and F. Gill, Eds.), National Academy of Sciences, Philadelphia, Pennsylvania.

Mowbray, T. B., F. Cooke, and B. Ganter. 2000. Snow Goose. *In* The Birds of North America, # 514 (A. Poole and F. Gill, Eds.), National Academy of Sciences, Philadelphia, Pennsylvania.

Mowbray, T. B., C. R. Ely, J. S. Sedinger, and R. E. Trost. 2002. Canada Goose. *In* The Birds of North America, # 682 (A. Poole and F. Gill, Eds.), National Academy of Sciences, Philadelphia, Pennsylvania.

Mueller, H. 1999. Common Snipe. *In* The Birds of North America, # 417 (A. Poole and F. Gill, Eds.), National Academy of Sciences, Philadelphia, Pennsylvania.

Muller, M. J., and R. W. Storer. 1999. Pied-billed Grebe. *In* The Birds of North America, # 410 (A. Poole and F. Gill, Eds.), National Academy of Sciences, Philadelphia, Pennsylvania.

Murphy, M. T. 1996. Eastern Kingbird. *In* The Birds of North America, # 253 (A. Poole and F. Gill, Eds.), National Academy of Sciences, Philadelphia, Pennsylvania.

NatureServe. 2005. NatureServe Explorer Species Database. http://www.natureserve.org/explorer/index.htm.

Naugler, C. T. 1993. American Tree Sparrow. *In* The Birds of North America, # 37 (A. Poole, and F. Gill, Eds.), National Academy of Sciences, Philadelphia, Pennsylvania.

Nettleship, D. N. 2000. Ruddy Turnstone. *In* The Birds of North America, # 537 (A. Poole and F. Gill, Eds.), National Academy of Sciences, Philadelphia, Pennsylvania.

Nice, M. M. 1937. Studies in the life history of the Song Sparrow. Pt. I. Trans. Linnean Soc. New York 4:1–247.

Nice, M. M. 1943. Studies in the life history of the Song Sparrow. Pt. II. Trans. Linnean Soc. New York 6:1–329.

Nisbet, I. C. T. 2002. Common Tern. *In* The Birds of North America, # 618 (A. Poole and F. Gill, Eds.), National Academy of Sciences, Philadelphia, Pennsylvania.

Nol, E., and M. S. Blanken. 1999. Semipalmated Plover. *In* The Birds of North America, # 444 (A. Poole and F. Gill, Eds.), National Academy of Sciences, Philadelphia, Pennsylvania.

Nol, E., and R. C. Humphrey. 1994. American Oystercatcher. *In* The Birds of North America, # 82 (A. Poole and F. Gill, Eds.), National Academy of Sciences, Philadelphia, Pennsylvania.

Nolan, V., Jr. 1978. The ecology and behavior of the Prairie Warbler *Dendroica discolor*. Ornithological Monographs 26:1–595.

Nolan, V., Jr., E. D. Ketterson, and C. A. Buerkle. 1999. Prairie Warbler. *In* The Birds of North America, # 455 (A. Poole and F. Gill, Eds.), National Academy of Sciences, Philadelphia, Pennsylvania.

Nolan, V., Jr., E. D. Ketterson, D. A. Cristol, C. M. Rogers, E. D. Clotfelter, R. C. Titus, S. J. Schoech, and E. Snajdr. 2002. Dark-eyed Junco. *In* The Birds of North America, # 716 (A. Poole and F. Gill, Eds.), National Academy of Sciences, Philadelphia, Pennsylvania.

Norris, R. A. 1958. Comparative biosystematics and life history of the nuthatches, *Sitta pygmaea* and *Sitta pusilla*. University of California Publications in Zoology 56:119–300.

Noss, R. F., E. T. LaRoe, and J. M. Scott. 1995. Endangered ecosystems of the United States: A preliminary assessment of loss and degradation. U. S. Department of the Interior, National Biological Service. Biological Report no. 28, Washington, D. C.

Nova Scotia Museum of Natural History. 2005. Common Black-headed Gull. *In* Birds of Nova Scotia. http://museum.gov.ns.ca/mnh/nature/nsbirds/bns0171.htm

Nowak, R. M. 1999. Walker's mammals of the world. Sixth Edition. Johns Hopkins University Press, Baltimore, Maryland.

Ogden, L. J. E., and B. J. Stutchbury. 1994. Hooded Warbler. *In* The Birds of North America, # 110 (A. Poole and F. Gill, Eds.), National Academy of Sciences, Philadelphia, Pennsylvania.

Orians, G. H. 1985. Blackbirds of the Americas. University of Washington Press, Seattle, Washington.

Oring, L. W., E. M. Gray, and J. M. Reed. 1997. Spotted Sandpiper. *In* The Birds of North America, # 289 (A. Poole and F. Gill, Eds.), National Academy of Sciences, Philadelphia, Pennsylvania.

Owens, J. G. 1984. *Sorex fumeus*. Mammalian Species # 215. American Society of Mammalogists, http://www.science.smith.edu/departments/Biology/VHAYSSEN/msi/.

Pague, C. A., J. C. Mitchell, and G. Hammerson. 2005. Cow Knob Salamander. NatureServe Explorer Species Database. http://www.natureserve.org/explorer/index.htm.

Paludan, K. 1952. Contributions to the breeding biology of *Larus argentatus* and *Larus fuscus*. Videns Medd Dansk Naturhist Foren 114:1–128.

Parker, J. W. 1999. Mississippi Kite. *In* The Birds of North

America, # 402 (A. Poole and F. Gill, Eds.), National Academy of Sciences, Philadelphia, Pennsylvania.

Parmelee, D. F. 1992. Snowy Owl. *In* The Birds of North America, # 10 (A. Poole, P. Stettenheim, and F. Gill, Eds.), National Academy of Sciences, Philadelphia, Pennsylvania.

Parmelee, D. F. 1992. White-rumped Sandpiper. *In* The Birds of North America, # 29 (A. Poole, P. Stettenheim, and F. Gill, Eds.), National Academy of Sciences, Philadelphia, Pennsylvania.

Parnell, J. F., R. M. Erwin, and K. C. Molina. 1995. Gull-billed Tern. *In* The Birds of North America, # 140 (A. Poole and F. Gill, Eds.), National Academy of Sciences, Philadelphia, Pennsylvania.

Parsons, K. C., and T. L. Master. 2000. Snowy Egret. *In* The Birds of North America, # 489 (A. Poole and F. Gill, Eds.), National Academy of Sciences, Philadelphia, Pennsylvania.

Pauley, T. K. 2005a. West Virginia salamanders. Marshall University herpetology website. http://www.marshall.edu/herp/salamanders.htm.

Pauley, T. K. 2005b. Frogs and Toads of West Virginia. Marshall University herpetology website. http://www.marshall.edu/herp/ANURANS.HTM.

Pauley, T. K. 2005c. Turtles of West Virginia. Marshall University herpetology website. http://www.marshall.edu/herp/pages/TURTLES.HTM

Paulson, D. R. 1995. Black-bellied Plover. *In* The Birds of North America, # 186 (A. Poole, and F. Gill, Eds.), National Academy of Sciences, Philadelphia, Pennsylvania.

Payne, L. X., and E. Pierce. 2002. Purple Sandpiper. *In* The Birds of North America, # 706 (A. Poole and F. Gill, Eds.), National Academy of Sciences, Philadelphia, Pennsylvania.

Payne, R. B. 1992. Indigo Bunting. *In* The Birds of North America, # 4 (A. Poole, P. Stettenheim, and F. Gill, Eds.), National Academy of Sciences, Philadelphia, Pennsylvania.

Peer, B. D., and E. K. Bollinger. Common Grackle. 1997. *In* The Birds of North America, # 271 (A. Poole and F. Gill, Eds.), National Academy of Sciences, Philadelphia, Pennsylvania.

Peterson, K. E., and T. L. Yates. 1980. *Condylura cristata*. Mammalian Species # 129. American Society of Mammalogists, http://www.science.smith.edu/departments/Biology/VHAYSSEN/msi/.

Petit, L. J. 1999. Prothonotary Warbler. *In* The Birds of North America, # 408 (A. Poole and F. Gill, Eds.), National Academy of Sciences, Philadelphia, Pennsylvania.

Petranka, J. W. 1998. Salamanders of the United States and Canada. Smithsonian Institution Press, Washington, D. C.

Pierotti, R. J., and T. P. Good. 1994. Herring Gull. *In* The Birds of North America, # 124 (A. Poole and F. Gill, Eds.), National Academy of Sciences, Philadelphia, Pennsylvania.

Pitocchelli, J. 1993. Mourning Warbler. *In* The Birds of

North America, # 72 (A. Poole, P. Stettenheim, and F. Gill, Eds.), National Academy of Sciences, Philadelphia, Pennsylvania.

Pitocchelli, J., J. Bouchie, and D. Jones. 1997. Connecticut Warbler. *In* The Birds of North America, # 320 (A. Poole and F. Gill, Eds.), National Academy of Sciences, Philadelphia, Pennsylvania.

Platz, J. E. 1989. Speciation within the chorus frog *Pseudacris triseriata*: morphometric and mating call analyses of the boreal and western subspecies. Copeia 1989:704–712.

Poole, A. F., R. O. Bierregaard, and M. S. Martell. 2002. Osprey. *In* The Birds of North America, # 683 (A. Poole and F. Gill, Eds.), National Academy of Sciences, Philadelphia, Pennsylvania.

Poole, A., and F. Gill (Eds.). 1992–2002. The birds of North America. National Academy of Sciences, Philadelphia, Pennsylvania.

Post, W., and J. S. Greenlaw. 1994. Seaside Sparrow. *In* The Birds of North America, # 127 (A. Poole and F. Gill, Eds.), National Academy of Sciences, Philadelphia, Pennsylvania.

Post, W., J. P. Poston, and G. T. Bancroft. 1996. Boat-tailed Grackle. *In* The Birds of North America, # 207 (A. Poole and F. Gill, Eds.), National Academy of Sciences, Philadelphia, Pennsylvania.

Poulin, R. G., S. D. Grindal, and R. M. Brigham. 1996. Common Nighthawk. *In* The Birds of North America, # 213 (A. Poole and F. Gill, Eds.), National Academy of Sciences, Philadelphia, Pennsylvania.

Powell, G. V. N. 1980. Migrant participation in Neotropical mixed species flocks. Pp. 477–483 *in* Migrant birds in the Neotropics (A. Keast and E. S. Morton, Eds.), Smithsonian Institution Press, Washington, D. C.

Powell, G. V. N., and J. H. Rappole. 1986. The Hooded Warbler. Pp. 827–853 *In* Audubon wildlife report (L. Di Silvestro, Ed.), National Audubon Society, New York.

Powell, R. A. 1981. *Martes pennanti*. Mammalian Species # 156. American Society of Mammalogists, http://www.science.smith.edu/departments/Biology/VHAYSSEN/msi/.

Pravosudov, V. V., and T. C. Grubb, Jr. 1993. White-breasted Nuthatch. *In* The Birds of North America, # 54 (A. Poole, P. Stettenheim, and F. Gill, Eds.), National Academy of Sciences, Philadelphia, Pennsylvania.

Preston, C. R., and R. D. Beane. 1993. Red-tailed Hawk. *In* The Birds of North America, # 52 (A. Poole, P. Stettenheim, and F. Gill, Eds.), National Academy of Sciences, Philadelphia, Pennsylvania.

Rappole, J. H. 1982. Management possibilities for beach-nesting shorebirds. Pp. 114–126 *In* nongame and endangered wildlife symposium (R. Odom and W. Guthrie, Eds.) Georgia Department of Natural Resources, Atlanta, Georgia.

Rappole, J. H. 1999. Ruby-throated Hummingbird. Pp. 671–672 *In* Handbook of the birds of the world, Vol. 5. Barn-owls to Hummingbirds (K.-L. Schuchmann, Ed.), Lynx Edicions, Barcelona, Spain.

Rappole, J. H., and M. V. MacDonald. 1994. Cause and

effect in population declines of migratory birds. Auk 111:652–660.

Rappole, J. H., and M. V. McDonald. 1998. A response to Latta and Baltz. Auk 115: 246–251.

Rappole, J. H., M. A. Ramos, and K. Winker. 1989. Wintering Wood Thrush and mortality in southern Veracruz. Auk 106:402–410.

Rappole, J. H., and D. W. Warner. 1980. Ecological aspects of avian migrant behavior in Veracruz, Mexico. Pp. 353–393 *In* Migrant birds in the Neotropics (A. Keast and E. S. Morton, Eds.) Smithsonian Institution Press, Washington, D. C.

Reed, A., D. H. Ward, D. V. Derksen, and J. S. Sedinger. 1998. Brant. *In* The Birds of North America, # 337 (A. Poole and F. Gill, Eds.), National Academy of Sciences, Philadelphia, Pennsylvania.

Reich, L. M. 1981. *Microtus pennsylvanicus*. Mammalian Species # 159. American Society of Mammalogists, http://www.science.smith.edu/departments/Biology/VHAYSSEN/msi/.

Richardson, M., and D. W. Brauning. 1995. Chestnut-sided Warbler. *In* The Birds of North America, # 190 (A. Poole and F. Gill, Eds.), National Academy of Sciences, Philadelphia, Pennsylvania.

Rimmer, C. C., and K. P. McFarland. 1998. Tennessee Warbler. *In* The Birds of North America, # 350 (A. Poole and F. Gill, Eds.), National Academy of Sciences, Philadelphia, Pennsylvania.

Rimmer, C. C., K. P. McFarland, W. G. Ellison, and J. E. Goetz. 2001. Bicknell's Thrush. *In* The Birds of North America, # 592 (A. Poole and F. Gill, Eds.), National Academy of Sciences, Philadelphia, Pennsylvania.

Rising, J. D., and N. J. Flood. 1998. Baltimore Oriole. *In* The Birds of North America, # 384 (A. Poole and F. Gill, Eds.), National Academy of Sciences, Philadelphia, Pennsylvania.

Robbins, C. S., D. K. Dawson, and B. A. Dowell. 1989. Habitat area requirements of breeding forest birds of the Middle Atlantic States. Wildlife Monographs 103:1–34.

Robertson, G. J., and R. I. Goudie. 1999. Harlequin Duck. *In* The Birds of North America, # 466 (A. Poole and F. Gill, Eds.), National Academy of Sciences, Philadelphia, Pennsylvania.

Robertson, G. J., and J.-P. L. Savard. 2002. Long-tailed Duck. *In* The Birds of North America, # 651 (A. Poole and F. Gill, Eds.), National Academy of Sciences, Philadelphia, Pennsylvania.

Robertson, R. J., B. J. Stutchbury, and R. R. Cohen. 1992. Tree Swallow. *In* The Birds of North America, # 11 (A. Poole, P. Stettenheim, and F. Gill, Eds.), National Academy of Sciences, Philadelphia, Pennsylvania.

Robinson, J. A., L. W. Oring, J. P. Skorupa, and R. Boettcher. 1997. American Avocet. *In* The Birds of North America, # 275 (A. Poole and F. Gill, Eds.), National Academy of Sciences, Philadelphia, Pennsylvania.

Robinson, J. A., J. M. Reed, J. P. Skorupa, and L. W. Oring. 1999. Black-necked Stilt. *In* The Birds of North America, # 449 (A. Poole and F. Gill, Eds.), National Academy of Sciences, Philadelphia, Pennsylvania.

Robinson, T. R., R. R. Sargent, and M. B. Sargent. 1996. Ruby-throated Hummingbird. *In* The Birds of North America, # 204 (A. Poole and F. Gill, Eds.), National Academy of Sciences, Philadelphia, Pennsylvania.

Robinson, W. D. 1995. Louisiana Waterthrush. *In* The Birds of North America, # 151 (A. Poole and F. Gill, Eds.), National Academy of Sciences, Philadelphia, Pennsylvania.

Robinson, W. D. 1996. Summer Tanager. *In* The Birds of North America, # 248 (A. Poole and F. Gill, Eds.), National Academy of Sciences, Philadelphia, Pennsylvania.

Roble, S. M. 2003. Natural heritage resources of Virginia: rare animal species. Natural Heritage Technical Report 03–04. Virginia Department of Conservation and Recreation, Division of Natural Heritage, Richmond, Virginia.

Rodewald, P. G., and R. D. James. 1996. Yellow-throated Vireo. *In* The Birds of North America, # 247 (A. Poole and F. Gill, Eds.), National Academy of Sciences, Philadelphia, Pennsylvania.

Rodewald, P. G., J. H. Withgott, and K. G. Smith. 1999. Pine Warbler. *In* The Birds of North America, # 438 (A. Poole and F. Gill, Eds.), National Academy of Sciences, Philadelphia, Pennsylvania.

Rodgers, J. A., Jr., and H. T. Smith. 1995. Little Blue Heron. *In* The Birds of North America, # 145 (A. Poole and F. Gill, Eds.), National Academy of Sciences, Philadelphia, Pennsylvania.

Rohwer, F. C., W. P. Johnson, and E. R. Loos. 2002. Blue-winged Teal. *In* The Birds of North America, # 625 (A. Poole and F. Gill, Eds.), National Academy of Sciences, Philadelphia, Pennsylvania.

Rosenberg, D., and T. Rothe. 1994. Swans. Alaska Dept. Fish and Game. www.adfg.state.ak.us.

Rosenfield, R. N., and J. Bielefeldt. 1993. Cooper's Hawk. *In* The Birds of North America, # 75 (A. Poole, P. Stettenheim, and F. Gill, Eds.), National Academy of Sciences, Philadelphia, Pennsylvania.

Roth, R. R., M. S. Johnson, and T. J. Underwood. 1996. Wood Thrush. *In* The Birds of North America, # 246 (A. Poole and F. Gill, Eds.), National Academy of Sciences, Philadelphia, Pennsylvania.

Rubega, M. A., D. Schamel, and D. M. Tracy. 2000. Red-necked Phalarope. *In* The Birds of North America, # 538 (A. Poole and F. Gill, Eds.), National Academy of Sciences, Philadelphia, Pennsylvania.

Rusch, D. H., S. Destefano, M. C. Reynolds, and D. Lauten. 2000. Ruffed Grouse. *In* The Birds of North America, # 515 (A. Poole and F. Gill, Eds.), National Academy of Sciences, Philadelphia, Pennsylvania.

Ryder, J. P. 1993. Ring-billed Gull. *In* The Birds of North America, # 33 (A. Poole, P. Stettenheim, and F. Gill, Eds.), National Academy of Sciences, Philadelphia, Pennsylvania.

Sallabanks, R., and F. C. James. 1999. American Robin. *In* The Birds of North America, # 462 (A. Poole and F. Gill, Eds.), National Academy of Sciences, Philadelphia, Pennsylvania.

Savard, J.-P. L., D. Bordage, and A. Reed. 1998. Surf Scoter. *In* The Birds of North America, # 363 (A. Poole and F. Gill, Eds.), National Academy of Sciences, Philadelphia, Pennsylvania.

Scharf, W. C., and J. Kren. 1996. Orchard Oriole. *In* The Birds of North America, # 255 (A. Poole and F. Gill, Eds.), National Academy of Sciences, Philadelphia, Pennsylvania.

Sedgwick, J. A. 2000. Willow Flycatcher. *In* The Birds of North America, # 533 (A. Poole and F. Gill, Eds.), National Academy of Sciences, Philadelphia, Pennsylvania.

Seidel, M. E. 1994. Morphometric analysis and taxonomy of cooter and red-bellied turtles in the North American genus *Pseudemys* (Emydidae). Chelonia Conservation Biology 1:117–130.

Senseman, R. 2002. *Cervus elaphus*. Animal Diversity Web, University of Michigan Museum of Zoology, http:// animaldiversity.ummz.umich.edu/site/accounts/ information/ Cervus_elaphus.html.

Sevon, W. D., and G. M. Fleeger. 1999. Pennsylvania and the Ice Age. Second Ed. Pennsylvania Geological Survey, 4th Ser., Educational Series 6.

Shackleford, C. E., R. E. Brown, and R. N. Conner. 2000. Red-bellied Woodpecker. *In* The Birds of North America, # 500 (A. Poole and F. Gill, Eds.), National Academy of Sciences, Philadelphia, Pennsylvania.

Shaffer, L. L. 1999. Pennsylvania amphibians and reptiles. Pennsylvania Fish and Boat Commission, Harrisburg, Pennsylvania.

Shealer, D. 1999. Sandwich Tern. *In* The Birds of North America, # 405 (A. Poole and F. Gill, Eds.), National Academy of Sciences, Philadelphia, Pennsylvania.

Sheffield, S. R., and C. M. King. 1994. *Mustela nivalis*. Mammalian Species # 454. American Society of Mammalogists, http://www.science.smith.edu/ departments/Biology/VHAYSSEN/msi/.

Sheffield, S. R., and H. H. Thomas. 1997. *Mustela frenata*. Mammalian Species # 570.

Sherry, T. W., and R. T. Holmes. 1997. American Redstart. *In* The Birds of North America, # 277 (A. Poole and F. Gill, Eds.), National Academy of Sciences, Philadelphia, Pennsylvania.

Shields, M. 2001. Brown Pelican. *In* The Birds of North America, # 609 (A. Poole and F. Gill, Eds.), National Academy of Sciences, Philadelphia, Pennsylvania.

Shump, K. A., Jr., and A. U. Shump. 1982a. *Lasiurus borealis*. Mammalian Species # 183. American Society of Mammalogists, http://www.science.smith.edu/ departments/Biology/VHAYSSEN/msi/.

Shump, K. A., Jr., and A. U. Shump. 1982b. *Lasiurus cinereus*. Mammalian Species # 185. American Society of Mammalogists, http://www.science.smith.edu/ departments/Biology/VHAYSSEN/msi/.

Skeel, M. A., and E. P. Mallory. 1996. Whimbrel. *In* The Birds of North America, # 219 (A. Poole and F. Gill, Eds.), National Academy of Sciences, Philadelphia, Pennsylvania.

Smallwood, J. A., and D. M. Bird. 2001. American Kestrel. *In* The Birds of North America, # 602 (A. Poole and F. Gill, Eds.), National Academy of Sciences, Philadelphia, Pennsylvania.

Smith, K. G., J. H. Withgott, and P. G. Rodewald. 2000. Red-headed Woodpecker. *In* The Birds of North America, # 518 (A. Poole and F. Gill, Eds.), National Academy of Sciences, Philadelphia, Pennsylvania.

Smith, S. M. 1993. Black-capped Chickadee. *In* The Birds of North America, # 39 (A. Poole, P. Stettenheim, and F. Gill, Eds.), National Academy of Sciences, Philadelphia, Pennsylvania.

Smith, W. P. 1991. *Odocoileus virginianus*. Mammalian Species # 388. American Society of Mammalogists, http://www.science.smith.edu/departments/Biology/ VHAYSSEN/msi/.

Smolen, M. J. 1981. *Microtus pinetorum*. Mammalian Species # 147. American Society of Mammalogists, http://www.science.smith.edu/departments/Biology/ VHAYSSEN/msi/.

Snell, R. R. 2002. Iceland Gull. *In* The Birds of North America, # 699 (A. Poole and F. Gill, Eds.), National Academy of Sciences, Philadelphia, Pennsylvania.

Snyder, D. P. 1982. *Tamias striatus*. Mammalian Species # 168. American Society of Mammalogists, http://www.science.smith.edu/departments/Biology/ VHAYSSEN/msi/.

Sodhi, N. S., L. W. Oliphant, P. C. James, and I. G. Warkentin. 1993. Merlin. *In* The Birds of North America, # 44 (A. Poole, P. Stettenheim, and F. Gill, Eds.), National Academy of Sciences, Philadelphia, Pennsylvania.

Sogge, M. K., W. M. Gilbert, and C. Van Riper, III. 1994. Orange-crowned Warbler. *In* The Birds of North America, # 101 (A. Poole and F. Gill, Eds.), National Academy of Sciences, Philadelphia, Pennsylvania.

Soule, J., and G. Hammerson. 2005. Wood Turtle. NatureServe Explorer Species Database. http://www. natureserve.org/explorer/index.htm.

Squires, J. R., and R. T. Reynolds. 1997. Northern Goshawk. *In* The Birds of North America, # 298 (A. Poole and F. Gill, Eds.), National Academy of Sciences, Philadelphia, Pennsylvania.

Stager, K. 1964. The role of olfaction in food location by the Turkey Vulture (Cathartes aura). Los Angeles County Museum Contributions in Science 81:1–63.

Stalheim, S. 1974. Behavior and ecology of the Yellow Rail (*Coturnicops noveboracensis*). Unpubl. M. S. Thesis, University of Minnesota, Minneapolis, Minnesota.

Stalling, D. T. 1997. *Reithrodontomys humulis*. Mammalian Species # 565. American Society of Mammalogists, http://www.science.smith.edu/departments/Biology/ VHAYSSEN/msi/.

Stedman, S. J. 2000. Horned Grebe. *In* The Birds of North America, # 505 (A. Poole and F. Gill, Eds.), National Academy of Sciences, Philadelphia, Pennsylvania.

Steele, M. A. 1998. *Tamiasciurus hudsonicus*. Mammalian Species # 586. American Society of Mammalogists, http://www.science.smith.edu/departments/Biology/ VHAYSSEN/msi/.

Stout, B. E., and G. L. Nuechterlein. 1999. Red-necked Grebe. *In* The Birds of North America, # 465 (A. Poole and F. Gill, Eds.), National Academy of Sciences, Philadelphia, Pennsylvania.

Straight, C. A., and R. J. Cooper. 2000. Chuck-will's-widow. *In* The Birds of North America, # 499 (A. Poole and F. Gill, Eds.), National Academy of Sciences, Philadelphia, Pennsylvania.

Stuart, J. N. 2002. Review of amphibian and reptile nomenclature lists by Crother (2000) and Collins and Taggart (2002). Bulletin of the Chicago Herpetological Society 37(11):197– 199.

Suydam, R. S. 2000. King Eider. *In* The Birds of North America, # 491 (A. Poole and F. Gill, Eds.), National Academy of Sciences, Philadelphia, Pennsylvania.

Takewa, J. Y., and N. Warnock. 2000. Long-billed Dowitcher. *In* The Birds of North America, # 493 (A. Poole and F. Gill, Eds.), National Academy of Sciences, Philadelphia, Pennsylvania.

Talbot, W. 1672. The discoveries of John Lederer, in three several marches from Virginia, to the west of Carolina, and other parts of the continent: begun in March 1669, and ended in September 1670, together with a general map of the whole territory which he traversed. Printed by F. C. for Samuel Heyrick at Grays-Inne Gate in Holborn, London. Reprinted by George P. Humphrey, Rochester, New York, in 1902.

Tarvin, K. A., and G. E. Woolfenden. 1999. Blue Jay. *In* The Birds of North America, # 469 (A. Poole and F. Gill, Eds.), National Academy of Sciences, Philadelphia, Pennsylvania.

Telfair, R. C., II. 1994. Cattle Egret. *In* The Birds of North America, # 113 (A. Poole and F. Gill, Eds.), National Academy of Sciences, Philadelphia, Pennsylvania.

Temple, S. A. 2002. Dickcissel. *In* The Birds of North America, # 703 (A. Poole and F. Gill, Eds.), National Academy of Sciences, Philadelphia, Pennsylvania.

Thompson, B. C., J. A. Jackson, J. Burger, L. A. Hill, E. M. Kirsch, and J. L. Atwood. 1997. Least Tern. *In* The Birds of North America, # 290 (A. Poole and F. Gill, Eds.), National Academy of Sciences, Philadelphia, Pennsylvania.

Thomson, C. E. 1982. *Myotis sodalis*. Mammalian Species # 163. American Society of Mammalogists, http://www.science.smith.edu/departments/Biology/VHAYSSEN/msi/.

Tibbitts, T. L., and W. Moskoff. 1999. Lesser Yellowlegs. *In* The Birds of North America, # 427 (A. Poole and F. Gill, Eds.), National Academy of Sciences, Philadelphia, Pennsylvania.

Tilley, S. G., and M. J. Mahoney. 1996. Patterns of genetic differentiation in salamanders of the *Desmognathus ochrophaeus* complex (Amphibia: Plethontidae). Herpetological Monographs 10:1–41.

Titman, R. D. 1999. Red-breasted Merganser. *In* The Birds of North America, # 443 (A. Poole, and F. Gill, Eds.), National Academy of Sciences, Philadelphia, Pennsylvania.

Towson University Biological Sciences Department. 2005.

Amphibians of Maryland. http://wwwnew.towson.edu/herpetology/Amphibians.htm.

Tracy, D. M., D. Schmael, and J. Dale. 2002. Red Phalarope. *In* The Birds of North America, # 698 (A. Poole and F. Gill, Eds.), National Academy of Sciences, Philadelphia, Pennsylvania.

Tuck, L. M. 1972. The snipes: a study of the genus *Capella*. Canadian Wildlife Monograph Series No. 5.

Tuttle, M. D. 1979. Status, causes of decline, and management of endangered gray bats. Journal of Wildlife Management 43:1–17.

Twedt, D. J., and R. D. Crawford. 1995. Yellow-headed Blackbird. *In* The Birds of North America, # 192 (A. Poole and F. Gill, Eds.), National Academy of Sciences, Philadelphia, Pennsylvania.

U. S. Geological Survey. 2005. Amphibians and reptiles of the southeastern United States and the U. S. Virgin Islands. USGS CARS website. http://cars.er.usgs.gov/herps/

Van Dam, B., and G. Hammerson. 2005a. Spotted Turtle. NatureServe Explorer Species Database. http://www.natureserve.org/explorer/index.htm.

Van Dam, B., and G. Hammerson. 2005b. Kirtland's Snake. NatureServe Explorer Species Database. http://www.natureserve.org/explorer/index.htm.

Van Horn, M. A., and T. M. Donovan. 1994. Ovenbird. *In* The Birds of North America, # 88 (A. Poole and F. Gill, Eds.), National Academy of Sciences, Philadelphia, Pennsylvania.

Vega Rivera, J. H., W. J. McShea, and J. H. Rappole. 2003. Comparison of breeding and post- breeding movements and habitat requirements for the Scarlet Tanager. Auk 120:632–644.

Vega Rivera, J. H., J. H. Rappole, W. J. McShea, and C. A. Haas. 1998. Wood Thrush post- fledging movements and habitat use in northern Virginia. Condor 100:69–78.

Veitch, D. 2005. *Felis catus*. Global Invasive Species Database. http://www.issg.org/database/species/ecology.asp?si=24&fr=1&sts=

Verbeek, N. A. M., and C. Caffrey. 2002. American Crow. *In* The Birds of North America, # 647 (A. Poole and F. Gill, Eds.), National Academy of Sciences, Philadelphia, Pennsylvania.

Verbeek, N. A. M., and P. Hendricks. 1994. American Pipit. *In* The Birds of North America, # 95 (A. Poole and F. Gill, Eds.), National Academy of Sciences, Philadelphia, Pennsylvania.

Vickery, P. D. 1996. Grasshopper Sparrow. *In* The Birds of North America, # 239 (A. Poole, and F. Gill, Eds.), National Academy of Sciences, Philadelphia, Pennsylvania.

Virginia Department of Game and Inland Fisheries. 2004. Virginia's wildlife. http://www.dgif.state.va.us/wildlife/va_wildlife/index.html

Wade-Smith, J., and B. J. Verts. 1978. *Mephitis mephitis*. Mammalian Species # 173. American Society of Mammalogists, http://www.science.smith.edu/departments/Biology/VHAYSSEN/msi/.

Walters, E. L., E. H. Miller, and P. E. Lowther. 2002. Yellow-bellied Sapsucker. *In* The Birds of North America, # 662 (A. Poole and F. Gill, Eds.), National Academy of Sciences, Philadelphia, Pennsylvania.

Warnock, N. D., and R. E. Gill. 1996. Dunlin. *In* The Birds of North America, # 203 (A. Poole, and F. Gill, Eds.), National Academy of Sciences, Philadelphia, Pennsylvania.

Watkins, L. C. 1972. *Nycticeius humeralis*. Mammalian Species # 23. American Society of Mammalogists, http://www.science.smith.edu/departments/Biology/VHAYSSEN/msi/.

Watts, B. D. 1995. Yellow-crowned Night-Heron. *In* The Birds of North America, # 161 (A. Poole and F. Gill, Eds.), National Academy of Sciences, Philadelphia, Pennsylvania.

Webster, W. D., J. K. Jones, Jr., and R. J. Baker. 1980. *Lasiurus intermedius*. Mammalian Species # 132.

Webster, W. D., J. F. Parnell, and W. C. Biggs, Jr. 1985. Mammals of the Carolinas, Virginia, and Maryland. University of North Carolina Press, Chapel Hill, North Carolina.

Weckstein, J. D., D. E. Kroodsma, and R. C. Faucett. 2002. Fox Sparrow. *In* The Birds of North America, # 715 (A. Poole and F. Gill, Eds.), National Academy of Sciences, Philadelphia, Pennsylvania.

Weeks, H. P., Jr. 1994. Eastern Phoebe. *In* The Birds of North America, # 94 (A. Poole and F. Gill, Eds.), National Academy of Sciences, Philadelphia, Pennsylvania.

Wells-Gosling, N., and L. R. Haney. 1984. *Glaucomys sabrinus*. Mammalian Species # 229. American Society of Mammalogists, http://www.science.smith.edu/departments/Biology/VHAYSSEN/msi/.

West, R. L., and G. K. Hess. 2002. Purple Gallinule. *In* The Birds of North America, # 626 (A. Poole and F. Gill, Eds.), National Academy of Sciences, Philadelphia, Pennsylvania.

Wheelwright, N. T., and J. D. Rising. 1993. Savannah Sparrow. *In* The Birds of North America, # 45 (A. Poole, P. Stettenheim, and F. Gill, Eds.), National Academy of Sciences, Philadelphia, Pennsylvania.

Whitaker, J. O., Jr. 1974. *Cryptotis parva*. Mammalian Species # 43. American Society of Mammalogists, http://www.science.smith.edu/departments/Biology/VHAYSSEN/msi/.

Whitaker, J. O., Jr. 1972. *Zapus hudsonius*. Mammalian Species # 11. American Society of Mammalogists, http://www.science.smith.edu/departments/Biology/VHAYSSEN/msi/.

Whitaker, J. O., Jr., and R. E. Wrigley. 1972. *Napaeozapus insignis*. Mammalian Species # 14. American Society of Mammalogists, http://www.science.smith.edu/departments/Biology/VHAYSSEN/msi/.

White, C. M., N. J. Clum, T. J. Cade, and W. G. Hunt. 2002. Peregrine Falcon. *In* The Birds of North America, # 660 (A. Poole and F. Gill, Eds.), National Academy of Sciences, Philadelphia, Pennsylvania.

White, J. F., Jr., and A. W. White. 2002. Amphibians and reptiles of Delmarva. Tidewater Publishers, Centreville, Maryland.

Whitehead, D. R., and T. Taylor. 2002. Acadian Flycatcher. *In* The Birds of North America, # 614 (A. Poole and F. Gill, Eds.), National Academy of Sciences, Philadelphia, Pennsylvania.

Wiley, R. H., and D. S. Lee. 1999. Parasitic Jaeger. *In* The Birds of North America, # 445 (A. Poole and F. Gill, Eds.), National Academy of Sciences, Philadelphia, Pennsylvania.

Wiley, R. W. 1980. *Neotoma floridana*. Mammalian Species # 139. American Society of Mammalogists, http://www.science.smith.edu/departments/Biology/VHAYSSEN/msi/.

Williams, J. M. 1996. Nashville Warbler. *In* The Birds of North America, # 205 (A. Poole, and F. Gill, Eds.), National Academy of Sciences, Philadelphia, Pennsylvania.

Williams, J. M. 1996. Bay-breasted Warbler. *In* The Birds of North America, # 206 (A. Poole, and F. Gill, Eds.), National Academy of Sciences, Philadelphia, Pennsylvania.

Willner, G. R., G. A. Feldhamer, E. E. Zucker, and J. A. Chapman. 1980. *Ondatra zibethicus*. Mammalian Species # 141. American Society of Mammalogists, http://www.science.smith.edu/departments/Biology/VHAYSSEN/msi/.

Wilson, A. 1808–1831. American ornithology of the birds of the United States. Caxton Press of Sherman and Co., Philadelphia, Pennsylvania.

Wilson, W. H. 1994. Western Sandpiper. *In* The Birds of North America, # 90 (A. Poole and F. Gill, Eds.), National Academy of Sciences, Philadelphia, Pennsylvania.

Wilson, W. H. 1996. Palm Warbler. *In* The Birds of North America, # 238 (A. Poole and F. Gill, Eds.), National Academy of Sciences, Philadelphia, Pennsylvania.

Withgott, J. H., and K. G. Smith. 1998. Brown-headed Nuthatch. *In* The Birds of North America, # 349 (A. Poole and F. Gill, Eds.), National Academy of Sciences, Philadelphia, Pennsylvania.

Witmer, M. C., D. J. Mountjoy, and L. Elliot. 1997. Cedar Waxwing. *In* The Birds of North America, # 308 (A. Poole and F. Gill, Eds.), National Academy of Sciences, Philadelphia, Pennsylvania.

Wolfe, J. L. 1982. *Oryzomys palustris*. Mammalian Species # 176. American Society of Mammalogists, http://www.science.smith.edu/departments/Biology/VHAYSSEN/msi/.

Wolfe, J. L., and A. V. Linzey. 1977. *Peromyscus gossypinus*. Mammalian Species # 70. American Society of Mammalogists, http://www.science.smith.edu/departments/Biology/VHAYSSEN/msi/.

Woodin, M. C., and T. C. Michot. 2002. Redhead. *In* The Birds of North America, # 695 (A. Poole and F. Gill, Eds.), National Academy of Sciences, Philadelphia, Pennsylvania.

Woods, C. A. 1973. *Erethizon dorsatum*. Mammalian Species

29. American Society of Mammalogists, http://www.science.smith.edu/departments/Biology/VHAYSSEN/msi/.

Woods, C. A., L. Contreras, G. Willner-Chapman, and H. P. Whidden. 1992. *Myocastor coypus*. Mammalian Species # 398. American Society of Mammalogists, http://www.science.smith.edu/departments/Biology/VHAYSSEN/msi/.

Wootton, J. T. 1996. Purple Finch. *In* The Birds of North America, # 208 (A. Poole and F. Gill, Eds.), National Academy of Sciences, Philadelphia, Pennsylvania.

Wyatt, V. E., and C. M. Francis. 2002. Rose-breasted Grosbeak. *In* The Birds of North America, # 692 (A. Poole and F. Gill, Eds.), National Academy of Sciences, Philadelphia, Pennsylvania.

Yasukawa, K., and W. A. Searcy. 1995. Red-winged Blackbird. *In* The Birds of North America, # 184 (A. Poole and F. Gill, Eds.), National Academy of Sciences, Philadelphia, Pennsylvania.

Yates, T. L., and D. J. Schmidly. 1978. *Scalopus aquaticus*. Mammalian Species # 105. American Society of Mammalogists, http://www.science.smith.edu/departments/ Biology/VHAYSSEN/msi/.

Yosef, R. 1996. Loggerhead Shrike. *In* The Birds of North America, # 231 (A. Poole and F. Gill, Eds.), National Academy of Sciences, Philadelphia, Pennsylvania.

Zug, G. R., L. J. Vitt and J. P. Caldwell. 1993 Herpetology: An introductory biology of amphibians and reptiles. Academic Press, New York.

INDEX

ACKNOWLEDGMENTS

I thank Jonathan Ballou and Scott Miller at the Smithsonian National Zoological Park's Conservation and Research Center for their support. Robert Lockhart, History Editor for University of Pennsylvania Press, was extremely helpful in overseeing details related to publication. Rosalyn Alexander provided a number of her excellent line drawings for the work, as did Jackson Shedd. Photographers Barth Schorre, Vernon Grove, and David Parmelee donated many of the outstanding photos that accompany the species accounts. Dick DeGraaf, senior author of *New England Wildlife*, was instrumental in initiating the project. He also provided a thorough review of the completed manuscript, which was very helpful. Richard Abel, Director of the University Press of New England, kindly provided many illustrations from *New England Wildlife*. I thank Donald W. Linzey for review of the mammal section, and an anonymous Pennsylvania herpetologist for review of the reptile and amphibian sections. Their comments improved the manuscript, although any remaining errors are my responsibility.

Most drawings from Richard M. DeGraaf and Mariko Yamasaki, *New England Wildlife: Habitat, Natural History, and Distribution* (Hanover, N. H.: University Press of New England, 2001). © 2001 University Press of New England, reprinted by permission.

Mammal drawings initialed R.A. reprinted by permission of R.A. Alexander, BSc., D.V.M.